Progress in Nonlinear Differential Equations and Their Applications
Volume 40

Recent Trends in Nonlinear Analysis

Festschrift Dedicated to Alfonso Vignoli on the Occasion of His Sixtieth Birthday

Jürgen Appell
Editor

Springer Basel AG

Jürgen Appell
Department of Mathematics
University of Würzburg
Am Hubland
97074 Würzburg
Germany
e-mail: appell@mathematik.uni-wuerzburg.de

1991 Mathematics Subject Classification 00B30 (primary); 28A80, 34B15, 35J05, 35J60, 35P05, 35Q72, 45G10, 46E30, 46E35, 47A10, 47A53, 47H04, 47H10, 47H11, 47H30, 49J35, 55M25, 55N91, 58Z05, 70H20, 90A09, 90A14 (secondary)

A CIP catalogue record for this book is available from the Library of Congress, Washington D.C., USA

Deutsche Bibliothek Cataloging-in-Publication Data

Recent trends in nonlinear analysis: Festschrift dedicated to Alfonso Vignoli on the occasion of his sixtieth birthday / Jürgen Appell, ed. – Basel ; Boston ; Berlin : Birkhäuser, 2000
(Progress in nonlinear differential equations and their applications ; Vol. 40)
ISBN 978-3-0348-9556-9 ISBN 978-3-0348-8411-2 (eBook)
DOI 10.1007/978-3-0348-8411-2

Originally published by Birkhäuser Verlag, Basel, Switzerland in 2000
Softcover reprint of the hardcover 1st edition 2000
Printed on acid-free paper produced of chlorine-free pulp. TCF ∞

ISBN 978-3-0348-9556-9

9 8 7 6 5 4 3 2 1

Dedication

This collection of 21 research papers is dedicated to Alfonso Vignoli on the occasion of his sixtieth birthday.

Alfonso's personal qualities and his incredible energy leave lasting impressions on his friends and colleagues. His influential research contributions cover a wide range of topics and illustrate his broad view of mathematical analysis as a whole.

This broad view is also reflected in the contributions of this *Festschrift* through which the authors express their appreciation, gratitude, and friendship.

Alfonso Vignoli (with Angela)

Contents

Progress in Nonlinear Differential Equations
and Their Applications, Vol. 40

Alfonso Vignoli – the Researcher, Teacher, and Friend

JÜRGEN APPELL, MASSIMO FURI, JORGE IZE

Alfonso Vignoli was born on February 1, 1940, in Florence, one of the most beautiful cities in the world; this explains perhaps his extraordinary eye for esthetical beauty. His academic and personal curriculum is really impressive. His father, a traditional communist in the best sense of the word, decided to send him to the Soviet Union, the "paradise" of the working class, to study mathematics. So after 6 years of study Alfonso graduated from the Faculty of Mathematics and Mechanics of the highly regarded Lomonossov State University of Moscow obtaining a Masters Degree in Physics in 1967. The corresponding Italian Diploma (laurea) in Physics was conferred on him in 1968 at the University of Florence.

Alfonso's academic career bears all the characteristics of a widely-travelled and open-minded *cittadino del mondo.* He lived 10 years in Argentina, 7 years in the Soviet Union, and more than 1 year in the United States. He held positions at various Italian Universities, among them Genoa, Florence, L'Aquila, Cosenza and, now for more than 20 years, Rome. Moreover, he was visiting professor, sometimes for a longer period, in Mexico, Germany, Switzerland, and the United States, and he speaks fluently English, French, Spanish, and Russian (and started learning German and Portuguese). So it is not surprising that he has colleagues and friends all over the world who consider themselves lucky to know him as a researcher, teacher, and friend.

Throughout his life, Alfonso has been interested in many fields of mathematics. Without aiming for a complete coverage, let us mention the following topics:[1]

- ill-posed problems and regularization [1-3,7,9],
- fixed point theorems and applications [4-6,8,10,12,14],
- nonlinear operator theory [13-15,18-21,23,24,27,29,32,33,38,39,51,55],
- extremal problems and variational calculus [7,9,11,16,49,50],
- nonlinear spectral and eigenvalue theory [17,22,25,26,30,52,54],
- bifurcation and implicit function theorems [25,30,31,33-37,45-47],
- topological degree theory [28,37,40,42,48,53],
- equivariant and related maps [40-42,44,48,53].

[1]The number in square brackets refer to the list of references at the end.

In what follows we will try to shed some light on Alfonso's life and personality, mainly from the viewpoint of common mathematical interests, with a particular emphasis on equivariant degree theory, noncompact fixed point theory, and nonlinear spectral theory, but also with regard to some personal reminiscences.

In 1968, Alfonso, who was again living in Florence at that time, met Massimo at the University, where he went to speak with Roberto Conti and Gaetano Villari about the possibility of a recognition of the degree obtained in Russia. Since then, an intense mathematical collaboration and a deep, ever lasting friendship began. As an immediate concrete result, after a few days of pleasant and fruitful collaboration, the first of a long line of joint papers was born [3]. Before the end of the year, Mario Martelli, who was a very close friend of Massimo since the time when they were both freshmen at the University of Florence, joined the other two to form the group of the "three musketeers", as they were called by some common friends. From that moment on, a fruitful mathematical cooperation started, and a profound and ever lasting friendship among the three was born. They were three different personalities with the same common interests, three young fellows moved by the same wind of enthusiasm. They are still close friends living in different cities, and only distance and ever growing academic duties separate them.

Jorge met Alfonso first, in 1973, in a month long meeting in Montréal which has been quite important since many of us started, at that time, a long friendship. Jorge was still a graduate student, taking careful notes at every talk and proud that his adviser, Louis Nirenberg, had included some of the results of his thesis in his mini-course. Alfonso was with a lively group of Italians, in particular the other two musketeers,[2] who used to sing, laugh and talk about politics until very late at night. They made a latin group with the participation of some Germans and even a native Hawaiian.

After a few casual encounters at meetings, Alfonso spent a month in 1976 at the *Universidad Nacional Autónoma de México* (UNAM), where he lectured on the nonlinear spectrum [22,23] and stayed at Jorge's home. One of Jorge's students wrote his bachelor thesis on this topic.

The next main get-together with Jorge was in Cosenza in the spring of 1982: Jorge had spent a sabbatical year in France and, before finishing it in Germany, he was invited by Alfonso to spend a month there and give some lectures on bifurcation. In Nice, he had been working on the relationship between extensions of maps and the non-triviality of bifurcation invariants for several parameters. Alfonso, Ivar Massabò and Jacobo Pejsachowicz had been playing with the idea of "complementing maps"[3] applied to different classes of maps. They had given a very nice argument, based on Zorn's lemma, to simplify the Alexander-Antman proof of the dimension of "bifurcating surfaces". This combination of ideas and previous work led them to begin a collaboration which is still going.

In Cosenza the whole "gang of four"[4] worked very intensely, in the house on campus. So intensely that, in order to maintain his ideas clear, Alfonso had to smoke *Gauloises*,

[2]Massimo Furi and Mario Martelli, see above; the three musketeers wrote many joint papers [3-11,17,19-26,29-32] between 1969 and 1983.

[3]The idea of complementing maps consists in "filling up" the dimension of the range, in rather the same way as the implicit function theorem may be proved by means of the inverse function theorem.

[4]Alfonso, Ivar, Jacobo and Jorge; as a result, they published the joint papers [34] and [35].

drink lots of *Cirò* wine, and keep up with the other three of the gang. This first step was followed by a whole year that Alfonso spent in Mexico and where he fell in love with – the country.

In the two papers [34,35] written by the gang of four, they took the idea of essential maps, or of 0-epi maps, and extended it to the situation where one has an arbitrary set S and an arbitrary (not necessarily bounded) open subset U of a Banach space E. One says that a continuous map $g : U \to G$, where G is another Banach space, is *0-epi* on $S\&U$ if the following holds:

1. $g^{-1}(0) \cap S$ is bounded and away from ∂U;
2. the equation $g(x) = h(x)$ has a solution $x \in S \cap U$ for any compact map $h : E \to G$, with bounded support contained in U.

0-epi maps have strong persistence properties with respect to the sets U and S and to a certain class of homotopies (for instance, compact or k-set contractions). In particular, if g is 0-epi on $S\&U$, then either $S \cap \partial V \neq \emptyset$ or $g(S \cap \overline{V}) = G$, for any open and bounded subset V of U which contains $g^{-1}(0) \cap S$.

This type of purely topological results leads to global continuation and global bifurcation for maps f such that $S = f^{-1}(0)$ and g is an auxiliary function, such as a section in the parameter space or a "complementing map".

The main result on dimension of these papers was the following: If S is closed in U and g is 0-epi on $S\&U$, proper and bounded on bounded and closed subsets of $S \cap U$, then there is a minimal subset Σ of $S \cap U$ such that:

1. g is 0-epi on Σ, $g^{-1}(0) \cap \Sigma \neq \emptyset$, Σ is either unbounded or $\overline{\Sigma}$ meets ∂U, $\dim(\Sigma \cap V) \geq \dim G$, and $\dim(\Sigma \cap \partial V) \geq \dim G - 1$ for any V as above;
2. if $\Sigma = \Sigma_1 \cup \Sigma_2$, proper subsets, then $\dim(\Sigma_1 \cap \Sigma_2) \geq \dim G - 1$; in particular, Σ is connected and has dimension at least $\dim G$ at each point.

This result relies on simple arguments of general topology and dimension theory. With it one recovers all the similar results given for all classes of maps where some sort of degree theory is defined. As a matter of fact, the gang of four provided a simple version of it in the Berkeley meeting proceedings [35] (for compact maps), but in the paper [34] the intention was to extract the essential ingredients of those arguments and to unify a whole zoo of different situations.[5]

When applied to a bifurcation situation with several parameters one obtains "surfaces" bifurcating from singular sets which also have a large dimension. These surfaces cover complete components of regular points.

Clearly, if there is an associated degree theory then one gets 0-epi maps. On the contrary, if one is not using the algebraic properties of a degree, one gets most of the results on existence, continuation and global alternative from the simpler, more versatile and primitive notion of 0-epi maps.

[5]The gang of four was maybe too successful on that ground, since, although the paper [34] has still the best results in that direction, few people have had the patience of going through its details.

The next paper (without Jacobo) appeared in the proceedings [36] of the Maratea meeting in 1985 and was due to several reciprocal visits, including one of Alfonso and Ivar to Utah where Jorge was spending a year and where they manufactured a splendid pasta from scratch. In that study the previous results were extended to the case where the set S may be the zero set of an equivariant map f, i.e., such that $f(\gamma x) = \tilde{\gamma} f(x)$, for all γ in a compact Lie group Γ, and f a map between two representations E and $\tilde{E}$ of Γ. One defines then the notion of a Γ-*epi map* g from E into another representation G, where now the sets are invariant and the maps equivariant. One obtains very similar results for sets in the orbit space, with dimension in terms of G^{Γ}. In that paper these ideas were applied to the case of $\mathbb{S}^1$-maps and the Hopf bifurcation problem for periodic solutions of autonomous differential equations recovering, in particular, via a trick Jorge had used previously, the Fuller degree and the Chow, Mallet-Paret and Yorke bifurcation results.

At the Maratea meeting, Kazimierz Gęba talked about a way of constructing a degree via a compactification of the space.[6] This idea rang a familiar bell, in view of Jorge's previous work on bifurcation with several parameters and obstruction and the complementing map construction. This led to the following simpler construction of a general degree for equivariant maps: Assume that B and E are Banach spaces, and $F : U \subset B \to E$ is a map which is non-zero on the boundary of a bounded open subset $U \subset B$.[7] The map F may have some compactness properties if one wishes to arrive at a non void theory but this is not essential. Since U is bounded, take a large ball B_R, centered at the origin and containing U. Extend F to $\tilde{F}$ to B_R (equivariantly and with the additional properties). Construct a neighborhood N of ∂U such that $\tilde{F}$ is non-zero on N (some compactness is required there), and a Urysohn function $\varphi(x)$ with value 0 in U and 1 outside $U \cup N$. Suspend the map $\tilde{F}$ on $I \times B$, with $I = [0, 1]$, by the map

$$\hat{F}(t, x) = (2t + 2\varphi(x) - 1, \tilde{F}(x)).$$

It is easy to see that $\hat{F}(t, x) = 0$ if and only if $x \in U$, $F(x) = 0$, and $t = 1/2$; in particular, $\hat{F}$ maps $\partial(I \times B_R)$ into $\mathbb{R} \times (E \setminus \{0\})$. Hence F defines an element of the group of homotopy classes of maps between the two topological spheres $\partial(I \times B_R)$ and $\mathbb{R} \times (E \setminus \{0\})$. This element will be defined as the degree of F with respect to U. If one deals with a class of maps with a special property, for instance equivariant maps, then the homotopy classes will have to preserve this property.

This degree extends the Leray-Schauder degree[8] and has all the properties of the degree (up to one suspension for the additivity). However, it is not in general a single integer but an element of the corresponding homotopy group of spheres (equivariant if there is a group action).

In the paper [40], Alfonso, Ivar and Jorge derived the abstract properties of this degree and applied it to the case of a semi-free action of $\mathbb{S}^1$, exploring the unstable case, recovering the Fuller degree and computing the index of an isolated orbit with application to the Hopf bifurcation problem. The computations relied on ideas of extension

[6]He elaborated on this idea with Alfonso and Ivar in the paper [41].

[7]If these are representations of a group Γ then U is invariant and F equivariant.

[8]The Leray-Schauder degree is recovered in case $B = E$ and one has compact perturbations of the identity.

of mappings and were quite explicit. Alfonso, Ivar and Jorge were confident, in particular in discussions at the ELAM meeting in Rio de Janeiro, that the abstract results on equivariant homotopy groups of spheres would be readily available in the topology literature.[9]

The second paper of this series was on the study of a general $\mathbb{S}^1$-action, where one has to do a step by step extension process on "fundamental cells" on each isotropy subspace in a very explicit form. One obtains obstructions to this extension which will constitute the elements of the $\mathbb{S}^1$-degree. Alfonso, Ivar and Jorge computed the equivariant homotopy groups for different group actions and when the difference of the dimensions of the isotropy subspaces is at most 2. In particular, when $B = \mathbb{R} \times E$, they proved that

$$\Pi_{\mathbb{S}^1} = \Pi_{k+1}(\mathbb{S}^k) \times \mathbb{Z} \times \ldots \times \mathbb{Z},$$

where the first term corresponds to the invariant part and one has one $\mathbb{Z}$ for each nontrivial isotropy subgroup. Extensions to infinite dimensions, applications to global continuation and global bifurcation and other properties were included, like the computation of the index of an isolated orbit or an isolated loop of stationary points, with applications to autonomous differential equations, for instance Hopf bifurcation and period doubling. A first version of the paper was written while Jorge was on sabbatical in Rome and sent to the Transactions of the AMS, but the editor suggested to have the work published as a Memoir of the AMS. So Alfonso and Jorge took advantage of this suggestion to expand the paper by adding the study of ordinary differential equations with fixed period and those which have a first integral,[10] as well as examples of symmetry breaking. Later they added a last chapter on virtual periods and orbits index in order to relate their results to those of Yorke and Fiedler, resulting in another "small" paper of 180 pages [42].

It was clear, by that time, that the proposed program would take much longer than expected and that one should study first the easier (and more explicit) case of actions of abelian groups. This was the matter of the next three articles [44,48,53], all published in the same journal (but without Ivar occupied in more interesting activities). The first part [44] was devoted to the topological part, i.e., the extension process (on fundamental cells), the obstruction and the homotopy groups of Γ-maps, where the action may be different on the two spaces and the dimension of the spaces are different. The groups are stratified according to the dimension of the Weyl groups. The arguments were based on extension of equivariant maps (the Hopf problem) by a careful analysis of the fundamental cells. In particular, Alfonso and Jorge proved that if $B = E$, then one has that Π_Γ is a product of $\mathbb{Z}$'s, one for each isotropy subgroup with finite Weyl group. On the other hand, if $B = \mathbb{R} \times E$, one has that $\Pi_\Gamma = \Pi_{\Gamma/\mathbb{T}^n} \times \mathbb{Z} \times \ldots \times \mathbb{Z}$, where the first term corresponds to the finite group $\Gamma/\mathbb{T}^n$ on the fixed point subspace of the maximal torus $\mathbb{T}^n$, and one $\mathbb{Z}$ for each isotropy subgroup with a one-dimensional Weyl group. Explicit generators were given for these groups, in particular for an explicit construction of $\Pi_{\Gamma/\mathbb{T}^n}$, a group which is made of finite groups according to the presentations of Γ/H, H an isotropy subgroup.

[9] They were completely wrong as they soon realized.

[10] Something had been done previously by Jorge in Australia.

The second part [48] dealt with the index computations, their relationship with Poincaré's sections, the index of an isolated orbit, in particular of twisted orbits for ordinary differential equations. Alfonso and Jorge gave also general results of Borsuk-Ulam type, in order to explore when one has equivariant extensions of maps between spheres of the same dimension but with different group actions. Finally, they computed the index of an isolated loop of stationary orbits and looked at the symmetry breaking process, products, composition, and applications to differential equations.

The third part [53], just published in 1999, looks at two classes of equivariant maps which have the additional property of being gradients or orthogonal to the infinitesimal generator of the action of the torus part of the action. This includes Hamiltonian systems. One obtains a degree which is an element of the corresponding group of orthogonal equivariant maps, with one $\mathbb{Z}$ for each isotropy subgroup (i.e., a much richer structure than that for the non-orthogonal case). One has also symmetry breaking and products. The index of an isolated orbit is computed, giving rise to bifurcation results where a change of a Morse number will give bifurcation from a certain torus to a higher dimensional torus. A section is devoted to the problem of bifurcation with several parameters, where the complex Bott periodicity theorem appears. For Hamiltonian systems, one has a complete study of the index of either a stationary orbit or a non-stationary one, with a relation to Maslov index.

At this point (and at this moment), it is clear that one has a rather complete picture for the case of an abelian action. Thus, Alfonso and Jorge decided that it was time to collect all these results and ideas in a single book, where one would be able to explain in more detail (and through examples) how this degree works. This will be the forthcoming monograph [C].

Another important field of interest of Alfonso for many years was nonlinear spectral theory. A basic ingredient is the notion of stably solvable maps which was introduced by the three musketeers in the paper [23]. A continuous map f between two Banach spaces E and F is called *stably solvable* if the equation $f(x) = h(x)$ has a solution $x \in E$ for any continuous compact map $h : E \to F$ satisfying

$$|h| = \limsup_{||x|| \to \infty} \frac{||h(x)||}{||x||} = 0.$$

Choosing, in particular, $h(x) \equiv y$ one sees that every stably solvable map is onto; the converse is true for linear maps.

A stably solvable map f is called *regular* if both

$$d(f) = \liminf_{||x|| \to \infty} \frac{||f(x)||}{||x||} > 0$$

and

$$\beta(f) = \sup \{k > 0 : \alpha(f(A)) \geq k\alpha(A) \text{ for any bounded } A \subset E\} > 0,$$

where $\alpha(A)$ denotes the (Kuratowski) measure of noncompactness of A. In spite of the rather technical definition, it turned out that this class of maps plays an important and natural role in various problems of nonlinear analysis. Thus, in the linear case a map $L \in \mathfrak{L}(E, F)$ is regular if and only if it is an isomorphism.[11] Moreover, a nonlinear map $f : \mathbb{R}^n \to \mathbb{R}^m$ is regular if and only if its homotopy class is nontrivial (i.e. contains nonconstant maps).

There is also an interesting connection between regular maps and essential vector fields.[12] Let E be an infinite dimensional Banach space, E_0 a finite codimensional closed subspace of E, and $S(E)$ the unit sphere in E. Recall that a map $\Phi : S(E) \to E_0 \setminus \{0\}$ is called *compact essential vector field* if $(I - \Phi)(S(E))$ is precompact, and every extension of Φ to the unit ball in E vanishes at some point.

Now, to each compact vector field $\Phi : S(E) \to E_0 \setminus \{0\}$ one may associate its "homogeneization" $\hat{\Phi} : E \to E_0$ defined by

$$\hat{\Phi}(x) = \begin{cases} \|x\|\Phi\left(\frac{x}{\|x\|}\right) & \text{if} \quad x \neq 0, \\ 0 & \text{if} \quad x = 0. \end{cases}$$

Clearly, $\hat{\Phi}$ is positively homogeneous and coincides with Φ on $S(E)$. In [26] the authors proved that the vector field Φ is essential if and only if its homogeneization $\hat{\Phi}$ is regular in the sense defined above.

Given a continuous nonlinear map $f : E \to E$, the spectrum[13] of f consists, by definition, of all scalars $\lambda \in \mathbb{K}$ such that $\lambda - f$ is not regular. So one has

$$\sigma(f) = \sigma_\delta(f) \cup \Sigma(f) \cup \sigma_\beta(f),$$

where $\lambda \in \sigma_\delta(f)$ if $\lambda - f$ is not stably solvable, $\lambda \in \Sigma(f)$ if $d(\lambda - f) = 0$, and $\lambda \in \sigma_\beta(f)$ if $\beta(\lambda - f) = 0$. In the context of bounded linear maps $L \in \mathfrak{L}(E, E)$, these subspectra have a natural meaning. For example, $\sigma_\delta(L)$ is the approximate defect spectrum, and $\Sigma(L) \cup \sigma_\beta(L)$ is the approximate point spectrum of L. In the nonlinear case one could say, loosely speaking, that the scalars $\lambda \in \sigma_\delta(f)$ are characterized by some lack of surjectivity of $\lambda - f$, the scalars $\lambda \in \Sigma(f)$ by some lack of coercivity $\lambda - f$, and the scalars $\lambda \in \sigma_\beta(f)$ by some lack of compactness of $\lambda - f$.

The subspectrum $\Sigma(f)$ has particularly interesting topological properties if E is infinite dimensional and f is compact:

1. $0 \notin \Sigma(f)$ implies that the connected component of $\mathbb{K} \setminus \Sigma(f)$ containing zero lies entirely in $\sigma_\delta(f)$;

2. $\Sigma(f)$ contains all asymptotic bifurcation points of f;

[11] This explains the importance of this definition of regularity in nonlinear spectral theory; in fact, for linear maps the spectrum defined by Furi, Martelli and Vignoli coincides with the familiar spectrum.

[12] Compact essential vector fields have been introduced and studied by Andrzej Granas in 1962. We also mention that Louis Nirenberg has found a characterization of essential vector fields using arguments of stable homotopy theory, and exhibited some surprisingly interesting applications to partial differential equations.

[13] Now there are many other definitions of spectra for nonlinear maps in Banach spaces. The spectrum $\sigma(f)$ is therefore called *Furi-Martelli-Vignoli spectrum* (or *FMV-spectrum*, for short) in the recent literature.

3. if f is positively homogeneous, then $\Sigma(f) \setminus \{0\}$ coincides with the set of all eigenvalues[14] of f.

In particular, the last property explains the importance of the FMV-spectrum in applications. Recall that a point $\lambda \in \mathbb{K}$ is called *asymptotic bifurcation point* of a continuous map $f : E \to E$ if there exist sequences $(\lambda_n)_n$ in $\mathbb{K}$ and $(x_n)_n$ in E such that $\lambda_n \to \lambda$, $||x_n|| \to \infty$, and $f(x_n) = \lambda_n x_n$. In other words, asymptotic bifurcation points are accumulation points of sequences of eigenvalues with a corresponding unbounded sequence of eigenvectors.

In [26], the authors proved the following results:

1. The set $B(f)$ of asymptotic bifurcation points of f is a closed subset of $\Sigma(f)$;
2. if E is infinite dimensional and f is compact with $0 < d(f) \leq |f| < \infty$, then $B(f) \neq \emptyset$;
3. if $\lambda, \mu \in \mathbb{K}\setminus[\Sigma(f)\cap\sigma_\beta(f)]$ are such that $\lambda \notin \sigma(f)$ and $\mu \in \sigma(f)$, then $B(f)\cap\sigma_\beta(f)$ separates λ and μ, i.e. λ and μ belong to different components of $\mathbb{K}\setminus[B(f)\cap\sigma_\beta(f)]$;
4. in case $E = \mathbb{R}^{2n+1}$ and $|f| < \infty$, the set $B(f)$ is always nonempty.

These important results lead to some quite interesting existence theorems for boundary value problems for differential equations; for details we refer to the recent paper [52] and the survey article [26].

Surprisingly, nonlinear spectral theory has not found much attention between 1978 and 1998. Motivated by some promising new results on nonlinear spectra obtained by Jürgen and his group in Würzburg, Alfonso became again interested in nonlinear spectral theory; so Alfonso and Jürgen decided to put their efforts together. As a result, two joint papers on the subject [52,54] were written, but the theory is far from being complete. Since there is no account of nonlinear spectral theory in book form, but only a wealth of special results scattered over many research journals, it seems reasonable to write a monograph on the theory and applications of nonlinear spectral theory; this will be the forthcoming book [D], with Espedito De Pascale from Calabria as third author. So, in a certain sense the circle was closed and Alfonso turned back to one of his earliest fields of interest. Or, as the French saying goes: *On revient toujours à son premier amour.*

This was more or less the state of the art till 1995. Then came the 12th of January, 1996, probably one of the darkest days in Alfonso's life. Being on a trip to Siena, he felt a strong indisposition which became worse and worse, and in the evening of that day he fell into coma and was taken to the intensive-care unit of a hospital. It turned out that he had sustained an extremely serious cerebral haemorrhage, and the medical prognosis was quite pessimistic. After several days at hospital in Rome he awoke to new life on the 1st of February – his birthday, in an incredibly significant sense. Since then a real and

[14] Here an *eigenvalue* of f is, as in the linear case, a scalar λ such that the equation $f(x) = \lambda x$ has a nontrivial solution $x \in E$. We point out, however, that this is not the only possible (and, as a matter of fact, not the most natural) definition of eigenvalue in the nonlinear case. For example, the "Vignoli seagull" $f(x) = \sqrt{|x|}$ on the real line has every non-zero real number as eigenvalue, but its FMV-spectrum is $\sigma(f) = \{0\}$ which is even disjoint from the eigenvalues.

unexpected *miracolo* happened: He recovered completely from his terrible illness and regained all his previous physical and mental capacities, including his mathematical sagacity, his sophisticated eloquence and, last but not least, his fine sense of humour.

Fortunately, he is now again as before in many aspects – always prepared to meet friends, to have a good glass of wine (or two), to make adventurous plans for the future, and to "philosophize" on God and the world. So in a certain sense, with this Festschrift we are also celebrating the fourth anniversary of Alfonso's new life.

Appreciating the mathematician Alfonso Vignoli would be incomplete without adding some remarks on his teaching abilities. We all know him as a brilliant lecturer who attaches great importance to conveying *ideas*, rather than enumerating definitions and theorems. From the earliest days of his career as professor Alfonso has shown devotion to, and ability in, attracting young people to his research interests. As a result, many of his immediate students with higher degrees are working successfully at various Italian and foreign Universities. The scientific enthusiasm and optimism that they have acquired over the years, from contacts to Alfonso and work done jointly with him, have inspired them for a long time, and probably will inspire them also in the future.

On behalf of all contributors to this Festschrift, we offer him our warmest congratulations on his sixtieth birthday, wish him all the best for his future mathematical activity, and sincerely hope that he will enjoy health and happiness for still many years to come:

Auguri Alfonso!

List of Publications of Alfonso Vignoli

1. Books

A. Schiaffino A., Vignoli A.: *Introduzione all'Analisi matematica, Vol. I*, Aracne Ed., Roma 1998.

B. Schiaffino A., Vignoli A.: *Introduzione all'Analisi matematica, Vol. II*, Aracne Ed., Roma 1999.

C. Ize J., Vignoli A.: *Equivariant Degree Theory*, in preparation.

D. Appell J., De Pascale E., Vignoli A.: *Nonlinear Spectral Theory*, in preparation.

2. Papers

1. Budak B.M., Gaponenko Ju. L., Vignoli A.: *On a method of regularization of the extremal problem for a continuous convex functional* [in Russian], Dokl. Akad. Nauk, SSSR **184, 1** (1969), 12-15.

2. Budak B.M., Gaponenko Ju. L., Vignoli, A.: *On a method of regularization for a continuous convex functional* [in Russian], Zhurn. Vychisl. Mat. Mat. Fiz. **9, 5** (1969), 1046-1056.

3. Furi M., Vignoli A.: *On the regularization of a nonlinear ill-posed problem in Banach spaces*, J. Optim. Theory Appl. **4, 3** (1969), 206-209.

4. Furi M., Vignoli A.: *A fixed point theorem in complete metric spaces*, Boll. Unione Mat. Ital. **2, 4** (1969), 505-509.

5. Furi M., Vignoli A.: *Fixed point theorems for densifying mappings*, Accad. Naz. Lincei **47, 6** (1969), 465-467.

6. Furi M., Vignoli A.: *On α-nonexpansive mappings and fixed points*, Accad. Naz. Lincei **48, 2** (1970), 195-198.

7. Furi M., Vignoli A.: *About well-posed optimization problems for functionals in metric spaces*, J. Optim. Theory Appl. **5, 3** (1970), 225-229.

8. Furi M., Vignoli A.: *A remark about some fixed point theorems*, Boll. Unione Mat. Ital. **3, 2** (1970), 197-200.

9. Furi M., Vignoli A.: *A characterization of well posed minimum problems in a complete metric space*, J. Optim. Theory Appl. **5, 6** (1970), 452-461.

10. Furi M., Vignoli A.: *On a property of the unit sphere in a linear normed space*, Bull. Acad. Pol. Sci. **18, 6** (1970), 333-334.

11. FURI M., MARTELLI M., VIGNOLI A.: *On minimum problems for families of functionals*, Annali Mat. Pura Appl. **86** (1970), 181-188.

12. VIGNOLI A.: *An intersection theorem in Banach spaces*, Accad. Naz. Lincei **49, 3** (1970), 180-183.

13. VIGNOLI A.: *On quasibounded mappings and nonlinear functional equations*, Accad. Naz. Lincei **50, 2** (1971), 114-117.

14. MONTAGNANA M., VIGNOLI A.: *On quasiconvex mappings and fixed point theorems*, Boll. Unione Mat. Ital. **4, 6** (1971), 870-878.

15. VIGNOLI A.: *On α-contractions and surjectivity*, Boll. Unione Mat. Ital. **4, 3** (1971), 446-455.

16. STAMPACCHIA G., VIGNOLI A.: *A remark on variational inequalities for a second order nonlinear differential operator with non Lipschitz obstacles*, Boll. Unione Mat. Ital. **5, 1** (1972), 123-131.

17. MARTELLI M., VIGNOLI A.: *Eigenvectors and surjectivity for α-Lipschitz mappings in Banach spaces*, Annali Mat. Pura Appl. **94** (1972), 1-9.

18. PEJSACHOWICZ J., VIGNOLI A.: *On differentiability and surjectivity of α-Lipschitz mappings*, Annali Mat. Pura Appl. **101** (1974), 49-63.

19. MARTELLI M., VIGNOLI A.: *Some surjectivity results for non compact multivalued maps*, Rend. Accad. Sci. Fis. Mat. Napoli **41** (1974), 1-12.

20. MARTELLI M., VIGNOLI A.: *On differentiability of multivalued maps*, Boll. Unione Mat. Ital. **10, 3** (1974), 701-712.

21. MARTELLI M., VIGNOLI A.: *A generalized Leray-Schauder condition*, Accad. Naz. Lincei **57, 5** (1974), 374-379.

22. FURI M., VIGNOLI A.: *A non-linear spectral approach to surjectivity in Banach spaces*, J. Funct. Anal. **20, 4** (1975), 304-318.

23. FURI M., MARTELLI M., VIGNOLI A.: *Stably solvable operators in Banach spaces*, Accad. Naz. Lincei **60, 1** (1976), 21-26.

24. FURI M., VIGNOLI A.: *On surjectivity for nonlinear maps in Banach spaces*, Annali Mat. Pura Appl. **112** (1977), 205-216.

25. FURI M., VIGNOLI A.: *Spectrum for nonlinear maps and bifurcation in the nondifferentiable case*, Annali Mat. Pura Appl. **113** (1977), 265-285.

26. FURI M., MARTELLI M., VIGNOLI A.: *Contributions to the spectral theory for nonlinear operators in Banach spaces*, Annali Mat. Pura Appl. **118** (1978), 229-294.

27. VIGNOLI A.: *Some topological methods for solving nonlinear operator equations*, Proc. Funct. Anal. Appl. (1979), Kumasi, Ghana, 60-134.

28. PEJSACHOWICZ J., VIGNOLI A.: *On the topological coincidence degree for perturbations of Fredholm operators*, Boll. Unione Mat. Ital. **17-B** (1980), 1457-1466.

29. FURI M., MARTELLI M., VIGNOLI A.: *On the solvability of nonlinear operators equations in normed spaces*, Annali Mat. Pura Appl. **124** (1980), 321-343.

30. FURI M., VIGNOLI A.: *Unbounded nontrivial branches of eigenfunctions for nonlinear equations*, Nonlin. Anal. TMA **6** (1982), 1267-1280.

31. FURI M., PERA M.P., VIGNOLI A.: *Components of positive solutions for nonlinear equations with several parameters*, Boll. Unione Mat. Ital. **1-C** (1982), 285-302.

32. MARTELLI M., VIGNOLI A.: *On the structure of the solution set of nonlinear equations*, Nonlin. Anal. TMA **7** (1983), 685-693.

33. MASSABÒ I., VIGNOLI A.: *Irreducible continua of solutions for Birkhoff-Kellogg type equations*, Nonlin. Anal. TMA **7** (1983), 453-461.

34. IZE J., MASSABÒ I., PEJSACHOWICZ J., VIGNOLI A.: *Structure and dimension of global branches of solutions to multiparameter equations*, Trans. Amer. Math. Soc. **291** (1985), 383-435.

35. IZE J., MASSABÒ I., PEJSACHOWICZ J., VIGNOLI A.: *Nonlinear multiparametric equations; structure and dimension of global branches of solutions*, Proc. Symp. Pure Math. **45** (1986), 529-540.

36. IZE J., MASSABÒ I., VIGNOLI A.: *Global results on continuation and bifurcation for equivariant maps*, NATO ASI, Series C, Mathematical und Physical Sciences **173** (1986), 75-111.

37. GĘBA K., MASSABÒ I., VIGNOLI A.: *Generalized topological degree and bifurcation*, NATO ASI, Series C, Mathematical und Physical Sciences **173** (1986), 55-73.

38. APPELL J., MASSABÒ I., VIGNOLI A., ZABREJKO P.P.: *Lipschitz and Darbo conditions for the superposition operator in ideal spaces*, Annali Mat. Pura Appl. **152** (1988), 123-137.

39. MASSABÒ I., VIGNOLI A.: *On the théorème fondamental of J. Leray and J. Schauder*, Colloq. Math. **57** (1989), 265-272.

40. IZE J., MASSABÒ I., VIGNOLI A.: *Degree theory for equivariant maps*, Trans. Amer. Math. Soc. **315** (1989), 433-510.

41. GĘBA K., MASSABÒ I., VIGNOLI A.: *On the Euler characteristic of equivariant gradient vector fields*, Boll. Unione Mat. Ital. **4-A** (1990), 243-252.

42. IZE J., MASSABÒ I., VIGNOLI A.: *Degree theory for equivariant maps: the general S^1 action*, Memoirs Amer. Math. Soc. **481** (1992), 1-179.

43. VIGNOLI A.: *L'analisi non lineare nella teoria della biforcazione*, in: Enciclopedia delle Scienze Fisiche dell'Istituto dell'Enciclopedia Italiana, Vol. **1** (1992), 134-144.

44. IZE J., VIGNOLI A.: *Equivariant degree for abelian actions. Part I: equivariant homotopy groups*, Topol. Methods Nonlin. Anal. **2, 2** (1993), 367-413.

45. APPELL J., MOROZ V.B., VIGNOLI A., ZABREJKO P.P.: *On the application of Kielhöfer's bifurcation theorem to Hammerstein equations with potential nonlinearity*, Boll. Unione Mat. Ital. **8-B** (1994), 833-850.

46. VIGNOLI A., ZABREJKO P.P.: *Some remarks on the Hildebrandt-Graves theorem*, Zeitschr. Anal. Anw. **14** (1995), 89-93; Erratum: ibidem **14** (1995), 653.

47. APPELL J., VIGNOLI A., ZABREJKO P.P.: *Implicit function theorems and nonlinear integral equations*, Expositiones Math. **14** (1996), 385-424.

48. IZE J., VIGNOLI A.: *Equivariant degree for abelian actions. Part II: index computations*, Topol. Methods Nonlin. Anal. **7, 2** (1996), 369-430.

49. VIGNOLI A., ZABREJKO P.P., MOROZ V.B.: *Critical values of lower-bounded functionals* [in Russian], Dokl. Akad. Nauk Belarusi **41, 2** (1997), 16-21.

50. MOROZ V.B., VIGNOLI A., ZABREJKO P.P.: *On the three critical points theorem*, Topol. Methods Nonlin. Anal. **11, 1** (1998), 103-115.

51. APPELL J., VÄTH M., VIGNOLI A.: *Compactness and existence results for ordinary differential equations in Banach spaces*, Zeitschr. Anal. Anw. **18, 3** (1999), 569-584.

52. APPELL J., CONTI G., VIGNOLI A.: *Teoria spettrale, punti di biforcazione e problemi nonlineari al contorno*, Atti Sem. Mat. Fis. Univ. Modena **47** (1999), 383-389.

53. IZE J., VIGNOLI A.: *Equivariant degree for abelian actions. Part III: orthogonal maps*, Topol. Methods Nonlin. Anal. **13, 1** (1999), 105-146.

54. APPELL J., DE PASCALE E., VIGNOLI A.: *A comparison of different spectra for nonlinear operators*, Nonlin. Anal. TMA (to appear).

55. FURI M., VIGNOLI A.: *An elementary topological approach to solvability of nonlinear equations* (submitted).

Progress in Nonlinear Differential Equations
and Their Applications, Vol. 40

How to Make Use of the Solution Set to Solve Boundary Value Problems

GIUSEPPE ANICHINI, GIUSEPPE CONTI

Friendly dedicated to Alfonso Vignoli

Summary: The set of solutions of a differential problem can be conveniently used as a tool to get existence results for boundary value problems. In this paper a survey of results, methods, and applications concerning such a tool is given.

Keywords: Boundary value problem, nonlinear condition, solution set, fixed point, multivalued operator, acyclic set.

Classification: 34B15, 34A60, 45G10, 47H04, 47H10, 47H15, 47H30.

Acknowledgement: The authors wish to thank the referee for some improvements.

1. Introduction. In this paper we consider a boundary value problem for a first order system of differential equations of the type

$$\text{(BV)} \qquad \begin{cases} x'(t) = f(t, x(t)), & t \in J \subset \mathbb{R}, \\ x \in S \end{cases}$$

where $f : J \times \mathbb{R}^n \to \mathbb{R}^n$ is a continuous (or sometimes a Carathéodory) function defined on J, and S is a suitable subset of $C(J, \mathbb{R}^n)$, the space of all continuous functions from J to $\mathbb{R}^n$.

The condition $x \in S$ is referred to informally as the "boundary condition" associated to the differential system of equations given above. Sometimes it is written as $Lx = r$ where $L : Dom(L) \to \mathbb{R}^n$ is a linear operator defined on a subspace $Dom(L)$ of $C(J, \mathbb{R}^n)$ and $r \in \mathbb{R}^n$; in this case S is a linear manifold in $C(J, \mathbb{R}^n)$.

The real interval J can be compact or, more frequently, noncompact. As a matter of fact, some interesting problems, mainly arising in applied sciences, imply the noncompactness requirement e.g. when we are looking for the asymptotic behaviour of the solutions of an ordinary differential equation.

Any interested reader can find in Chapter XII of the book [27] a detailed study of boundary value problems, related definitions, and classical results; but it will be chiefly in the survey paper [16] that a detailed exposition of differential problems with both linear and nonlinear boundary conditions, in the scalar or vector case, can be found. It is firm to say that essentially all the papers we quoted in the references owe something to the survey [16].

It is well known that problem (BV) can be very often reduced, by some kind of linearization technique, to a fixed point equation of the type

$$x = Tx,$$

or

$$x \in \Sigma(x),$$

according to the "uniqueness" or "nonuniqueness" property of the solutions of the differential problem. Here uniqueness means that a set of assumptions ensuring that the operator T is singlevalued is at our disposal; when this is not the case a multivalued nonlinear operator arises, and the approach is no longer the same. The above operator Σ is then a multivalued operator defined on a subset of a suitable Banach space with the property that its (possible) fixed points are solutions of the boundary value problem (BV).

As a matter of fact, in the first case a classical fixed point theorem (e.g. Brouwer, Banach-Caccioppoli, Schauder, etc.) can be conveniently used, while in the second case some results from multivalued operator theory must be singled out. In this paper we shall consider a number of results concerning the second case.

A procedure often used in trying to prove the existence of one solution of the boundary value problem (BV) is the following:

- Consider a map $g : J \times \mathbb{R}^n \times \mathbb{R}^n \to \mathbb{R}^n$ with the "diagonal" property

  $$g(t, \alpha, \alpha) = f(t, \alpha),$$

  for all $\alpha \in \mathbb{R}^n$.

- Study the so called "linearized" boundary value problem

  $$(\mathrm{BV})_q \qquad \begin{cases} x' = g(t, x, q), & t \in J \subset \mathbb{R}, \\ x \in S. \end{cases}$$

- Find the possible fixed points of the operator Σ defined by

  $$q \mapsto \Sigma(q) = \{\text{the set of solutions of } (\mathrm{BV})_q\}$$

  for $q \in Q$ where Q is a suitable subset of $C(J, \mathbb{R}^n)$.

  Clearly any fixed point of Σ is a solution of the problem (BV).

This approach has been developed by many authors; among them we want to recall [1,3,13-15,17,29,30,33,41].

In some cases a standard linearization of the problem allows us to get a multivalued operator with nonempty, compact and convex values: so the Ky-Fan, Tychonov, or similar fixed point theorem for multivalued functions is used as main tool in order to get existence of solutions. The reader is referred to the papers [1,13-15,17,29] to find applications of this method.

The previous method can also be extended in several directions. In particular, the regularity assumption on Σ – usually continuity – can be replaced by the assumption that the mapping is upper semicontinuous; but, apart from this regularity hypothesis, the "geometry" of the images has also to be taken into consideration. As a matter of fact, it is well known that, if these images are convex (and nonempty and compact), then it is not hard to get fixed points for the operator Σ; some problems arise when the values of the operator fail to be convex. In this case a different approach has to be taken into account.

Already in the forties, in [8], a detailed study of the set of the solutions of the Cauchy problem for ordinary differential equations caused a deeper insight into a concept coming from algebraic topology, namely acyclicity, a concept which is weaker than convexity but stronger than connectedness and simple connectedness.

We refer to [42] for the following definition of an acyclic set:

Definition. Let X be a topological space and let $\check{H}^p(X,\mathbb{Z})$ denote the reduced Cěch cohomology group of X in dimension p with coefficients in $\mathbb{Z}$. Then X is *acyclic* if $\check{H}^p(X,\mathbb{Z}) = 0$ for every $p \geq 0$.

Remark. One of the most important examples of acyclic sets are R_δ sets (see [30] for a reference). Also contractible (and hence also convex) sets are acyclic. Concerning a topological property of acyclic sets it is known that they are connected and even simply connected.

The papers [9-12,22-24,26,30,35,36,38-40,43-46] deal with the problem how to overcome the lack of convexity. There are also cases in which the set of solutions is neither convex nor acyclic, for instance, when a finite number of solutions arises [2,6,34].

We will end with some results concerning the use of the set of solutions as a useful tool to get the existence of solutions of nonlinear boundary value problems in Banach spaces. This argument is also widely considered and a number of results is obtained in the papers [18-20,25,28,34,37].

Finally, in order to show some results and the assumptions needed to obtain them, we shall introduce a series of papers each of which will be concerned with the topic we just deal with; in these papers different approaches and various tools useful to accomplish them will be presented. The main peculiarity of these papers will be the use of the properties of the set of solutions of a (hopefully) easier "linearized" boundary value problem as a tool to get the existence of solutions of a more general nonlinear boundary value problem.

2. Some results. The first result comes from [2].

Theorem 1. *Consider the boundary value problem*

$$\text{(BV)} \qquad \begin{cases} \dot{x} = f(t,x), & t \in J \subset \mathbb{R}, \quad x \in \mathbb{R}^n, \\ x \in S, \end{cases}$$

where $f \in C(J \times \mathbb{R}^n, \mathbb{R}^n)$ *and* $S \subset C(J, \mathbb{R}^n)$. *Assume that a continuous function* $g : J \times \mathbb{R}^n \times \mathbb{R}^n \to \mathbb{R}^n$ *is given such that* $g(t,\alpha,\alpha) = f(t,\alpha)$ *for all* $(t,\alpha) \in J \times \mathbb{R}^n$

and such that, for some closed and convex set $Q \subset C(J, \mathbb{R}^n)$ and for some closed and bounded subset $S_1 \subset Q \cap S$, the boundary value problem

$$(\mathrm{BV})_q \qquad \begin{cases} \dot{x} = g(t, x, q), & t \in J \subset \mathbb{R}, \quad x \in \mathbb{R}^n, \\ x \in S_1 \end{cases}$$

has, for each $q \in Q$, an acyclic set of solutions. Then the boundary value problem (BV) *has at least one solution.*

As mentioned above the property of the solution set of a suitably "linearized" boundary value problem to be acyclic is the assumption we need in order to get existence results for the general nonlinear boundary value problem.

In this paper we are also concerned with the case when the values of Σ are sets with a finite number of elements, and an application to the existence of periodic solutions for a Liénard equation is given as well. The main tool used here is a direct application of the fixed point theorem of Eilenberg-Montgomery (see [29] for a reference).

The following results can be found in [4].

Theorem 2. *Consider the Neumann problem*

$$(\mathrm{N}) \qquad \begin{cases} x'' + g(u) = h(t, u, u'), & t \in [0, \pi], \quad u \in \mathbb{R}, \\ u'(0) = u'(\pi) = 0. \end{cases}$$

Assume the following conditions hold:

(i) *h is a continuous and bounded function such that for all $q \in C^1([0, \pi])$ we have*

$$\frac{1}{\pi} \int_0^\pi h(s, q(s), q'(s)\, ds \in Int(Im(g)),$$

where $Int(Im(g))$ denotes the (relative) interior of the image of the function g.

(ii) *g is a continuous, bounded, increasing function such that $ug(u) > 0$ for $u \neq 0$, and*

$$\sup \{|g(u)| : u \in \mathbb{R}\} > \sup \{|h(t, u, v)| : (t, u, v) \in [0, 1] \times \mathbb{R}^2\}.$$

Then the problem (N) *has at least one solution.*

In this paper an analogous result concerning the Dirichlet problem is also obtained. In [31] a similar result is given in a suitable Banach space X (we refer to [34] for definitions and results concerning index, degree, and measures of noncompactness).

Theorem 3. *Consider the boundary value problem*

$$(\mathrm{BV}) \qquad \begin{cases} \dot{x} = f(t, x), & t \in J \subset \mathbb{R}, \quad x \in X, \\ x \in S, \end{cases}$$

where $f \in C(J \times X, X)$ and $S \subset C(J, X)$. Assume that there exist a closed, bounded and convex set $Q \subset C(J, X)$ and a closed subset $S_1 \subset Q \cap S$ such that the solutions of the integral equation

$$x = K(x, q)$$

are also solutions of the "linearized" boundary value problem

$$(\mathrm{BV})_q \qquad \begin{cases} \dot{x} = g(t, x, q), & t \in J \subset \mathbb{R}, \quad x \in X, \\ x \in S_1 \end{cases}$$

for any $q \in Q$. Suppose that $K : \Omega \times Q \to C(J, X)$ is a condensing map in the first variable with respect to a monotone and regular measure of noncompactness, and Ω is an open, bounded and convex subset of X such that $\operatorname{ind}(K(\cdot, q), \Omega) \neq 0$ *for some (and hence for all) $q \in Q$, and the function $g : J \times X^2 \to X$ is continuous and such that $g(t, \alpha, \alpha) = f(t, \alpha)$ for all $(t, \alpha) \in J \times X$.*
Let $\Sigma : Q \to Q$ be the operator which associates to each $q \in Q$ the set of solutions of $(\mathrm{BV})_q$. *Assume that, for each $q \in Q$, the set $\Sigma(q)$ is discrete. Then the boundary value problem* (BV) *has at least one solution.*

In the paper [5] the Carathéodory case is taken into consideration:

Theorem 4. *Consider the boundary value problem*

$$(\mathrm{BV}) \qquad \begin{cases} x'' = f(t, x, x') & t \in I = [a, b] \subset \mathbb{R}, \\ x \in S, \end{cases}$$

where $f \in Car(I \times \mathbb{R}^n \times \mathbb{R}^n, \mathbb{R}^n)$ and $S \subset C^1(I, \mathbb{R}^n)$.
Let us assume that the following conditions hold:
(i) *There is a function $g \in Car(I \times \mathbb{R}^n \times \mathbb{R}^n \times \mathbb{R}^n \times \mathbb{R}^n, \mathbb{R}^n)$ such that $g(t, c, c, d, d) = f(t, c, d)$ for almost all $t \in I$ and for all $(c, d) \in \mathbb{R}^n \times \mathbb{R}^n$.*
(ii) *For a given closed and convex set $Q \subset C^1(I, \mathbb{R}^n)$ and a given closed and bounded $S_1 \subset Q \cap S$, the boundary value problem*

$$(\mathrm{BV})_q \qquad \begin{cases} x'' = g(t, x, q, x', q') \text{ for almost all } t \in I, \\ x \in S_1 \end{cases}$$

has, for every $q \in Q$, an acyclic set of solutions. Then the boundary value problem (BV) *has at least one solution.*

In the paper [6] an integral equation with an exact number of solutions is studied in order to show the existence of solutions of some boundary value problem. Here a useful tool is the concept of w-map in Darbo's sense (we refer to [21] for precise definitions).

Theorem 5. *Consider the boundary value problems*

$$(\mathrm{BV}) \qquad \begin{cases} x' = f(t, x) \text{ for almost all } t \in I = [a, b] \subset \mathbb{R}, \quad x \in \mathbb{R}^n, \\ x \in S, \end{cases}$$

where $S \subset AC(I, \mathbb{R}^n)$, and

$$(\mathrm{BV})_q \qquad \begin{cases} x' = g(t, x, q) \text{ for almost all } t \in I, \\ x \in S_1, \end{cases}$$

where $f \in Car(I \times \mathbb{R}^n, \mathbb{R}^n)$ *is such that* $g(t,c,c) = f(t,c)$ *for almost all* $c \in \mathbb{R}^n$, *with a suitable function* $g \in Car(I \times \mathbb{R}^n \times \mathbb{R}^n, \mathbb{R}^n)$. *We assume that there exist a bounded, closed and convex set* $Q \subset AC(I, \mathbb{R}^n)$ *and a closed set* $S_1 \subset Q \cap S$ *such that the boundary value problem* $(\mathrm{BV})_q$ *is equivalent to some "integral equation"*

$$\text{(IE)} \qquad x = A(x,q) = A_q(x)$$

for all $q \in Q$, *where* $A : \Omega \times Q \to S_1 \subset AC(I, \mathbb{R}^n)$ *is a compact operator for each* $q \in Q$ *such that* $\deg(I - A_q, \Omega, 0) \neq 0$ *for some (and hence for all)* $q \in Q$ *and some open and convex set* $\Omega \subset AC(I, \mathbb{R}^n)$.
Let Σ *be the multivalued operator which associates to each* $q \in Q$ *the set of solutions of the integral equation* (IE). *Then, if the set* $\Sigma(q)$ *is discrete for each* $q \in Q$, *the boundary value problem* (BV) *has at least one solution.*

As an application of the previous result consider the two-point boundary value problem

$$\text{(P)} \qquad \begin{cases} -u'' = g(t,u) + h(t)u & t \in [0,\pi], \\ u(0) = u(\pi) = 0. \end{cases}$$

The following conditions on the function $g : \Omega \times \mathbb{R} \to \mathbb{R}$ are here imposed:
(i) g is a Carathéodory function such that the limits

$$\lim_{x \to +\infty} \frac{g(t,x)}{x} = g(+), \qquad \lim_{x \to -\infty} \frac{g(t,x)}{x} = g(-)$$

exist for almost all $t \in (0,\pi)$, and the closed interval $[g(+), g(-)]$ (or $[g(-), g(+)]$) contains the first (simple) eigenvalue of the problem

$$\text{(PO)} \qquad \begin{cases} -v'' = \lambda v, \\ v(0) = v(\pi) = 0 \end{cases}$$

with $v(t) \geq 0$ in $(0,\pi)$.
(ii) The function $t \mapsto \sup_{x \in \mathbb{R}} \left| \frac{g(t,x)}{x} \right|$ is (essentially) bounded.
(iii) The function $g(t,\cdot)$ is strictly convex and increasing.
Let Q denote the subset of $L^2((0,\pi))$ defined by

$$Q = \{u \in L^2((0,\pi)) : ||u|| + ||u'|| \leq \overline{M},\ u(0) = u(\pi) = 0\}.$$

It is easy to see that Q is closed and convex. Moreover, if ϕ is a normalized eigenfunction of problem (PO), let us denote by $\phi^\perp$ the orthogonal complement of ϕ in $L^2((0,\pi))$.

Theorem 6. *Assume that* (i), (ii), (iii) *hold true. Then, for almost all* $t \in (0,\pi)$, *for each* $h_0 \in \phi^\perp$ *and for each function* $h \in L^2((0,\pi))$ *such that* $hq = h_0 + r\phi$ ($r \in \mathbb{R}$, $q \in Q$) *there exists* $a \in \mathbb{R}$ *such that the boundary value problem*

$$\text{(P)}_q \qquad \begin{cases} -u'' = g(t,u) + h(t)q(t), \\ u(0) = u(\pi) = 0 \end{cases}$$

has exactly one solution if $r = a$ *and exactly two solutions if* $r < a$.

Finally, we consider some boundary value problems which are treated by using methods connected to implicit differential equations argument. In this case tools from the theory of multivalued mappings and differential inclusions are widely used. Here we want also to stress the use of a similar argument by focusing it through the use of a function $\mu(X)$, defined by K. Kuratowski [7] on the family of all bounded sets, which is some kind of "measure of noncompactness". The measures we are dealing with in this paper is generated by a derivative operator we shall introduce in the "nonsingular case" i.e., roughly speaking, when the boundary operator is invertible. In this paper, by using the theory of boundary value problems due to R. Conti [16], the (Penrose) generalized inverse of a matrix, measures of noncompactness, and a fixed point result for condensing multivalued mappings as crucial tools, we may solve the given boundary value problem in the singular case.

So we are dealing with boundary value problems for differential equations of the type

$$(\mathrm{BV}) \qquad \begin{cases} x'(t) = A(t)x(t) + g(t, x(t), x'(t)), & t \in I = [t_1, t_2] \subset \mathbb{R}, \\ Lx = r, \end{cases}$$

where g is a continuous function from $[t_1, t_2] \times \mathbb{R}^{2n}$ into $\mathbb{R}^n$ such that

(i) $g(t, x, \cdot)$ is a Lipschitz function, i.e.

$$|g(t, x, y_1) - g(t, x, y_2)| \le k|y_1 - y_2|, \qquad k > 0.$$

(ii) $A(t)$ is a continuous $n \times n$-matrix-valued function defined on I; it is known [16] that there exists a unique continuous matrix function $(t, s) \mapsto U(t, s)$ such that

$$U(t, s) = I + \int_s^t A(\tau)U(\tau, s)\, d\tau.$$

(iii) L is supposed to be a linear operator from $C([t_1, t_2], \mathbb{R}^n)$ into $\mathbb{R}^n$ describing the boundary conditions we want to associate to our differential problem; here we specialize the boundary condition by saying that any function which is a (possible) solution of the differential problem is required to satisfy a (boundary) condition of the type

$$Lx = M_1 x(t_1) + M_2 x(t_2) = r \in \mathbb{R}^n,$$

where M_1 and M_2 are constant $n \times n$-matrices.

We consider now two boundary value problems. The first one is a general boundary value problem we are investigating in order to obtain the existence of at least one solution; the second one is a "linear type" boundary value problem we need for solving the first one by using a fixed point theorem.

The general boundary value problem we already introduced is

$$(\mathrm{BV}) \qquad \begin{cases} x'(t) = A(t)x(t) + g(t, x(t), x'(t)), & t \in I = [t_1, t_2] \subset \mathbb{R}, \\ Lx = 0. \end{cases}$$

The "linear type" boundary value problem has the form

$$(\mathrm{BV})_q \qquad \begin{cases} x' = A(t)x + g(t, x, q'), & t \in I = [t_1, t_2] \subset \mathbb{R}, \\ Lx = 0, \end{cases}$$

where q belongs to some subset $Q \subset C^1(I, \mathbb{R}^n)$. The following hypothesis will be used in the sequel:

(H0): Assume that

$$\ker W = \ker [M_1 + M_2 U(t_2, t_1)] \neq \{0\}$$

and suppose that g, A and L satisfy the conditions (i), (ii) and (iii) with some $k \in [0, 1)$. Moreover, assume that the linear boundary value problem

$$\begin{cases} x'(t) = A(t)x(t) + g(t, w(t), q'(t)), & t \in I = [t_1, t_2] \\ Lx = 0 \end{cases}$$

has a solution for each $w \in C(I, \mathbb{R}^n)$ such that $Lw = 0$.

Now we are ready to state our main result:

Theorem 7. *Consider the boundary value problem* (BV). *Assume that the condition* (H0) *holds. Then, if the function* g *satisfies* $|g(t, x, y)| \leq \gamma$ *for some* $\gamma > 0$ *and for all* $(t, x, y) \in [t_1, t_2] \times \mathbb{R}^n \times \mathbb{R}^n$, *the boundary value problem* (BV) *has at least one solution.*

References

[1] G. Anichini: *Nonlinear problems for systems of differential equations,* Nonlin. Anal. TMA **1, 6** (1976/77), 691-699

[2] G. Anichini, G. Conti, P. Zecca: *Using solution sets for solving boundary value problems for ordinary differential equations,* Nonlin. Anal. TMA **15, 5** (1991), 465-472

[3] G. Anichini, G. Conti: *Boundary value problem with nonlinear boundary conditions,* Nonlinearity **1** (1988), 531-540

[4] G. Anichini, G. Conti: *Existence of solutions of a boundary value problem through the solution map of a linearized type problem,* Rend. Sem. Mat. Univ. Pol. Torino **48, 2** (1990), 149-159

[5] G. Anichini, G. Conti: *A direct approach to the existence of solutions of a boundary value problem for a second order differential system,* Diff. Equ. Dynam. Systems **3, 1** (1995), 23-34

[6] G. Anichini, G. Conti: *About the existence of solutions of a boundary value problem for a Carathéodory differential system,* Zeitschr. Anal. Anw. **16, 3** (1997), 621-630

[7] G. Anichini, G. Conti: *Boundary value problems for implicit ODE's in a singular case,* Diff. Equ. Dynam. Systems, to appear

[8] M. Aronszajn: *Le correspondent topologique de l'unicité dans la théorie des équations differentielles,* Ann. Math. **43** (1942), 730-738

[9] J. Bebernes, M. Martelli: *On the structure of the solution set for periodic boundary value problems,* Nonlin. Anal. TMA **4, 4** (1980), 821-830

[10] D. Bielawski, T. Pruszko: *On the structure of the set of solutions of a functional equation with application to boundary value problems,* Ann. Pol. Math. **53, 3** (1991), 201-209

[11] A. V. Bogatyrev: *Fixed points and properties of solutions of differential inclusions* [in Russian], Izv. Akad. Nauk SSSR **47, 4** (1983), 895-909

[12] A. I. Bulgakov, L. N. Lyapin: *Certain properties of the set of solutions of a Volterra-Hammerstein integral inclusion* [in Russian], Diff. Uravn. **14, 8** (1978), 1465-1472

[13] A. Carbone, G. Conti, G. Marino: *A nonlinear boundary value problem for multivalued differential systems,* Atti Sem. Mat. Fis. Univ. Modena **38** (1990), 493-509

[14] M. Cecchi, M. Furi, M. Marini: *About the solvability of ordinary differential equations with asymptotic boundary conditions*, Boll. Unione Mat. Ital. **4-C** (1985), 329-345

[15] M. Cecchi, M. Furi, M. Marini: *On continuity and compactness of some nonlinear operators associated with differential equations in noncompact intervals,* Nonlin. Anal. TMA **9, 2** (1985), 171-180

[16] R. Conti: *Recent trends in the theory of boundary value problems for ordinary differential equations,* Boll. Unione Mat. Ital. **22, 3** (1967), 135-178

[17] G. Conti, R. Iannaci: *Nonzero solutions of nonlinear systems of differential equations via fixed point theorems for multivalued maps,* Nonlin. Anal. TMA **6, 5** (1982), 415-421

[18] G. Conti, P. Nistri, P. Zecca: *Nonconvex set valued systems in Banach spaces,* Funk. Ekvac. **37, 1** (1994), 101-114

[19] G. Conti, P. Nistri, P. Zecca: *Systems of set valued equations in Banach spaces,* Lect. Notes Math. **1475**, Springer, Berlin 1990

[20] G. Conti, V. Obukhovskij, P. Zecca: *On the topological structure of the solution set for a semilinear functional differential inclusion in a Banach space,* in: Topology in Nonlinear Analysis, Banach Center Publ. **35**, Warszawa 1996, p. 159-169

[21] G. Darbo: *Estensione alle mappe ponderate del teorema di Lefschetz sui punti fissi*, Rend. Sem. Mat. Univ. Padova **31** (1961), 46-57

[22] J. L. Davy: *Properties of the solution set of a generalized differential equation,* Bull. Austral. Math. Soc. **6, 3** (1972), 379-398

[23] F. S. De Blasi, J. Myjak: *On the solution sets for differential inclusions,* Bull. Pol. Acad. Sci. **33** (1985), 17-23

[24] K. Deimling, M. Rao: *On solution sets of multivalued differential equations,* Appl. Anal. **30, 8** (1988), 129-135

[25] R. Dragoni, J. W. Macki, P. Nistri, P. Zecca: *Solution Sets of Differential Equations in Banach Spaces,* Pitman Research Notes Math. **342**, Longman, Harlow 1996

[26] L. Górniewicz: *On the solution sets of differential inclusions,* J. Math. Anal. Appl. **113** (1986), 235-244

[27] P. Hartman: *Ordinary Differential Equations*, J. Wiley & Sons, New York 1973

[28] M. Kamenskij, V. Obukhovskij, P. Zecca: *Methods of solution sets for a quasilinear functional-differential inclusion in a Banach space,* Diff. Equ. Dynam. Systems **4, 3-4** (1996), 339-350

[29] A. G. Kartsatos: *Nonzero solutions to boundary value problems for nonlinear systems,* Pacific J. Math. **53** (1974), 425-433

[30] J. Lasry, R. Robert: *Analyse nonlinéaire multivoque*, UER Math. de la decision **249**, Ed. Univ. Paris Dauphine, Paris 1979

[31] A. Margheri, P. Zecca: *Solution sets and boundary value problems in Banach spaces,* Topol. Methods Nonlin. Anal. **2, 1** (1993), 179-188

[32] A. Margheri, P. Zecca: *Solution sets of multivalued Sturm-Liouville problems in Banach Spaces,* Atti Acc. Naz. Lincei Cl. Sci. Fis. Mat. Nat. **5, 2** (1994), 161-166

[33] G. Marino: *Nonlinear boundary value problems for multivalued differential equations in Banach spaces,* Nonlin. Anal. TMA **14, 7** (1990), 545-558

[34] I. Massabò, P. Nistri, J. Pejsachowicz: *On the solvability of nonlinear equations in Banach spaces,* Lect. Notes Math. **886**, Springer, Berlin 1981

[35] J. J. Nieto: *Structure of the solution set for semilinear elliptic equations,* Coll. Math. Soc. Janos Bolyai **47** (1987), 799-807

[36] J. J. Nieto: *Aronszajn's theorem for some nonlinear Dirichlet problem,* Proc. Edinburgh Math. Soc. **31** (1988), 345-351

[37] P. Nistri, V. Obukhovskij, P. Zecca: *On the solvability of systems of inclusions involving noncompact operators,* Trans. Amer. Math. Soc. **342, 2** (1994), 543-562

[38] N. S. Papageorgiou: *On the solution set of differential inclusions in Banach spaces,* Appl. Anal. **25, 4** (1987), 319-329

[39] N. S. Papageorgiou: *On the solution set of differential inclusions with state constraints,* Appl. Anal. **31, 4** (1989), 279-289

[40] S. Plaskacz: *On the solution sets for differential inclusions,* Boll. Unione Mat. Ital. **6-A** (1992), 387-394

[41] E. ROVDEROVÁ: *Existence of monotone solution of a nonlinear equations,* J. Math. Anal. Appl. **192** (1995), 1-15

[42] E. SPANIER: *Algebraic Topology,* McGraw Hill, New York 1966

[43] S. SZUFLA: *Some properties of the solution set of ordinary differential equations,* Bull. Acad. Polon. Sci. **22, 7** (1974), 675-678

[44] S. SZUFLA: *On the structure of solution set of nonlinear equations,* in: Proc. Conf. Diff. Equ. Optimal Control, Zielona Gora 1989, p. 33-39

[45] A. A. TOLSTONOGOV: *On the structure of the solution set for differential inclusions in a Banach space* [in Russian], Mat. Sbornik **46** (1983), 1-15

[46] G. VIDOSSICH: *On the structure of the set of solutions of nonlinear equations,* J. Math. Anal. Appl. **34** (1971), 602-617

GIUSEPPE ANICHINI, Università di Firenze, Dipartimento di Matematica Applicata "G. Sansone", Via S. Marta 3, I-50139 Firenze, Italy; `anichini@dma.unifi.it`

GIUSEPPE CONTI, Università di Firenze, Istituto di Matematica, Via dell'Agnolo 14, I-50122 Firenze, Italy; `gconti@cesit1.unifi.it`

Progress in Nonlinear Differential Equations
and Their Applications, Vol. 40

On the Unique Solvability of Hammerstein Integral Equations with Non-Symmetric Kernels

JÜRGEN APPELL, ESPEDITO DE PASCALE, PETR P. ZABREJKO

Al caro amico Alfonso, con l'affetto di sempre (anche di più)

Summary: Some elementary results about Hammerstein integral equations with non-symmetric kernels are proved by means of Minty's monotonicity principle.

Key words: Hammerstein integral equation, Minty monotonicity principle, non-symmetric kernel, monotonicity constant, positivity constant, superposition operator, Orlicz space, ideal space.

Classification: 45G10, 47H30, 47H05, 47H17, 46E30.

Acknowledgement: This paper was written in the framework of a DFG project (Gz. AP 40-15/1). Financial support by the Deutsche Forschungsgemeinschaft is gratefully acknowledged.

The aim of this article is to prove a new result on the unique solvability of the Hammerstein integral equation

$$x(t) = \int_{\Omega} k(t,s)\, f(s, x(s))\, ds + g(t). \tag{1}$$

Here Ω is an arbitrary set, ds is a σ-additive and σ-finite measure on a σ-algebra of subsets of Ω, $k : \Omega \times \Omega \to \mathbb{R}$ is a kernel which defines a bounded linear integral operator

$$Kx(t) = \int_{\Omega} k(t,s) x(s)\, ds \tag{2}$$

in the space $L_2 = L_2(\Omega)$, and $f : \Omega \times \mathbb{R} \to \mathbb{R}$ is a Carathéodory function, i.e. $f(\cdot, u)$ is measurable on Ω for all $u \in \mathbb{R}$ and $f(s, \cdot)$ is continuous on $\mathbb{R}$ for almost all $s \in \Omega$. The function g is a given measurable function, and the function x is unknown.

The first results about the unique solvability of the equation (1) have been obtained by Hammerstein [4], see also [3]. In these results, the integral equation (1) was considered under the assumption that the kernel k is symmetric and positive definite, and the nonlinearity f is either Lipschitz continuous or monotone with respect to u on the whole axis $\mathbb{R}$. Later, these results have been generalized building on Minty's important theory of monotone operators in Hilbert spaces ([11], see also [10,13]). The symmetry of the kernel k was a standard assumption in these articles; the case of non-symmetric kernels was studied later, for example, in [1,5]. Our Theorem 1 below is similar to some results from [1,5] but seems to be somewhat simpler and more natural.

In this article we are going to show that Minty's approach applies as well to Hammerstein integral equations with non-symmetric kernels. For simplicity, we formulate our results in the setting of Lebesgue spaces $L_p = L_p(\Omega)$ $(1 \leq p \leq \infty)$; it is possible without essential changes to use Orlicz spaces instead, or even abstract ideal spaces which have been considered by many authors.

Below we denote the scalar product in L_2 by $\langle \cdot, \cdot \rangle$, i.e.

$$\langle h_1, h_2 \rangle = \int_\Omega h_1(s) h_2(s)\, ds.$$

Suppose that the operator (2) acts not only in the space L_2 but also from $L_{p'}$ into L_p, where $2 \leq p \leq \infty$ and $p' = p/(p-1)$ as usual, including $\infty' = 1$. Let

$$A = \frac{1}{2}(K + K^*) \tag{3}$$

denote the selfadjoint part of K, and assume that this operator is positive definite. Similarly, let

$$B = \frac{1}{2}(K - K^*) \tag{4}$$

be the skew-adjoint part of K. Under our assumptions, both A and B act from the space $L_{p'}$ into the space L_p. It is well-known that the operator A can be represented in the form $A = CC^*$, where $C = A^{1/2}$ is the square root of A acting from L_2 into L_p, and the adjoint operator C^* acts from $L_{p'}$ into L_2.

We call the operator (2) *$\mathcal{P}$-positive* if the operator $C^{-1}K(C^*)^{-1}$ exists and is bounded in L_2, and *$\mathcal{S}$-positive* if the operator $K(C^*)^{-1}$ exists and is bounded in L_2. It is evident that each $\mathcal{P}$-positive operator is $\mathcal{S}$-positive but not vice versa. Observe that the operator (2) is $\mathcal{P}$-positive if and only if $C^{-1}B(C^*)^{-1}$ is bounded in L_2, and $\mathcal{S}$-positive if and only if $B(C^*)^{-1}$ is bounded in L_2.

The class of $\mathcal{P}$-positive operators was apparently introduced by Petryshyn, that of $\mathcal{S}$-positive operators by Samarskij (see [12]). We point out that $\mathcal{P}$-positivity is equivalent to *angle-boundedness* of K, i.e.

$$|\langle Kh_1, h_2 \rangle - \langle h_1, Kh_2 \rangle| \leq a \sqrt{\langle Kh_1, h_1 \rangle} \sqrt{\langle Kh_2, h_2 \rangle} \quad (h_1, h_2 \in L_2).$$

All these operator classes have been studied in detail in [10], see also [16].

Below by M and N we denote the closure of the operators $C^{-1}K(C^*)^{-1}$ and $K(C^*)^{-1}$, respectively, in L_2. Both operators M and N are defined on the closure (in L_2) of the range of $C = A^{1/2}$; this closure in our situation coincides with L_2. It is essential for us that, for $\mathcal{P}$-positive operators K, the decompositions

$$K = CMC^*, \qquad K = NC^* \tag{5}$$

hold. Moreover, K, M, and N are related by the the equalities

$$N = CM, \qquad N^* = M^*C^*, \tag{6}$$

and we have $\langle Mh, h\rangle = ||h||^2$ for all $h \in L_2$. The latter equality implies, in particular, that both operators M and M^* have a trivial nullspace.

In what follows the number

$$\mu(K) = ||N||^2$$

will be important. It is easy to see that $\mu(K)$ is the *positivity constant* of K in the sense of Krasnosel'skij [10], i.e. the smallest number μ for which

$$\langle Kh, Kh\rangle \leq \mu \langle Kh, h\rangle \qquad (h \in L_2).$$

Assume that the nonlinearity f in (1) generates a superposition operator

$$Fx(s) = f(s, x(s)) \tag{7}$$

from L_p into $L_{p'}$. Due to the famous Krasnosel'skij theorem [7] this is, in case $1 \leq p < \infty$, equivalent to the fact that the function f satisfies a growth estimate

$$|f(s,u)| \leq a(s) + b|u|^{p-1} \qquad (s \in \Omega,\ u \in \mathbb{R}) \tag{8}$$

with some $a \in L_{p'}$ and $b \geq 0$. In case $p = \infty$ the estimate (8) has to be replaced by

$$|f(s,u)| \leq a_n(s) \qquad (s \in \Omega,\ |u| \leq n),$$

where $a_n \in L_1$ may depend on n. We suppose in addition that there exists a number c such that

$$(f(s,u) - f(s,v))(u-v) \leq c(u-v)^2 \qquad (s \in \Omega,\ u \in \mathbb{R}). \tag{9}$$

The condition (9) is equivalent to the the fact that the function $u \mapsto cu - f(s,u)$ is monotonically increasing for almost all $s \in \Omega$. The smallest number c for which (9) holds will be denoted by $\gamma(f)$ and called the *monotonicity constant* of f in the sequel.

Theorem 1. *Suppose that K is $\mathcal{P}$-positive in L_2, f satisfies the Carathéodory conditions with $f(s,0) = 0$, and the inequality*

$$\gamma(f)\,\mu(K) < 1 \tag{10}$$

holds. Then the Hammerstein integral equation (1) *has, for arbitrary $g \in N(L_2)$, a solution $x_* \in N(L_2)$. If $g = Nk$ for $k \in L_2$, then $x_* = Nh_*$ for some $h_* \in L_2$ with*

$$||h_*|| \leq \frac{||k||}{1 - \gamma(f)\,\mu(K)}; \tag{11}$$

moreover, this solution is unique in the space L_p.

Proof. Consider the auxiliary operator equation

$$\Phi h = 0 \tag{12}$$

in L_2, where

$$\Phi h = M^*h - N^*FNh - M^*k. \tag{13}$$

If h_* is a solution of (12), i.e. $M^*h_* = N^*FNh_* + M^*k$, then $M^*(h_* - C^*FNh_* - k) = 0$, by (6), hence $h_* = C^*FNh_* + k$, since M^* has a trivial nullspace. Applying N to the last equality we get

$$Nh_* = NC^*FNh_* + Nk = KFNh_* + g,$$

by (6). Consequently, $x_* = Nh_*$ is a solution of (1).

Thus, in order to prove the existence of solution for the equation (1) under the hypotheses of Theorem 1, it suffices to study the equation (12) with Φ given by (13). We claim that the operator (13) is monotone in Minty's sense. In fact, for all $h_1, h_2 \in L_2$ we have

$$\begin{aligned}\langle \Phi h_1 - \Phi h_2, h_1 - h_2\rangle &= \langle M^*h_1 - M^*h_2 - N^*FNh_1 + N^*FNh_2, h_1 - h_2\rangle \\ &= \langle M^*(h_1 - h_2), h_1 - h_2\rangle - \langle FNh_1 - FNh_2, Nh_1 - Nh_2\rangle \\ &\geq ||h_1 - h_2||^2 - \gamma(f)\,\langle Nh_1 - Nh_2, Nh_1 - Nh_2\rangle \\ &\geq ||h_1 - h_2||^2 - \gamma(f)\,\mu(K)\,||h_1 - h_2||^2.\end{aligned}$$

Thus,

$$\langle \Phi h_1 - \Phi h_2, h_1 - h_2\rangle \geq (1 - \gamma(f)\,\mu(K))\,||h_1 - h_2||^2.$$

Furthermore, for all $h \in L_2$ on the sphere $||h|| = r$ we have, since $\Phi 0 = -k$,

$$\langle \Phi h, h\rangle = \langle \Phi h - \Phi 0, h - 0\rangle + \langle \Phi 0, h\rangle \geq$$

$$\geq (1 - \gamma(f)\,\mu(K))\,||h||^2 - ||k||\,||h|| = (1 - \gamma(f)\,\mu(K))\,r^2 - ||k||\,r.$$

Consequently, $\langle \Phi h, h\rangle \geq 0$ on the sphere $||h|| = r$, provided we choose

$$r \geq \frac{||k||}{1 - \gamma(f)\,\mu(K)}.$$

Now, due to Minty's existence principle [10,11,13], equation (12) has a unique solution $h_* \in L_2$. By what we have shown above, $x_* = Nh_*$ is then a solution of the Hammerstein equation (1), and so we have proved existence.

Concerning uniqueness, suppose that x_* and x_{**} are two solutions of equation (1) with $g = Nk$ for some $k \in L_2$. Let

$$h_* = C^*Fx_* + k, \qquad h_{**} = C^*Fx_{**} + k.$$

It is easily seen that both elements h_* and h_{**} belong to L_2, by (6), and that

$$h_* = C^*FNh_* + k, \qquad h_{**} = C^*FNh_{**} + k.$$

By (13), this means that $\Phi h_* = \Phi h_{**} = 0$. Since equation (12) has only one solution, we conclude that $h_* = h_{**}$, hence $x_* = x_{**}$. □

In Theorem 1 we supposed that the selfadjoint part (3) of the linear integral operator K is positive definite. However, the methods described above carry over to some classes of operators which are not necessarily positive definite. We restrict ourselves to the case when A is *quasi-positive definite*; this means that A has at most a finite number of negative eigenvalues of finite multiplicity.

Assume again that the operator (2) acts not only in the space L_2 but also from $L_{p'}$ into L_p, where p and p' are as above, and let A and B be defined as in (3) and (4). Under our assumptions, both A and B act between $L_{p'}$ and L_p.

Suppose that the operator A is quasi-positive definite in L_2. In this case we can consider the finite-dimensional orthogonal projection P of L_2 onto the subspace of eigenvectors of A which correspond to negative eigenvalues of A. This projection commutes with A (but not necessarily with B); moreover, P acts both in L_p and $L_{p'}$. The operator

$$|A| = (I - 2P)A \tag{14}$$

is then positive definite. As the operator A in the previous section, the operator $|A|$ can be represented in the form $|A| = DD^*$, where $D = |A|^{1/2}$ acts from L_2 into L_p, and D^* acts from $L_{p'}$ into L_2.

Let us call an operator K *$\mathcal{P}$-quasi-positive* if the operator $D^{-1}K(D^*)^{-1}$ exists and is bounded in L_2, and *$\mathcal{S}$-quasi-positive* if the operator $K(D^*)^{-1}$ exists and is bounded in L_2. Below we denote by M and N again the closure (in L_2) of the bounded operators $D^{-1}K(D^*)^{-1}$ and $K(D^*)^{-1}$, respectively. Both operators are defined on the whole space L_2, and

$$K = DMD^*, \qquad K = ND^*. \tag{15}$$

Moreover, we have $N = DM$, $N^* = M^*D^*$, and $\langle Mh, h\rangle = ||h||^2 - 2||Ph||^2$ for all $h \in L_2$.

In what follows we use the number

$$\nu(K) = \sup\{\nu : \nu > 0,\ ||Nh|| \geq \sqrt{\nu}\,||Ph||\ (h \in L_2)\}. \tag{16}$$

It easy to see that, in the case of a selfadjoint operator K in L_2, the number (16) is just the absolute value of the largest negative eigenvalue of K. For an operator K which is not selfadjoint, the number (16) cannot be interpreted in such an easy way.

Suppose again that the function f satisfies the Carathéodory conditions and generates a superposition operator (7) between L_p and $L_{p'}$ satisfying the additional condition (9). As before, the smallest number c for which (9) holds will be denoted by $\gamma(f)$; we point out, however, that $\gamma(f)$ may now be negative.

Theorem 2. *Suppose that K is $\mathcal{P}$-quasi-positive in L_2, f satisfies the Carathéodory conditions with $f(s,0) = 0$, and the inequality*

$$\gamma(f)\,\nu(K) < -1 \tag{17}$$

holds. Then the Hammerstein integral equation (1) *has, for arbitrary $g \in N(L_2)$, a solution $x_* \in N(L_2)$. If $g = Nk$ for $k \in L_2$, then $x_* = Nh_*$ for some $h_* \in L_2$ with*

$$||h_*|| \leq -\frac{||k||}{1 + \gamma(f)\,\nu(K)}; \tag{18}$$

moreover, this solution is unique in the space L_p.

Proof. The proof is almost identical with that of Theorem 1, building on the analysis of the auxiliary equation (12) with Φ defined by (13). The essential difference consists in the use of Minty's monotonicity method.

In fact, for all $h_1, h_2 \in L_2$, we have now

$$\begin{aligned}\langle \Phi h_1 - \Phi h_2, h_1 - h_2\rangle &= \langle M^*h_1 - M^*h_2 - N^*FNh_1 + N^*FNh_2, h_1 - h_2\rangle \\ &= \langle M^*(h_1 - h_2), h_1 - h_2\rangle - \langle FNh_1 - FNh_2, Nh_1 - Nh_2\rangle \\ &\geq ||h_1 - h_2||^2 - 2||P(h_1 - h_2)||^2 - \gamma(f)\,\langle Nh_1 - Nh_2, Nh_1 - Nh_2\rangle \\ &\geq ||h_1 - h_2||^2 - 2||P(h_1 - h_2)||^2 - \gamma(f)\,||N(h_1 - h_2)||^2 \\ &\geq ||h_1 - h_2||^2 - 2||P(h_1 - h_2)||^2 - \gamma(f)\,\nu(K)\,||P(h_1 - h_2)||^2 \\ &= ||h_1 - h_2||^2 - (2 + \gamma(f)\,\nu(K))\,||P(h_1 - h_2)||^2 \geq -(1 + \gamma(f)\,\nu(K))\,||h_1 - h_2||^2.\end{aligned}$$

It remains to remark that for all $h \in L_2$ on the sphere $||h|| = r$ we get, since $\Phi 0 = -k$,

$$\begin{aligned}\langle \Phi h, h\rangle &= \langle \Phi h - \Phi 0, h - 0\rangle + \langle \Phi 0, h\rangle \\ &\geq -(1 + \gamma(f)\,\nu(K))\,||h||^2 - ||k||\,||h|| \\ &= -(1 + \gamma(f)\,\nu(K))\,r^2 - ||k||\,r.\end{aligned}$$

Consequently, $\langle \Phi h, h\rangle \geq 0$ on the sphere $||h|| = r$, provided we choose

$$r \geq -\frac{||k||}{1 + \gamma(f)\,\nu(K)}.$$

The last part of the proof literally repeats the corresponding part of the proof of Theorem 1. □

It is evident that one can repeat all ideas and constructions given above in the case when the superposition operator (7) acts from an Orlicz space L_M into its dual space L_{M^*}, where M and M^* are mutually conjugate Young functions, see e.g. [8,9]. In case of a finite atomic-free measure ds the use of Orlicz spaces makes it possible to consider Hammerstein integral equations with nonlinearities f of arbitrary growth with respect to u, but under the more restrictive assumption that k is the kernel of a linear integral operator (2) acting from L_{M^*} into L_M (see [8,9]).

In the same way one can also repeat the above reasoning in the case when the superposition operator (7) acts from a so-called ideal space E into its Köthe dual E', and k is the kernel of a linear integral operator (2) acting from E' into E (see e.g. [2,6,14,15]).

Finally, there are no difficulties in generalizing Theorems 1 and 2 to the case of systems of Hammerstein equations. Such systems can be written in the same form (1). However, in this case k is a matrix kernel (i.e. a function from $\Omega \times \Omega$ into $\mathbb{R}^{m\times m}$) which defines a bounded linear operator in the space L_2 of vector functions on Ω with values in $\mathbb{R}^m$, and f is a function from $\Omega \times \mathbb{R}^m$ into $\mathbb{R}^m$. Almost all constructions remain unchanged in this case; the only exception is that the scalar inequality (9) must be replaced by the condition

$$(f(s,u) - f(s,v), u - v) \leq c\,||u - v||^2 \qquad (s \in \Omega,\ u \in \mathbb{R}^m),$$

with $(\cdot,\cdot)$ denoting the scalar product in $\mathbb{R}^m$. Of course, all results for systems can be proved as well in the setting of Orlicz spaces or general ideal spaces.

References

[1] H. AMANN: *Ein Existenz- und Eindeutigkeitssatz für die Hammersteinsche Gleichung in Banachräumen*, Math. Zeitschr. **111** (1965), 175-190

[2] J. APPELL, P. P. ZABREJKO: *Nonlinear Superposition Operators*, Cambridge University Press, Cambridge 1990

[3] C. L. DOLPH, G. J. MINTY: *On nonlinear integral equations of Hammerstein type*, in: *Nonlinear Integral Equations* [Ed.: P. M. ANSELONE], Univ. of Wisconsin, Madison 1974, 99-154

[4] A. HAMMERSTEIN: *Nichtlineare Integralgleichungen nebst Anwendungen*, Acta Math. **54** (1930), 117-176

[5] P. HESS: *On nonlinear equations of Hammerstein type in Banach spaces*, Proc. Amer. Math. Soc. **30, 2** (1971), 308-312

[6] L. V. KANTOROVICH, G. P. AKILOV: *Functional Analysis* [in Russian], Fizmatgiz, Moscow 1978; Engl. transl.: Pergamon Press, Oxford 1982

[7] M. A. KRASNOSEL'SKIJ: *Topological Methods in the Theory of Nonlinear Integral Equations* [in Russian], Gostekhizdat, Moscow 1956; Engl. transl.: Macmillan, New York 1964

[8] M. A. KRASNOSEL'SKIJ, JA. B. RUTITSKIJ: *Convex Functions and Orlicz Spaces* [in Russian], Fizmatgiz, Moscow 1958; Engl. transl.: Noodhoff, Groningen 1961

[9] M. A. KRASNOSEL'SKIJ, JA. B. RUTITSKIJ: *Orlicz spaces and nonlinear integral equations* [in Russian], Trudy Moskov. Matem. Obshch. **7** (1958), 63-120

[10] M. A. KRASNOSEL'SKIJ, P. P. ZABREJKO: *Geometrical Methods of Nonlinear Analysis* [in Russian], Nauka, Moscow 1975; Engl. transl.: Springer, Berlin 1984

[11] G. MINTY: *Monotone nonlinear operators in Hilbert spaces*, Duke Math. J. **29** (1962), 341-346

[12] D. PASCALI, S. SBURLAN: *Nonlinear Mappings of Monotone Type*, Edit. Acad., Bucharest 1978

[13] M. M. VAJNBERG: *The Variational Method and Method of Monotone Operators in the Theory of Nonlinear Equations* [in Russian], Nauka, Moscow 1972; Engl. transl.: Halsted Press, Jerusalem 1973

[14] P. P. ZABREJKO, A. I. POVOLOTSKIJ: *Remarks on existence theorems for solutions of Hammerstein equations* [in Russian], Uchen. Zap. Leningrad. Gos. Ped. Inst. **404** (1971), 374-379

[15] P. P. Zabrejko, A. I. Povolotskij: *The Hammerstein operator and Orlicz spaces* [in Russian], Kachestv. Pribl. Metody Issled. Operat. Uravn. (Jaroslavl') **2** (1977), 39-51

[16] P. P. Zabrejko, A. I. Povolotskij, E. I. Smirnov: *On two classes of linear operators in Hilbert spaces* [in Russian], Kachestv. Pribl. Metody Issled. Operat. Uravn. (Jaroslavl') **7** (1982), 90-93

Jürgen Appell, Universität Würzburg, Mathematisches Institut, Am Hubland, D-97074 Würzburg, Germany; appell@mathematik.uni-wuerzburg.de

Espedito De Pascale, Università della Calabria, Dipartimento di Matematica, I-87036 Arcavacata di Rende (CS), Italy; depascal@pobox.unical.it

Petr P. Zabrejko, Belgosuniversitet, Matematicheskij Fakul'tet, pr. F. Skoriny 4, BY-220050 Minsk, Belorussia; zabreiko@mmf.bsu.unibel.by

Progress in Nonlinear Differential Equations
and Their Applications, Vol. 40

The Invariance of Domain for C^1 Fredholm Maps of Index Zero

PIERLUIGI BENEVIERI, MASSIMO FURI, MARIA PATRIZIA PERA

Dedicated to Alfonso Vignoli on the occasion of his 60th birthday

Summary: We give a version of the classical Invariance of Domain Theorem for nonlinear Fredholm maps of index zero between Banach spaces (and Banach manifolds). The proof is based on a finite dimensional reduction technique combined with a mod 2 degree argument for continuous maps between (finite dimensional) differentiable manifolds.

Keywords: Domain invariance, nonlinear Fredholm maps, mod 2 degree.

Classification: 58C25, 47H11.

1. Introduction. Is the one-to-one continuous image of an open set in $\mathbb{R}^n$ still open? This problem is the so called *invariance of domain* and, as H. Freudenthal pointed out in his essay on the history of topology [6] (see also [12]), it has been considered of a great importance at the beginning of this (20th) century "for it was essential for the justification of the Poincaré continuation method in the theory of automorphic functions and the uniformization (of analytic functions)."

As it is well-known, a positive answer to the problem has been given by Brouwer in 1912 (see [3]). This result is usually obtained as a consequence of the Odd Mapping Theorem, as it is shown, for instance, in [5]. A simpler proof can be done by using the Jordan Separation Theorem (see, e.g., [7] or [10]).

The extension of the Domain Invariance Theorem to Banach spaces is due to Schauder [9], who proved that if Ω is a bounded open subset of a Banach space E and $f : \overline{\Omega} \to E$ is a one-to-one map of the form $f = I - h$ with I the identity of E and $h : \overline{\Omega} \to E$ compact, then $f(\Omega)$ is open. Again this result can be deduced by the infinite dimensional version of the Odd Mapping Theorem. A proof displaying this connection can be found, for instance, in [2,5] or [7].

The result of Schauder is, clearly, a nonlinear counterpart of the well-known Fredholm alternative.

In this paper, we present a version of the Invariance of Domain Theorem for nonlinear Fredholm maps of index zero between Banach spaces (or, more generally, Banach manifolds). To prove our result, we use a finite dimensional reduction method for Fredholm maps that goes back to Caccioppoli (see [4]). In his paper, Caccioppoli defines a mod 2 degree for such type of maps, in order to give an answer to this question:

When does a point y_0 belong to the interior of the image of a Fredholm map of index zero?

His answer is the following:

Let E and F be Banach spaces, $f : \Omega \to F$ a Fredholm map of index zero defined on an open subset Ω of E, and $y_0 \in f(\Omega)$. If the mod 2 degree of f in Ω at y_0 is nonzero, then $f(\Omega)$ is a neighborhood of y_0.

However, it should be observed that Caccioppoli does not provide sufficient conditions for the degree to be different from zero. It is a matter of definition that, if f is one-to-one and y is a regular value of f, belonging to $f(\Omega)$ and sufficiently close to y_0, then the mod 2 degree of f in Ω at y_0 is equal to 1. On the other hand, it seems to be a nontrivial fact to prove the existence of a regular value belonging to the image of f, unless one preliminarily shows that such an image has an interior point. Actually, this is our aim here.

2. Preliminaries. Let E and F be two real Banach spaces. We recall that a bounded linear operator is said to be *Fredholm* if both $\operatorname{Ker} L$ and $\operatorname{coKer} L$ have finite dimension. In this case its *index* is the integer

$$\operatorname{ind} L = \dim \operatorname{Ker} L - \dim \operatorname{coKer} L.$$

A map $f : M \to N$ between real Banach manifolds is Fredholm of index zero (see [11]) if it is C^1 and its Fréchet derivative $f'(x)$, from the tangent space T_xM of M at x to the tangent space $T_{f(x)}N$ of N at $f(x)$, is Fredholm of index zero for any $x \in M$.

A map $f : M \to N$ between manifolds is said to be *proper* if $f^{-1}(K)$ is compact for any compact subset K of N. In particular, let us recall that Fredholm maps are locally proper (see [11]).

The proof of the Invariance of Domain Theorem we present here is based on a notion of mod 2 degree for continuous maps between (finite dimensional) differentiable manifolds. A classical version of such a degree can be found, for example, in the textbook of Milnor [8], where the author takes into account continuous maps between compact manifolds.

The notion of degree we use in this paper is a straightforward extension of the mod 2 degree in [8] and it is defined in the following context. Let $f : M \to N$ be a continuous map between two differentiable boundaryless manifolds of the same dimension. Given an open subset U of M and a point $y \in N$, the triple (f, U, y) is called *admissible for the mod 2 degree*, if $f^{-1}(y) \cap U$ is compact.

The mod 2 degree, denoted by $\deg_2$, is a function defined in the class of all admissible triples, taking values in $\mathbb{Z}_2$, and verifying the following properties:

(i) (*Normalization*) If f is a homeomorphism of M onto N, then

$$\deg_2(f, M, y) = 1,$$

for all $y \in N$.

(ii) (*Additivity*) If (f, M, y) is an admissible triple and U_1, U_2 are two open disjoint subsets of M such that $f^{-1}(y) \subseteq U_1 \cup U_2$, then

$$\deg_2(f, M, y) = \deg_2(f, U_1, y) + \deg_2(f, U_2, y).$$

(iii) (*Homotopy invariance*) Let $H : M \times [0,1] \to N$ be a continuous homotopy. Then, given any (continuous) path $y : [0,1] \to N$, such that the set $\{(x,t) \in M \times [0,1] : H(x,t) = y(t)\}$ is compact, $\deg_2(H(\cdot,t), M, y(t))$ does not depend on t.
(iv) (*Topological invariance*) Let (f, M, y) be admissible. Given two differentiable manifolds W and Z with two homeomorphisms $\phi : M \to W$ and $\psi : N \to Z$, one has

$$\deg_2(f, M, y) = \deg_2(\psi \circ f \circ \phi^{-1}, W, \psi(y)).$$

The following consequences of the additivity property will be used in the proof of our main result.
(v) (*Excision*) If (f, M, y) is admissible and U is an open neighborhood of $f^{-1}(y)$, then

$$\deg_2(f, M, y) = \deg_2(f, U, y).$$

(vi) (*Existence*) If (f, M, y) is admissible and $\deg_2(f, M, y) = 1$, then the equation $f(x) = y$ admits a solution in M.

3. The invariance of domain. The following is our version of the invariance of domain for Fredholm maps.

Theorem 1. *Let M and N be two real Banach manifolds and $f : M \to N$ be an injective Fredholm map of index zero. Then $f(M)$ is open in N.*

Proof. The above statement is clearly equivalent to the following. If E and F are real Banach spaces, U is an open subset of E and $f : U \to F$ is an injective Fredholm map of index zero, then $f(U)$ is open. Thus, we will prove our result directly in the context of Banach spaces.

Take $y_0 \in f(U)$ and denote $x_0 = f^{-1}(y_0)$. There exists an open neighborhood V of x_0 in E such that $\overline{V} \subset U$ and f is proper on $\overline{V}$ (recall that Fredholm maps are locally proper). Let F_0 be any finite dimensional subspace of F through y_0 such that $F_0 + \text{Range} f'(x_0) = F$, i.e. f is transverse to F_0 at x_0. Without loss of generality, we may assume that f is transverse to F_0 at any $x \in V$ (see e.g. [1]). Since $f(\partial V)$ is closed in F and does not contain y_0, there exists an open ball B centered at y_0 such that $\overline{B} \cap f(\partial V) = \emptyset$. We will show that B is contained in $f(U)$. Let $y \in B$ and denote $Y = \text{span}(F_0 \cup \{y\})$. The map f is still transverse to the subspace Y at any $x \in V$ and, consequently, the set $f^{-1}(Y) \cap V$ is a submanifold of U (containing x_0) of the same dimension as Y (again see e.g. [1]). Denote $W = f^{-1}(Y) \cap V \cap f^{-1}(B)$ and observe that the restriction $f|_W : W \to Y$ is a continuous (actually C^1) map between two manifolds of the same dimension with $(f|_W)^{-1}(y_0)$ compact, since it reduces to $\{x_0\}$. Therefore, as pointed out in the preliminaries, the mod 2 degree, $\deg_2(f|_W, W, y_0)$, is well defined.

Let us show that $\deg_2(f|_W, W, y_0) = 1$. By excision, to compute this degree it suffices to restrict f to a coordinate neighborhood W_1 of x_0 in W. Hence, using local charts, by the topological invariance, one can replace the triple $(f|_{W_1}, W_1, y_0)$ with a triple (f_1, Ω, η), where Ω is a bounded open subset of $\mathbb{R}^n$, $f_1 : \Omega \to \mathbb{R}^n$ is injective, and $\eta \in f_1(\Omega)$. In this situation, as it is well-known (see e.g. [7]), the oriented Brouwer degree is ± 1. Consequently, the mod 2 degree, $\deg_2(f_1, \Omega, \eta)$, is equal to 1 and, thus, $\deg_2(f|_W, W, y_0) = 1$ as well, as claimed.

Now, since B is a ball containing y_0 and y, the line segment Γ joining y_0 and y is contained in $B \cap Y$. Moreover, $(f|_W)^{-1}(\Gamma)$ is a compact subset of W (observe, in fact, that $\Gamma \cap f(\partial W) = \emptyset$). Thus, by the homotopy invariance of the degree, $\deg_2(f|_W, W, (1-t)y_0+ty)$ is well defined and independent of t. Consequently, $\deg_2(f|_W, W, y) = 1$, which implies, by the existence property, that $(f|_W)^{-1}(y)$ is nonempty. □

Obviously, as a consequence of Theorem 1, we get the following nonlinear Fredholm alternative:

If $f : M \to N$ is Fredholm of index zero, $x_0 \in M$, and the equation $f(x) = y$ has at most one solution (close to x_0) for any y in a convenient neighborhood of $y_0 = f(x_0)$, then $f(x) = y$ is (uniquely) solvable for y sufficiently close to y_0.

An easy improvement of our previous result is the following.

Theorem 2. *Let M be a real Banach manifold, F a real Banach space, $f : M \to F$ the sum of two maps, g and h, where g is Fredholm of index zero and h is continuous with $h(M)$ in a finite dimensional space. Assume that f is injective. Then $f(M)$ is open in F.*

Sketch of the proof. The proof is in the outline of that of Theorem 1, with only minor differences. More precisely, in the same notation as above, choose F_0 containing $\{y_0\} \cup h(M)$ and such that $F_0 + \operatorname{Range} g'(x_0) = F$ (observe that, F_0 being a vector space, $g(x_0) \in F_0$). In addition, consider the manifold $g^{-1}(Y) \cap V$ and take $W = g^{-1}(Y) \cap V \cap (f)^{-1}(B)$. □

We close the paper with a question whose answer we believe is affirmative, but, as far as we know, unknown.

Does the above domain invariance result hold if the perturbation h of g, instead of having finite dimensional image, is assumed to be only locally compact?

References

[1] R. Abraham, J. Robbin: *Transversal Mappings and Flows*, Benjamin, New York 1967

[2] M. Berger: *Nonlinearity and Functional Analysis*, Academic Press, New York 1977

[3] L. Brouwer: *Zur Invarianz des n-dimensionalen Gebietes*, Math. Ann. **71** (1912), 305-313

[4] R. Caccioppoli: *Sulle corrispondenze funzionali inverse diramate: teoria generale e applicazioni ad alcune equazioni funzionali non lineari e al problema di Plateau I/II*, Rend. Accad. Naz. Lincei **24** (1936), p. 258-263 and 416-421 [= Opere scelte, Vol. 2, Cremonese, Roma 1963, p. 157-177]

[5] K. Deimling: *Nonlinear Functional Analysis*, Springer, Berlin 1985

[6] H. Freudenthal: *Die Topologie in historischen Durchblicken*, in: Überblicke Mathematik [Ed.: D. Laugwitz] **4**, Bibliogr. Institut, Mannheim 1971, p. 7-24

[7] N. LLOYD: *Degree Theory*, Cambridge Univ. Press, Cambridge 1978

[8] J. MILNOR: *Topology from the Differentiable Viewpoint*, Univ. Virginia Press, Charlottesville 1965

[9] J. SCHAUDER: *Invarianz des Gebiets in Funktionalräumen*, Studia Math. **1** (1929), 123-139

[10] J. SCHWARTZ: *Nonlinear Functional Analysis*, Gordon & Breach, New York 1969

[11] S. SMALE: *An infinite dimensional version of Sard's theorem*, Amer. J. Math. **87** (1965), 861-866

[12] E. ZEIDLER: *Nonlinear Functional Analysis and its Applications I*, Springer, Berlin 1986

PIERLUIGI BENEVIERI, Dipartimento di Matematica Applicata, Università di Firenze, Via S. Marta 3, I-50139 Firenze, Italy; benevieri@dma.unifi.it

MASSIMO FURI, Dipartimento di Matematica Applicata, Università di Firenze, Via S. Marta 3, I-50139 Firenze, Italy; furi@dma.unifi.it

MARIA PATRIZIA PERA, Dipartimento di Matematica Applicata, Università di Firenze, Via S. Marta 3, I-50139 Firenze, Italy; pera@dma.unifi.it

Progress in Nonlinear Differential Equations
and Their Applications, Vol. 40

Positive Eigenfunctions for Some Unbounded Differential Operators

LUCIO BOCCARDO

Dedicated to Alfonso Vignoli on the occasion of his 60th birthday

... y no reconoció otra autoridad que la suya ni más servidumbre que la de su obsesión. Gabriel García Márquez

Summary: We study the nonlinear differential eigenvalue problem (3), where the nonlinear matrix function $M(x,s)$ can be unbounded and with degenerate coercivity with respect to the variable s.

Keywords: Nonlinear eigenvalue problem, nonlinear differential equation.

Classification: 35P30, 35J65.

1. Introduction. Twenty years ago, thanks to the "moral support" of Alfonso Vignoli, I proved the existence of positive eigenfunctions for some nonlinear differential eigenvalue problems. Here, I shall prove the same result in a more general setting.
Let Ω be a bounded domain of $\mathbb{R}^N$, $N \geq 1$, $M(x,s)$ a Caratheodory function defined in $\Omega \times \mathbb{R}$ with values in $\mathbb{R}^{N\times N}$ such that, for every $k > 0$,

$$(1) \qquad \begin{cases} M(x,s) = M^*(x,s), \\ \alpha_k \,|\xi|^2 \leq M(x,s)\xi \cdot \xi,\ \forall \xi \in \mathbb{R}^N,\ \forall s \in [0,k], \\ |M(x,s)| \leq \beta_k,\ \forall s \in [0,k], \end{cases}$$

where $\alpha_k, \beta_k > 0$, and define the differential operator on $W_0^{1,2}(\Omega)$

$$(2) \qquad Q(v) = -\operatorname{div}(M(x,v)\nabla v).$$

We point out that Q is not well defined on the whole $W_0^{1,2}(\Omega)$, since the function $x \mapsto M(x,v)\nabla v$ may not even belong to $L^1(\Omega)$.
Consider, for $r \in \mathbb{R}^+$ fixed, the following eigenvalue problem:

$$(3) \qquad \begin{cases} (\lambda_r, u_r) \in \mathbb{R}^+ \times W_0^{1,2}(\Omega) : \\ -\operatorname{div}(M(x,u_r)\nabla u_r) = \lambda_r\, u_r & \text{in } \Omega, \\ u_r > 0 & \text{on } \Omega, \\ u_r = 0 & \text{on } \partial\Omega, \\ \|u_r\|_{L^2(\Omega)} = r. \end{cases}$$

In [1], under the assumption

$$\alpha_k = \alpha, \quad \beta_k = \beta, \ \forall k > 0, \tag{4}$$

the existence of (λ_r, u_r) is proved, for any fixed $r \in \mathbb{R}^+$, thanks to the Schauder fixed point theorem.

Here we will adapt the method of [1] to the present framework: the main change will be the use of the $L^\infty(\Omega)$ norm instead of the $L^2(\Omega)$ norm in the normalization of the eigenfunctions.

2. Existence. Throughout this section, $r \in \mathbb{R}^+$ is fixed. Define the mapping $S : L^2(\Omega) \to L^2(\Omega)$, which associates to any fixed $w \in L^\infty(\Omega)$ the first eigenfunction $z_w \in W_0^{1,2}(\Omega)$ of the problem

$$\begin{cases} -\mathrm{div}\,(M(x, w(x))\nabla z_w) = \lambda_w z_w & \text{in } \Omega, \\ z_w > 0 & \text{on } \Omega, \\ \|z_w\|_{L^\infty(\Omega)} = r, \end{cases} \tag{5}$$

where λ_w is the first eigenvalue.

Since $w \in L^\infty(\Omega)$, $M(x, w(x))$ is uniformly elliptic and bounded even under our assumptions (1), with constants $\alpha_{\|w\|_{L^\infty(\Omega)}}$, $\beta_{\|w\|_{L^\infty(\Omega)}}$. It is well known [2] that λ_w is simple, that z_w belongs to $L^\infty(\Omega)$, that it can be chosen positive and (by linearity) with $L^\infty(\Omega)$ norm equal to r. It is clear that if $u = S(u)$, then u solves

$$\begin{cases} (\lambda_r, u_r) \in \mathbb{R}^+ \times W_0^{1,2}(\Omega) \cap L^\infty(\Omega) : \\ -\mathrm{div}\,(M(x, u_r)\nabla u_r) = \lambda_r\, u_r & \text{in } \Omega, \\ u_r > 0 & \text{on } \Omega, \\ u_r = 0 & \text{on } \partial\Omega, \\ \|u_r\|_{L^\infty(\Omega)} = r. \end{cases} \tag{6}$$

Remark the different choice of the norm (L^2 and L^∞) in (3) and (6). Moreover, recall that

$$\lambda_w = \inf_{v \in W_0^{1,2}(\Omega)} \frac{\int_\Omega M(x, w(x))\nabla v \nabla v}{\int_\Omega |v|^2},$$

so that

$$\alpha_r\, \mu = \alpha_r \inf_{v \in W_0^{1,2}(\Omega)} \frac{\int_\Omega |\nabla v|^2}{\int_\Omega |v|^2} \le \lambda_w \le \beta_r \inf_{v \in W_0^{1,2}(\Omega)} \frac{\int_\Omega |\nabla v|^2}{\int_\Omega |v|^2} = \beta_r\, \mu, \tag{7}$$

where μ is the first eigenvalue of the Laplacian with Dirichlet boundary conditions.

Theorem 1. *Under the assumptions* (1), *for any fixed* $r \in \mathbb{R}^+$, *there exists* (λ_r, u_r), *with* $\alpha_r\, \mu \le \lambda_r \le \beta_r\, \mu$, *solution of the nonlinear eigenvalue problem* (6).

Proof. By definition of $S : L^2(\Omega) \to L^2(\Omega)$ the set $B = \{v \in L^2(\Omega) : \|v\|_{L^\infty(\Omega)} \leq r\}$ is invariant. Furthermore, B is convex and closed with respect to the topology of $L^2(\Omega)$. The complete continuity of S between $L^2(\Omega)$ and $L^2(\Omega)$ has been proved in [1] under the assumptions α_k, β_k independent of k, for any $k > 0$. Under the assumption (1), we point out that on B we have $\alpha_k = \alpha_r$, $\beta_k = \beta_r$, for any $k > 0$; thus the quoted proof still holds.

The Schauder theorem then implies the existence of a fixed point u of S, that is a solution of (6), since the range of S is a subset of $W_0^{1,2}(\Omega) \cap L^\infty(\Omega)$. Moreover (7) implies that

$$\alpha_r\, \mu \leq \lambda_r \leq \beta_r\, \mu \tag{8}$$

as claimed. □

3. Examples. In this section we will discuss two model examples. In some sense, they show that (8) is sharp.

3.1. An unbounded operator. Let

$$M(x,s) = (1+|s|)^\theta I \quad (0 \leq \theta \leq 1),$$

that is,

$$Q(v) = -\operatorname{div}((1+|v|)^\theta \nabla v).$$

The estimate on λ_r given in Theorem 1 implies that now

$$\mu \leq \lambda_r \leq (1+|r|)^\theta \mu.$$

Proposition 1. *If $\ell \leq \mu$ and $0 \leq \theta \leq 1$, there are no solutions of the Dirichlet problem*

$$\begin{cases} u \in W_0^{1,2}(\Omega) \cap L^\infty(\Omega) : \\ -\operatorname{div}(M(x,u)\nabla u) = \ell\, u & \text{in } \Omega, \\ u > 0 & \text{on } \Omega, \\ u = 0 & \text{on } \partial\Omega. \end{cases} \tag{9}$$

Proof. If $\theta < 1$, use $\frac{(1+u)^{1-\theta}-1}{1-\theta}$ as test function in (9), and use $\log(1+u)$, if $\theta = 1$. Thus (if $\theta < 1$), by definition of μ,

$$\mu \int_\Omega |u|^2 \leq \int_\Omega |\nabla u|^2 = \frac{\ell}{1-\theta}\int_\Omega u\,[(1+u)^{1-\theta} - 1].$$

Remark that, if $t \in \mathbb{R}^+$, the inequality $(1+t)^{1-\theta} - 1 < (1-\theta)t$ holds, so that

$$\mu \int_\Omega |u|^2 \leq \frac{\ell}{1-\theta}\int_\Omega u\,[(1+u)^{1-\theta} - 1] < \mu \int_\Omega |u|^2,$$

a contradiction.

If $\theta = 1$ the inequalities are

$$\mu \int_\Omega |u|^2 \leq \ell \int_\Omega u \log(1+u) < \mu \int_\Omega |u|^2,$$

again a contradiction. □

3.2. A noncoercive operator. Let

$$M(x,s) = \frac{1}{(1+|s|)^\theta} I \quad (0 \leq \theta \leq 1),$$

that is,

$$Q(v) = -\mathrm{div}\left(\frac{\nabla v}{(1+|v|)^\theta}\right).$$

The estimate on λ_r given in Theorem 1 implies that now

$$\frac{\mu}{(1+|r|)^\theta} \leq \lambda_r \leq \mu.$$

Proposition 2. *If $\ell \geq \mu$ and $0 \leq \theta \leq 1$, there are no solutions of the Dirichlet problem*

$$(10) \qquad \begin{cases} u \in W_0^{1,2}(\Omega) \cap L^\infty(\Omega) : \\ -\mathrm{div}\,(M(x,u)\nabla u) = \ell\, u & \text{in } \Omega, \\ u > 0 & \text{on } \Omega, \\ u = 0 & \text{on } \partial\Omega. \end{cases}$$

Proof. We are not able to give a direct proof: we shall use the particular structure of the model problem. If $\theta < 1$, we perform the change of variable

$$w = \frac{(1+u)^{1-\theta} - 1}{1-\theta},$$

so that the boundary value problem (10) can be written as

$$w \in W_0^{1,2}(\Omega) \cap L^\infty(\Omega) : \quad -\Delta w = \ell\,\{[(1-\theta)w + 1]^{\frac{1}{1-\theta}} - 1\}$$

The use of the first eigenfuntion ψ of the Laplacian as test function implies that

$$\mu \int_\Omega w\psi = \ell \int_\Omega ([(1-\theta)w + 1]^{\frac{1}{1-\theta}} - 1)\psi.$$

The inequality $(1+t)^{1-\theta} - 1 < (1-\theta)t,\ t \in \mathbb{R}^+$ can be used again and it implies that

$$\mu \int_\Omega w\psi \geq \ell \int_\Omega \psi w > \mu \int_\Omega \psi w,$$

a contradiction. If $\theta = 1$, we perform the change of variable

$$w = \log(1+u),$$

so that the boundary value problem (10) can be written as

$$w \in W_0^{1,2}(\Omega) \cap L^\infty(\Omega) : \quad -\Delta w = \ell\,(e^w - 1).$$

The use of the first eigenfuntion ψ of the Laplacian as test function implies that

$$\mu \int_\Omega w\psi = \ell \int_\Omega (e^w - 1)\psi > \mu \int_\Omega \psi w,$$

again a contradiction. □

References

[1] L. BOCCARDO: *Positive eigenfunctions for a class of quasi-linear operators*, Boll. Unione Mat. Ital. **18-B** (1981), 951-959

[2] A. MANES, A. M. MICHELETTI: *Un'estensione della teoria variazionale classica degli autovalori per operatori ellittici del secondo ordine*, Boll. Unione Mat. Ital. **7** (1973), 285-301

LUCIO BOCCARDO, Università di Roma “La Sapienza”, Dipartimento di Matematica, Piazzale A. Moro 2, I-00185 Roma, Italy; boccardo@mat.uniroma1.it

Progress in Nonlinear Differential Equations
and Their Applications, Vol. 40

Some Geometrical Properties of Rearrangement Invariant Spaces

ILYA BRISKIN, EVGUENI M. SEMENOV

Dedicated to Professor Alfonso Vignoli on the occasion of his 60th birthday

Summary: We study the set of rearrangement invariant spaces with the property that the unit sphere generated by a pair of elements from such a space is a parallelogram, as well as characteristic properties of L_1 and L_∞ in the class of Lorentz spaces. We also investigate a monotonicity property of rearrangement invariant spaces.

Keywords: Rearrangement invariant space, Lorentz space, Orlicz space, geometry of Banach spaces.

Classification: 46E30, 47A57.

Acknowledgement: The second author's work is partly supported by RFBR (Russia), grant 98-01-00044, and by the Universities of Russia, grant 3667. The authors thank the referee for some useful remarks.

0. Introduction. The spaces L_1 and L_∞ contain two-dimensional subspaces whose unit sphere is a parallelogram. Each strictly normed space does not possess this property. In the setting of rearrangement invariant (r.i.) Banach function spaces the subspace structure has been studied by several authors (e.g. [2,3]). Denote by $\mathcal{K}$ the class of r.i. spaces E such that the subspace generated by some $x, y \in E$ is isometric to ℓ_1^2. If, in addition, all $\operatorname{supp} x$ and $\operatorname{supp} y$ are disjoint subsets, then this class of r.i. spaces is denoted by $\mathcal{K}_d$.

In this article we study the class $\mathcal{K}$. It is proved that $\mathcal{K}$ is not stable with respect to equivalent renorming. A characteristic property of L_1 and L_∞ in the class of Lorentz spaces is found, and a criterion for $L_M \in \mathcal{K}$, where L_M is an Orlicz space, is obtained. Moreover, we investigate a monotonicity property related to the class $\mathcal{K}_d$.

1. Definitions and notations. A Banach function space E on $[0, 1]$ with the Lebesgue measure is called *rearrangement invariant* (r.i.) or *symmetric* provided $x^*(t) \leq y^*(t)$ for every $t \in [0, 1]$ and $y \in E$, imply $x \in E$ and $\|x\|_E \leq \|y\|_E$, where $x^*(t)$ denotes the decreasing rearrangement of $|x(t)|$. We shall assume that the r.i. space E is separable or isometric to a conjugate space. It follows from the Calderón-Mityagin theorem that E is an interpolation space with constant 1 with respect to the pair (L_1, L_∞).

For every r.i. space E one has

(1) $$L_\infty \subseteq E \subseteq L_1.$$

For definiteness we shall assume that $\|\kappa_{[0,1]}\|_E = 1$, where $\kappa_e(t)$ denotes the characteristic function of a measurable set $e \subseteq [0,1]$. Then

(2) $$\|x\|_{L_1} \leq \|x\|_E \leq \|x\|_{L_\infty}$$

for every $x \in L_\infty$, the left inequality being true for each $x \in E$.

We will consider the *associate space* E' of all measurable functions $x(t)$ such that $\int_0^1 x(t)y(t)\,dt < \infty$ for every $y \in E$, equipped with the norm

$$\|x\|_{E'} = \sup\left\{\int_0^1 x(t)y(t)\,dt : \ \|y\|_E \leq 1\right\}.$$

For every r.i. space E, the embedding $E \subseteq E''$ is isometric. If E is separable, then E' coincides with the dual space E^*. Here and further the equality $E = F$ of two r.i. spaces means that the sets E and F coincide and $\|x\|_E = \|x\|_F$ for every $x \in E$. The space E' restores the norm of E in the sense that

(3) $$\|y\|_E = \sup\left\{\int_0^1 x(t)y(t)\,dt : \ \|x\|_{E'} \leq 1\right\}.$$

The function $\varphi_E(s) = \|\kappa_e\|_E$, where $\operatorname{mes} e = s$, is called the *fundamental function* of E. Each fundamental function φ_E is quasi-concave, i.e. $\varphi_E(t)$ and $t/\varphi_E(t)$ are increasing on $(0,1]$.

Let us recall some classical examples of r.i. spaces. Denote by Ω the set of increasing concave functions $\varphi(t)$ on $[0,1]$ with $\varphi(0) = 0$. Each $\varphi \in \Omega$ generates the *Lorentz space* $\Lambda(\varphi)$ equipped with the norm

$$\|x\|_{\Lambda(\varphi)} = \int_0^1 x^*(t)\,d\varphi(t)$$

and the *Marcinkiewicz space* $M(\varphi)$ equipped with the norm

$$\|x\|_{M(\varphi)} = \sup_{0<\tau\leq 1} \frac{1}{\varphi(\tau)} \int_0^\tau x^*(t)\,dt.$$

For every $\varphi \in \Omega$ one has $(\Lambda(\varphi))' = M(\varphi)$. If φ is continuous at 0, then $\Lambda(\varphi)$ is separable and $(\Lambda(\varphi))^* = M(\varphi)$. The assumption $\|\kappa_{[0,1]}\|_{\Lambda(\varphi)} = 1$ means that $\varphi(1) = 1$. Denote by $M_0(\varphi)$ the closure of step functions in $M(\varphi)$. If $\lim\limits_{t\to 0} t/\varphi(t) = 0$, then $M_0(\varphi)$ is a separable r.i. space and $M(\varphi)$ is not separable.

A convex even function $M(u)$ is called *N-function* if $M(0) = 0$. Each N-function $M(u)$ generates the *Orlicz space* L_M equipped with the norm

(4) $$\|x\|_{L_M} = \inf\{\lambda : \ \lambda > 0, \ \int_0^1 M\left(\frac{1}{\lambda}x(t)\right)\,dt \leq 1\}.$$

If E and F are isomorphic Banach spaces, then

$$\rho(E,F) = \inf \|T\|_{E\to F}\|T^{-1}\|_{F\to E},$$

where the infimum is taken over all invertible operators T from E into F, is called the *Banach-Mazur distance* of E and F. The subspace generated by two elements $x, y \in E$ is denoted by $[x, y]$.

For more information about all the above mentioned properties of r.i. spaces we refer to the monographs [1-3].

2. Two-dimensional sections of r.i. spaces. As we have mentioned in the Introduction, $\mathcal{K}$ is the class of r.i. spaces such that $[x, y]$ is isometric to ℓ_1^2 for some $x, y \in E$. The class $\mathcal{K}$ is not stable with respect to an equivalent r.i. renorming. This follows from Theorems 1 and 2 below.

Theorem 1. *Let E be a r.i. space. Then there exists an equivalent r.i. norm $\|\cdot\|_F$ such that $\rho(E, F) \leq 2$ and $F \in \mathcal{K}_d$.*

Proof. Consider the function

$$\psi(t) = \min(1, 2\varphi_E(t)).$$

Since $\varphi_E(t)$ is quasi-concave, $\psi(1/2) = 1$ and

$$\varphi_E(t) \leq \psi(t) \leq 2\varphi_E(t).$$

By [2, Theorem 2.5.8] one can introduce an equivalent norm $\|\cdot\|_F$ such that $\varphi_F(t) = \psi(t)$. It follows from the proof of that theorem that $\rho(E, F) \leq 2$. Since $\varphi_F(1/2) = \varphi_F(1) = 1$ we have

$$\|\kappa_{(0,1/2)}\|_F = \|\kappa_{(1/2,1)}\|_F = \|\kappa_{(0,1/2)} \pm \kappa_{(1/2,1)}\|_F = 1$$

as claimed. □

Theorem 2. *Let E be a r.i. space and $\varepsilon > 0$. Then there exists an equivalent r.i. norm $\|\cdot\|_G$ such that $\rho(E, G) \leq 1 + \varepsilon$ and $G \notin \mathcal{K}$.*

Proof. Each $E \in \mathcal{K}$ is not strictly normed. Therefore, it is sufficient to construct a strictly normed r.i. space G such that $\rho(E, G) \leq 1 + \varepsilon$. Put

$$\|x\|_G = \lambda\left(\|x\|_E + \varepsilon\|x\|_{L_M}\right),$$

where $M(u)$ is a strictly convex N-function, $M \in \Delta_2$ with $M(u) \leq u$ for every $u \geq 0$, and λ is chosen so that $\varphi_G(1) = 1$. Since $\|x\|_{L_M} \leq \|x\|_{L_1}$ and $\|x\|_{L_1} \leq \|x\|_E$ (see [2, 2.4.1]), we have

$$\|x\|_G \leq \lambda(1 + \varepsilon)\|x\|_E$$

for every $x \in E$. Clearly, $\|x\|_G \geq \lambda\|x\|_E$. It follows from the obtained inequalities that $\rho(E, G) \leq 1 + \varepsilon$. The strict convexity of $M \in \Delta_2$ implies that L_M is a strictly normed space [4]. Therefore, G is strictly normed, too. □

As a matter of fact, we have proved that G is a strictly normed space. It is well known that, up to equivalent renorming, each separable space is strictly normed. Theorem 2 shows that this statement is valid for every r.i. space.

Now we shall study the intersection of $\mathcal{K}$ with some classes of r.i. spaces. Clearly, $E \in \mathcal{K}$ if and only if there exist $u, v \in E$ such that $\|u\|_E = \|v\|_E = 1$ and $\|u+v\|_E = \|u-v\|_E = 2$.

Theorem 3. *Let $\varphi \in \Omega$. Then $\Lambda(\varphi) \in \mathcal{K}$ if and only if $\varphi(t) = \operatorname{sign} t$ or $\varphi(t) = bt$ for all $t \in [0, a]$ with $0 < a \le 1/b$.*

Proof. The sufficiency is obvious. In this case $E = L_1$ or $E = L_\infty$, up to equivalence. If $\Lambda(\varphi) \in \mathcal{K}$, then there exist $u, v \in \Lambda(\varphi)$ such that

$$\int_0^1 u^*(t)\,d\varphi(t) = \int_0^1 v^*(t)\,d\varphi(t) = 1 \tag{5}$$

and

$$\int_0^1 (u+v)^*(t)\,d\varphi(t) = \int_0^1 (u-v)^*(t)\,d\varphi(t) = 2. \tag{6}$$

Let c be the minimal number from $[0,1]$ such that $\varphi'(t) = 0$ for $t \in [c, 1]$. If $c = 0$, then $\varphi(t) = \operatorname{sign} t$, $E = L_\infty$, and the theorem is proved.

Suppose that $c > 0$. Without loss of generality, one can assume that $u = u^*$. Then $\operatorname{supp} u = [0, d]$ or $[0, d)$, where $0 < d \le c$. Suppose that

$$\operatorname{mes}\{t : 0 \le t \le d,\ v(t) > 0\} > 0. \tag{7}$$

There exist $\varepsilon > 0$ and $e \subseteq [0, d]$ such that $\operatorname{mes} e > 0$ and both $u(t) \ge \varepsilon$ and $v(t) \ge \varepsilon$ for $t \in e$. Put

$$u_1(t) = \begin{cases} u(t), & t \notin e, \\ u(t) - \varepsilon, & t \in e, \end{cases} \qquad v_1(t) = \begin{cases} v(t), & t \notin e, \\ v(t) - \varepsilon, & t \in e. \end{cases}$$

Then $\|u_1\|_{\Lambda(\varphi)} < \|u\|_{\Lambda(\varphi)} = 1$, $\|v_1\|_{\Lambda(\varphi)} \le \|v\|_{\Lambda(\varphi)} = 1$, and $u_1(t) - v_1(t) = u(t) - v(t)$ for $t \in [0,1]$. Hence,

$$2 = \|u - v\|_{\Lambda(\varphi)} = \|u_1 - v_1\|_{\Lambda(\varphi)} \le \|u_1\|_{\Lambda(\varphi)} + \|v_1\|_{\Lambda(\varphi)} < 2.$$

The obtained contradiction shows that the assumption (7) is false. By a similar reasoning one can prove that the assumption

$$\operatorname{mes}\{t : 0 \le t \le d,\ v(t) < 0\} > 0$$

is not true. Consequently, $\operatorname{supp} u \cap \operatorname{supp} v = \emptyset$.

Now we must prove that $\varphi(+0) = 0$. It follows from (5) and (6) that

$$\|u+v\|_{L_\infty}\varphi(+0) + \int_0^1 (u+v)^*(t)\varphi'(t)\,dt$$
$$= \|u\|_{L_\infty}\varphi(+0) + \int_0^1 u^*(t)\varphi'(t)\,dt + \|v\|_{L_\infty}\varphi(+0) + \int_0^1 v^*(t)\varphi'(t)\,dt.$$

Since

$$\int_0^1 (u+v)^*(t)\varphi'(t)\,dt \le \int_0^1 u^*(t)\varphi'(t)\,dt + \int_0^1 v^*(t)\varphi'(t)\,dt,$$

we obtain

$$\|u+v\|_{L_\infty}\varphi(+0) \geq (\|u\|_{L_\infty} + \|v\|_{L_\infty})\,\varphi(+0).$$

But $\|u+v\|_{L_\infty} = \max(\|u\|_{L_\infty}, \|v\|_{L_\infty})$ and $\|u\|_{L_\infty} > 0, \|v\|_{L_\infty} > 0$. Therefore $\varphi(+0) = 0$.

Denoting

$$n_x(\tau) = \text{mes }\{t : |x(t)| > \tau\},$$

we have

$$n_{u+v}(\tau) = n_u(\tau) + n_v(\tau)$$

for every $\tau > 0$. Using formula (2.5.2) from [2] and (6), we get

$$\int_0^\infty \varphi(n_u(\tau))\,d\tau + \int_0^\infty \varphi(n_v(\tau))\,d\tau = \int_0^\infty \varphi(n_u(\tau) + n_v(\tau))\,d\tau. \tag{8}$$

If $\varphi(t)/t$ is not a constant function on any interval $(0,\varepsilon)$, $\varepsilon > 0$, then $\varphi(t_1+t_2) < \varphi(t_1) + \varphi(t_2)$ for every $t_1, t_2 > 0$, and we have strict inequality instead of equality in (8). The obtained contradiction proves that $\varphi(t)/t$ is constant on some interval $(0,\varepsilon]$ with $\varepsilon > 0$. □

Theorem 3 shows that $\Lambda(\varphi) \in \mathcal{K}$ holds only for $\Lambda(\varphi) = L_\infty$ or $\Lambda(\varphi) = L_1$, up to equivalence.

If $\varphi \in \Omega$ and $\varphi(+0) = 0$, then $M(\varphi) \in \mathcal{K}_d$. Let us prove this statement. Given $s \in (0,1)$, we have

$$1 \geq \lim_{\tau\to 0} \|\varphi'\kappa_{(\tau,s)}\|_{M(\varphi)} \geq \lim_{\tau\to 0} \frac{1}{\varphi(s-\tau)} \int_\tau^s \varphi'(t)\,dt = \lim_{\tau\to 0} \frac{\varphi(s)-\varphi(\tau)}{\varphi(s-\tau)} = 1.$$

Therefore, there exists a sequence $\tau_k \downarrow 0$ such that

$$\lim_{k\to\infty} \|\varphi'\kappa_{(\tau_k,\tau_{k-1})}\|_{M(\varphi)} = 1.$$

Put

$$x(t) = \varphi'(t)\sum_{k=1}^\infty \kappa_{(\tau_{2k},\tau_{2k-1})}(t), \quad y(t) = \varphi'(t)\sum_{k=1}^\infty \kappa_{(\tau_{2k+1},\tau_{2k})}(t).$$

Then $\|x\|_{M(\varphi)} = \|y\|_{M(\varphi)} = \|x+y\|_{M(\varphi)} = 1$ and $\operatorname{supp} x \cap \operatorname{supp} y = \emptyset$. Consequently, $M(\varphi) \in \mathcal{K}_d$.

Theorem 4. *Let L_M be an Orlicz space. Then $L_M \in \mathcal{K}_d$ if and only if $M \notin \Delta_2$.*

Proof. Suppose $M \in \Delta_2$. By the Levi Lemma, the infimum in the definition (4) is attained. If $\|x\|_{L_M} = \|y\|_{L_M} = 1$ and $\operatorname{supp} x \cap \operatorname{supp} y = \emptyset$, then

$$\int_0^1 M(x(t))\,dt = \int_0^1 M(y(t))\,dt = 1.$$

Therefore

$$\int_0^1 M(x(t)+y(t))\,dt = 2.$$

Since the function

$$F(\lambda) = \int_0^1 M\left(\frac{1}{\lambda}[x(t)+y(t)]\right)dt$$

is finite for every $\lambda > 0$ and continuous, the solution of the equation $F(\lambda) = 1$ is bigger than 1. This means that $L_M \notin \mathcal{K}$.

Conversely, if $M \notin \Delta_2$, then there exists $x \in L_M$ such that $\operatorname{supp} x \subseteq (0, 1/2)$ and

$$\int_0^1 M(x(t))\, dt \leq 1/2, \qquad \int_0^1 M(\lambda x(t))\, dt = \infty$$

for every $\lambda > 1$. Consider the function

$$y(t) = \begin{cases} 0, & 0 \leq t \leq \frac{1}{2}, \\ x(t - \frac{1}{2}), & \frac{1}{2} < t \leq 1. \end{cases}$$

Then $\|x\|_{L_M} = \|y\|_{L_M} = \|x + y\|_{L_M} = 1$ and $\operatorname{supp} x \cap \operatorname{supp} y = \emptyset$. This completes the proof. □

3. A monotonicity property. Some r.i. spaces E have the following property: Let $0 < a < 1$, then

$$\mu(E, a) = \inf \{\|x + y\|_E : \ x, y \geq 0, \ \|x\|_E = 1, \ \|y\|_E = a\} > 1.$$

The Lorentz-Shimogaki Theorem (see [2, 2.3.1]) implies that the infimum in this definition is attained on pairs (x, y) with $\|x\|_E = 1$, $\|y\|_E = a$, and $\operatorname{supp} x \cap \operatorname{supp} y = \emptyset$. For example,

$$\mu(L_p, a) = (1 + a^p)^{1/p}.$$

If $\lim\limits_{\tau \to 0} \varphi(2t)/\varphi(t) = 1$, then

$$1 \leq \mu(\Lambda(\varphi), a) \leq \lim_{\tau \to 0} \left\| \frac{\kappa_{(0,\tau)}}{\varphi(\tau)} + \frac{\kappa_{(0,\tau)}}{\varphi(\tau)} \right\|_{\Lambda(\varphi)} = \lim_{\tau \to 0} \frac{\varphi(2\tau)}{\varphi(\tau)} = 1.$$

If $\varphi(+0) = 0$, then

$$\mu(M_0(\varphi), a) = \mu(M(\varphi), a) = 1.$$

Indeed, this follows from the fact that

$$\lim_{\tau \to 0} \|\varphi'(\tau)\kappa_{(\tau,1)}\|_{M(\varphi)} = 1$$

and

$$\frac{\varphi(\tau)}{\tau} \|\kappa_{(0,\tau)}\|_{M(\varphi)} = 1, \qquad \left\| \frac{\varphi(\tau)}{\tau} \kappa_{(0,\tau)} + \varphi'(\tau)\kappa_{(\tau,1)} \right\|_{M(\varphi)} = 1$$

for each $\tau \in (0, 1)$. However, non-uniform estimates are valid for $a = 1$ under some additional assumptions, as follows from Theorem 5 below.

Theorem 5. *Let E be a separable r.i. space and $\varphi_E(1/2) < 1$. If $x, y \in E$, $x, y \geq 0$, and $\|x\|_E = \|y\|_E = 1$, then $\|x + y\|_E > 1$.*

Proof. Since $E' = E^*$ (see [3, 2.a.3]), the supremum in (3) is attained. There exist $u, v \in E'$ such that $\|u\|_{E'} = \|v\|_{E'} = 1$ and

$$\int_0^1 x(t)u(t)\, dt = \int_0^1 y(t)v(t)\, dt = 1.$$

Without loss of generality one can assume that $x = x^*$. By formula (2.2.25) from [2] we have $u = u^*$.

Suppose that $\|x+y\|_E = 1$. Then

$$\int_0^1 (x+y)^*(t)u(t)\,dt \le \|x+y\|_E\|u\|_{E'} = 1.$$

Since $(x+y)^* \ge x^* = x$, we get

$$\int_0^1 (x+y)^*(t)u(t)\,dt \ge \int_0^1 x(t)u(t)\,dt = 1.$$

Hence

$$\int_0^1 (x+y)^*(t)u(t)\,dt = 1$$

and

$$\int_0^1 [(x+y)^*(t) - x(t)]\,u(t)\,dt = 0.$$

It follows from the obtained equality and $u = u^*$ that $(x+y)^*(t) = x(t)$ for every $t \in [0,b]$ for some $b \in (0,1)$. This means that

$$\inf_{t\in \operatorname{supp} x} x(t) \ge \sup_{t\in[0,1]} y(t).$$

Repeating this reasoning we get

$$\inf_{t\in \operatorname{supp} y} y(t) \ge \sup_{t\in[0,1]} x(t).$$

Consequently, there exist $e_1, e_2 \subseteq [0,1]$ and $c > 0$ such that $x(t) = c\kappa_{e_1}(t)$, $y = c\kappa_{e_2}(t)$ and $e_1 \cap e_2 = \emptyset$. Clearly, $\operatorname{mes} e_1 = \operatorname{mes} e_2 \le \frac{1}{2}$. Hence

$$c\varphi_E(2\operatorname{mes} e_1) = \|x+y\|_E = \|x\|_E = c\varphi_E(\operatorname{mes} e_1).$$

This equality shows that $\varphi_E(2\tau) = \varphi_E(\tau)$ for some $\tau \in (0,1/2]$. So $\varphi(t) = \text{const}$ for $t \in [\tau,1]$. The obtained contradiction proves our statement. □

Corollary. *Let $\psi \in \Omega$ and $\psi(+0) = \lim_{t\to 0} t/\psi(t) = 0$. Then $M_0(\psi) \notin \mathcal{K}_d$.*

Indeed, the concavity of $\psi(t)$ and the assumption $\lim_{t\to 0} t/\psi(t) = 0$ imply that $\psi(1/2) > 1/2$. Hence, $\varphi_{M(\psi)}(1/2) < 1$ and the statement follows from Theorem 5.

We have proved that $M(\psi) \in \mathcal{K}_d$. These examples illustrate the significant role of separability of r.i. spaces in studying the class $\mathcal{K}_d$.

References

[1] C. Bennett, R. Sharpley: *Interpolation of Operators*, Academic Press, Boston 1988

[2] S. G. Krein, Yu. I. Petunin, E. M. Semenov: *Interpolation of Linear Operators* [in Russian], Nauka, Moscow 1978; Engl. transl.: Amer. Math. Soc., Providence RI 1982

[3] J. Lindenstrauss, L. Tzafriri: *Classical Banach Spaces II. Function Spaces*, Springer, Berlin 1979

[4] M. M. Rao, Z. D. Ren: *Theory of Orlicz Spaces*, M. Dekker, New York 1991

Ilya Briskin, Center for Technological Education Holon, Department of Sciences, P. O. Box 305, Holon 58102, Israel; `gotlib@barley.cteh.ac.il`

Evgueni M. Semenov, Voronezh State University, Department of Mathematics, RUS-394693 Voronezh, Russian Federation; `root@func.vsu.ru`

Progress in Nonlinear Differential Equations
and Their Applications, Vol. 40

Strong Surjections and Nearness

ADRIANA BUICA

Dedicated to Alfonso Vignoli on the occasion of his 60th birthday

Summary: We show that the properties of being strongly surjective or stably solvable (in the sense of Furi-Martelli-Vignoli) carry over to maps which are "near" (in the sense of Campanato).

Keywords: Near mapping, strong surjection, stably solvable map, implicit equations.

Classification: 47H99, 47H15, 34A09, 34B15.

1. Introduction. M. Furi, M. Martelli and A. Vignoli in [3] introduced the notions of *strong surjection* and *stably solvable map* between two normed spaces E and F in order to define the spectrum for a nonlinear operator. Also, these concepts are related to that of *zero-epi map*, which is due to the same authors [4] and is very important in the study of solvability of nonlinear equations.

Near operators have been introduced by S. Campanato and also studied by A. Tarsia and S. Leonardi in [2,7,11,14] and have applications in nonlinear differential equations, too.

We prove that the property of being a strong surjection or stably solvable is preserved by nearness and notice that this can be used to prove existence results for differential equations in implicit form.

2. Main results. Let E be a normed space and F be a Banach space. A continuous map $f : E \to F$ is called a *strong surjection* if the equation $f(x) = h(x)$ has a solution for any continuous compact map $h : E \to F$.

A continuous map $f : E \to F$ is said to be *stably solvable* if the equation $f(x) = h(x)$ has a solution for any completely continuous map $h : E \to F$ with quasinorm $|h| = 0$. Recall that the *quasinorm* of a map h is defined by

$$|h| = \limsup_{||x|| \to \infty} \frac{||h(x)||}{||x||}.$$

We say that $g : E \to F$ is *near* $f : E \to F$ if there exist two positive constants α and k, with $k \in (0,1)$, such that for all $x_1, x_2 \in E$ we have

$$||f(x_1) - f(x_2) - \alpha[g(x_1) - g(x_2)]|| \leq k||f(x_1) - f(x_2)||. \tag{1}$$

In order to prove our main results we shall give two elementary lemmas.

Lemma 1. *Let $f : E \to F$ be continuous and suppose that $g : E \to F$ is near f. Then g is also continuous.*

Proof. Using that g is near f we obtain the estimate

$$\begin{aligned} &||g(x_1) - g(x_2)|| \\ &= \frac{1}{\alpha}||f(x_1) - f(x_2) - \alpha[g(x_1) - g(x_2)] - [f(x_1) - f(x_2)]|| \\ &\leq \left(\frac{k}{\alpha} + 1\right) ||f(x_1) - f(x_2)|| \end{aligned}$$

for all $x_1, x_2 \in E$. □

In what follows we shall denote by f_d^{-1} a right inverse for a surjective map f.

Lemma 2. *Let $f : E \to F$ be surjective and suppose that $g : E \to F$ is near f. Then the following statements are true:*

(a) *$f(x) = f(\hat{x})$ implies that $g(x) = g(\hat{x})$;*

(b) *the map $s = (f - \alpha g) \circ f_d^{-1} : F \to F$ is a contraction and does not depend on the choice of the right inverse of f.*

Proof. (a) Replacing $x_1 = x$ and $x_2 = \hat{x}$ in (1) we see that $f(x) = f(\hat{x})$ implies $g(x) = g(\hat{x})$. From this we can deduce that s does not depend on the choice of f_d^{-1}.

(b) Using (1) we get the estimate

$$\begin{aligned} &||(f - \alpha g)(f_d^{-1} y_1) - (f - \alpha g)(f_d^{-1} y_2)|| \\ &\leq k||f(f_d^{-1} y_1) - f(f_d^{-1} y_2)|| = k||y_1 - y_2|| \end{aligned}$$

for all $y_1, y_2 \in F$, which shows that s is a contraction. □

Remark. Relation (1) also shows that the map $f - \alpha g$ is a contraction with respect to f. For other considerations in this direction we refer to [1,6,13].

Theorem 1. *Let $g : E \to F$ be near $f : E \to F$. If f is a strong surjection, then g is also a strong surjection.*

Proof. By Lemma 1, the map g is continuous. Let $h : E \to F$ be continuous and compact. We know that $s = (f - \alpha g) \circ f_d^{-1} : F \to F$ is a contraction, by Lemma 2, where f_d^{-1} is any right inverse for f. Moreover, in this situation we have that $I - s$ is a homeomorphism.

The map $(I - s)^{-1} \circ \alpha h : E \to F$ is continuous and compact. So, it has a coincidence point $x \in E$ with the strong surjection f, i.e. $f(x) = (I - s)^{-1}(\alpha h(x))$. Denote $\hat{x} = f_d^{-1}(f(x))$ and notice that $f(x) = f(\hat{x})$ and $g(x) = g(\hat{x})$. Moreover, the implications

$$\begin{aligned} &f(x) = (I - s)^{-1}(\alpha h(x)) \Rightarrow (I - s)(f(x)) = \alpha h(x) \\ &\Rightarrow f(x) - (f - \alpha g)(\hat{x}) = \alpha h(x) \Rightarrow g(x) = h(x) \end{aligned}$$

are valid. This means that x is a coincidence point of g and h, and so g is a strong surjection. □

Theorem 2. *Let $g : E \to F$ be near $f : E \to F$. If f is stably solvable, then g is also stably solvable.*

Proof. Let $h : E \to F$ be completely continuous with quasinorm $|h| = 0$. The arguments follow like in the previous theorem, noticing that $(I - s)^{-1} \circ \alpha h$ is completely continuous with quasinorm $|(I - s)^{-1} \circ \alpha h| = 0$. □

3. An application. Let us consider two mappings $L, N : E \to F$ such that $L - N$ is a strong surjection. In applications (see [4,5,8,11]), usually, L is linear and bounded (in many cases, a differential operator) and N is completely continuous (e.g. a Nemytzkii operator from a "small" space into a "large" space). In the case that $||(L-N)(x)|| \to \infty$ as $||x|| \to \infty$ there are some relations between the theory of strong surjections and the theory of zero-epi maps, or degree theory, or the theory of essential compact fields (see [3,4,7,9]). Using this fact, we can find many examples of strong surjections of the form $L - N$. One of them which is due to M. Furi, M. Martelli, and A. Vignoli [4] is the following.

Let $C_0^2[0,1]$ be the space of C^2-functions such that $x(0) = x(1) = 0$ and $L, N : C_0^2[0,1] \to C[0,1]$ be defined by $Lx(t) = x''(t)$ and $N(x)(t) = x^3(t)$. Then $L - N$ is a strong surjection.

We use our main results in order to state that a map of the (implicit) form

$$g : E \to F, \quad g(x) = G(Lx, N(x))$$

is a strong surjection provided that $L - N$ is a strong surjection and $G : F \times F \to F$ satisfies the following relation for some $\alpha > 0$ and $k \in (0,1)$ and for all $y_1, y_2, z_1, z_2 \in F$

$$(2) \qquad ||y_1 - z_1 - y_2 + z_2 - \alpha[G(y_1, z_1) - G(y_2, z_2)]|| \leq k||y_1 - z_1 - y_2 + z_2||.$$

For example, $g : C_0^2[0,1] \to C[0,1]$ defined by $g(x)(t) = \tilde{g}(x''(t), x^3(t))$ is a strong surjection if $\tilde{g} : \mathbb{R} \times \mathbb{R} \to \mathbb{R}$ satisfies (2) for all $y_1, y_2, z_1, z_2 \in \mathbb{R}$.

References

[1] A. Buica: *Data dependence theorems on coincidence problems*, Studia UBB (Cluj) **41** (1996), 33-40

[2] S. Campanato: *Further contribution to the theory of near mappings*, Le Matematiche **48** (1993), 183-187

[3] M. Furi, M. Martelli, A. Vignoli: *Contributions to the spectral theory for nonlinear operators in Banach spaces*, Annali Mat. Pura Appl. **118** (1978), 229-294

[4] M. Furi, M. Martelli, A. Vignoli: *On the solvability of nonlinear operator equations in normed spaces*, Annali Mat. Pura Appl. **124** (1980), 321-343

[5] R. E. Gaines, J. Mawhin: *The Coincidence Degree and Nonlinear Differential Equations*, Lect. Notes Math. **568**, Springer, Berlin 1977

[6] K. Goebel: *A coincidence theorem*, Bull. Acad. Pol. Sci. **16** (1968), 733-735

[7] A. Granas: *The theory of compact vector fields and some of its applications to topology of functional spaces I*, Rozprawy Mat. **30**, Warsaw 1962

[8] A. Granas, R. B. Guenther, J. W. Lee: *Some general principles in the Carathéodory theory of nonlinear differential systems*, J. Math. Pures Appl. **70** (1991), 153-196

[9] D. H. Hyers, G. Isac, T. M. Rassias: *Topics in Nonlinear Analysis and Applications*, World Scientific, Singapore 1997

[10] S. Leonardi: *On the Campanato nearness condition*, Le Matematiche **48** (1993), 179-181

[11] R. Precup: *Existence theorems for nonlinear problems by continuation methods*, Nonlin. Anal. TMA **30, 12** (1997), 3313-3322

[12] I. A. Rus: *Some remarks on coincidence theory*, Pure Math. Manuscr. **9** (1990/91), 137-148

[13] A. Tarsia: *Some topological properties preserved by nearness between operators and applications to PDE's*, Czechosl. Math. J. **46** (121) (1996), 607-624

[14] A. Tarsia: *Differential equations and implicit functions: a generalization of the near operators theorem*, Topol. Methods Nonlin. Anal. **11** (1998), 115-133

Adriana Buica, Department of Applied Mathematics, Babeş-Bolyai University, Str. M. Kogălniceanu 1, RO-3400 Cluj-Napoca, Romania; abuica@math.ubbcluj.ro

Progress in Nonlinear Differential Equations
and Their Applications, Vol. 40

On the Vanishing Viscosity Approximation of a Time Dependent Hamilton-Jacobi Equation

ITALO CAPUZZO DOLCETTA, FABIANA LEONI

Dedicated to Alfonso Vignoli on the occasion of his 60th birthday

Caro Alfonso, che bella occasione per ricordare il piacevole periodo aquilano e l'entusiasmo che trasmettevi a tutti noi! Italo

Summary: We prove a new comparison result for monotone viscosity solutions of Hamilton-Jacobi equations. In combination with the classical vanishing viscosity method this allows us to prove existence and uniqueness of solution for Cauchy-Dirichlet problems.

Keywords: Viscosity solutions, Hamilton-Jacobi equations, vanishing viscosity method.

Classification: 49L25, 35F25, 70H20.

Acknowledgement: This work is partially supported by the TMR Network "Viscosity Solutions and Applications".

1. Introduction. The aim of this paper is to show how the viscosity solutions theory and vanishing viscosity method lead to the existence and uniqueness of generalized solutions of the following Cauchy-Dirichlet problem for Hamilton-Jacobi equations

$$(1) \qquad \begin{cases} u_t + H(x,t,Du) = 0 & \text{in } Q, \\ u(x,0) = u_0(x) & \text{in } \Omega, \\ u(x,t) = 0 & \text{on } \partial\Omega \times (0,T). \end{cases}$$

Here Q is the cylinder $\Omega \times (0,T)$, with Ω smooth, open bounded domain in $\mathbb{R}^N$ and T given in $(0,\infty)$, and $u(x,t)$ is the unknown function defined on $\overline{\Omega} \times [0,T)$, whose time derivative and spatial gradient are respectively denoted by u_t and Du. The initial datum $u_0(x)$ is given in $C^1(\Omega) \cap C_0(\overline{\Omega})$, where $C_0(\overline{\Omega})$ denotes the space of continuous functions in $\overline{\Omega}$ vanishing on $\partial\Omega$, the hamiltonian $H : \overline{Q} \times \mathbb{R}^N \to \mathbb{R}$ is a continuous function and they are assumed to satisfy the following assumptions:

(H1) $\displaystyle\lim_{|\xi|\to\infty} H(x,t,\xi) = \infty$ uniformly w.r. t. $(x,t) \in \overline{\Omega} \times [0,T]$,

(H2) $t \in [0,T] \mapsto H(x,t,\xi)$ is nonincreasing $\forall\, (x,\xi) \in \overline{\Omega} \times \mathbb{R}^N$,

(H3) $$H(x, 0, Du_0(x)) \leq 0 \text{ in } \Omega.$$

For problem (1) we will use the notion of generalized viscosity solution given in [3, 5], in which the boundary conditions are appropriately taken into account, consistently with the requirement imposed by the equation inside the domain.

Definition 1. An upper semicontinuous (u.s.c.) function $u : \overline{\Omega} \times [0, T) \to \mathbb{R}$ is a *viscosity subsolution* of problem (1) if for every $(x_0, t_0) \in \overline{\Omega} \times [0, T)$ and for every C^1 function $\varphi(x, t)$ such that $u - \varphi$ has a local maximum at (x_0, t_0), it results

$$\varphi_t(x_0, t_0) + H(x_0, t_0, D\varphi(x_0, t_0)) \leq 0 \quad \text{if } (x_0, t_0) \in \Omega \times (0, T),$$

$$\min\{u(x_0, 0) - u_0(x_0), \varphi_t(x_0, 0) + H(x_0, 0, D\varphi(x_0, 0))\} \leq 0 \quad \text{if } t_0 = 0,$$

$$\min\{u(x_0, t_0), \varphi_t(x_0, t_0) + H(x_0, t_0, D\varphi(x_0, t_0))\} \leq 0 \quad \text{if } x_0 \in \partial\Omega.$$

Viscosity supersolutions of (1) are defined in a similar way by replacing local maximum points with local minima, the inequality ≤ 0 with ≥ 0 and the min operation with max.

A function $u \in C(\overline{\Omega} \times [0, T))$ will be said a *viscosity solution* of problem (1) if it is both a sub- and a supersolution. □

In order to prove the existence of a solution for problem (1) we will use an approximation argument, consisting in regularizing the original equation, on the one hand, by adding the vanishing viscosity term $-\varepsilon\Delta u$ (with ε devoted to go to zero), and, on the other hand, by substituting H with a smooth, bounded approximating hamiltonian H_ε; thus, approximating also the initial datum u_0 with a smooth function $u_0^\varepsilon \in C_0^\infty(\Omega)$, we are led to the second order parabolic Dirichlet problem

$$\text{(2)} \qquad \begin{cases} u_t^\varepsilon - \varepsilon\Delta u^\varepsilon + H_\varepsilon(x, t, Du^\varepsilon) = 0 & \text{in } Q, \\ u^\varepsilon(x, 0) = u_0^\varepsilon(x) & \text{in } \Omega, \\ u^\varepsilon(x, t) = 0 & \text{on } \partial\Omega \times (0, T). \end{cases}$$

The viscosity solutions strategy, based on ideas introduced in [3], works as follows: assume that for every $\varepsilon > 0$ there exists a viscosity solution (taking pointwise the boundary data) $u^\varepsilon \in C(\overline{\Omega} \times [0, T))$ of problem (2) satisfying the uniform bound

$$\|u^\varepsilon\|_{L^\infty(Q)} \leq C\,;$$

define then the weak limits

$$\overline{u}(x, t) = \limsup_{\substack{y \to x \\ s \to t \\ \varepsilon \downarrow 0}} u^\varepsilon(y, s)\,, \qquad \underline{u}(x, t) = \liminf_{\substack{y \to x \\ s \to t \\ \varepsilon \downarrow 0}} u^\varepsilon(y, s)$$

which are in fact a subsolution and a supersolution, repsectively, of problem (1) in the sense of Definition 1.

If a comparison result holds for generalized semisolutions of problem (1), then we obtain

$$\underline{u} \geq \overline{u} \qquad \text{in } \overline{\Omega} \times [0,T),$$

from which it follows, since $\underline{u} \leq \overline{u}$ by definition, that

$$\underline{u} \equiv \overline{u} \qquad \text{in } \overline{\Omega} \times [0,T).$$

It then follows that the sequence $\{u^\varepsilon\}$ converges locally uniformly to the unique solution $u = \underline{u} = \overline{u}$ of problem (1).

The most delicate step in the previous procedure is the comparison between $\underline{u}$ and $\overline{u}$ which essentially reduces the existence question to that of uniqueness.

A comparison principle for problem (1) is given in [3], in which the hamiltonian $H(x,t,\xi)$ is in particular assumed to be uniformly continuous with respect to its last variable $\xi \in \mathbb{R}^N$. However, under the assumptions (H1), (H2) and (H3) made in the present paper, such a comparison result is not known in the literature.

Nevertheless, under only the coercivity assumption (H1), we will prove in Section 2 a new comparison principle for subsolutions and supersolutions of (1) which are monotone nondecreasing with respect to the time variable t. It will be obtained by adapting the arguments of the comparison principle for stationary equations having coercive hamiltonians (see [1, 2]) and the technical devices introduced in [3] to deal with generalized semisolutions.

The next step is then to show the existence, for every $\varepsilon > 0$, of a viscosity solution $u^\varepsilon(x,t)$ of problem (2) which is monotone nondecreasing with respect to t.

The Cauchy-Dirichlet problem for second order quasilinear parabolic equations in divergence form such as (2) has been extensively studied in the literature (see e.g. [7-9]) and well known results allow to establish the existence of "variational" solutions, that is non differentiable functions satisfying the weak formulation of (2) based on integration by parts formula. In section 2 we will show that each variational solution actually is, thanks to the smoothness of H_ε and u_0^ε and from the standard parabolic regularity results, a classical solution and therefore the only viscosity solution of (2); furthermore, using assumptions (H2) and (H3), this solution is proved to be monotone nondecreasing with respect to t.

Thus, we can conclude that problem (1) admits a unique generalized viscosity solution nondecreasing in t. More precisely, we have

Theorem 0. *Assume $u_0 \in C^1(\Omega) \cap C_0(\overline{\Omega})$ and that H is a continuous function satisfying* (H1), (H2), (H3). *Then there exists a unique monotone nondecreasing in t generalized viscosity solution $u \in \mathrm{Lip}(\overline{Q})$ of problem* (1).

We finally point out that the above result still holds for more general hamiltonians and more general boundary data; in particular, it allows functions of the form $H(x,t,u,\xi)$ which are measurable with respect to t (see [6, 11] for definitions of viscosity solutions in this framework), and for which there exists a $\gamma \in \mathbb{R}$ such that $u \in \mathbb{R} \mapsto H(x,t,u,\xi) + \gamma u$ is nondecreasing.

2. A comparison principle. In this section we are concerned with the comparison property for the solutions of the problem

$$(3)\qquad \begin{cases} u_t + H(x,t,Du) = 0 & \text{in } Q, \\ u(x,0) = u_0(x) & \text{in } \Omega, \\ u(x,t) = 0 & \text{on } \partial\Omega \times (0,T), \end{cases}$$

where the hamiltonian $H(x,t,\xi)$ will be assumed continuous in $\overline{Q} \times \mathbb{R}^N$.
By a subsolution (supersolution, solution) of (3) we will always mean a viscosity subsolution (supersolution, solution) in the sense of Definition 1.
First of all, we observe that, because of the structure of the equation, under the only assumption of continuity of H, the generalized condition imposed at time $t = 0$ is actually satisfied pointwise by the bounded semisolutions. Moreover, extending for $t = T$ the bounded from above subsolutions u and the bounded from below supersolutions v by setting

$$u(x,T) = \limsup_{\substack{y\to x\\ t\to T\\ t<T}} u(y,t)\,, \qquad v(x,T) = \liminf_{\substack{y\to x\\ t\to T\\ t<T}} v(y,t)$$

the equation holds automatically in the upper region $\Omega \times \{T\}$.
These properties are summarized in the following Proposition, the proof of which can be found in [3].

Proposition 1. *If $u(x,t)$ (respectively, $v(x,t)$) is a bounded from above (below) subsolution (supersolution) of problem* (3), *then it satisfies*

$$\begin{aligned} &u(x,0) \le u_0(x) && \text{in } \overline{\Omega}, \\ &u_t + H(x,t,Du) \le 0 && \text{in } \Omega\times]0,T], \\ &\min\{u, u_t + H(x,t,Du)\} \le 0 && \text{in } \partial\Omega \times \{T\} \end{aligned}$$

(*respectively,*

$$\begin{aligned} &v(x,0) \ge u_0(x) && \text{in } \overline{\Omega}, \\ &v_t + H(x,t,Dv) \ge 0 && \text{in } \Omega\times]0,T], \\ &\max\{v, v_t + H(x,t,Dv)\} \ge 0 && \text{in } \partial\Omega \times \{T\}). \end{aligned}$$

Assuming that H verifies also the coercivity hypothesis (H1), we can deduce a further property for the generalized subsolutions, which is the extension to the evolutive case of Lemma 7.1, section 7.1.2 of [2].

Proposition 2. *Let $H \in C(\overline{\Omega} \times [0,T] \times \mathbb{R}^N)$ satisfy* (H1). *If u is a bounded from above subsolution of problem* (3), *then*

$$u(x,t) \le 0 \qquad \text{in } \partial\Omega \times [0,T].$$

Proof. Let $(x_0,t_0) \in \partial\Omega \times [0,T]$ and assume, by contradiction, that $u(x_0,t_0) > 0$. Since $u(x_0,0) \le u_0(x_0) = 0$ by Proposition 1 it follows that $t_0 > 0$.

For fixed $\alpha, \beta, C > 0$ let us consider the function

$$\Psi(x,t) = u(x,t) - \frac{|x-x_0|^2}{2\alpha} - \frac{(t-t_0)^2}{2\beta} - Cd(x),$$

with $d(x) = \text{dist}(x, \partial\Omega)$ (distance function from $\partial\Omega$), which is a C^2 function in a neighborhood of $\partial\Omega$ if $\partial\Omega$ is a C^2 surface.
Let $(x^\star, t^\star) \in \overline{Q}$ be a maximum point for Ψ; since

$$\begin{aligned} u(x_0,t_0) &= \Psi(x_0,t_0) \leq \Psi(x^\star,t^\star) \\ &\leq u(x^\star,t^\star) - \frac{|x^\star-x_0|^2}{2\alpha} - \frac{(t^\star-t_0)^2}{2\beta}, \end{aligned}$$

it follows that

$$\frac{|x^\star-x_0|^2}{2\alpha} + \frac{(t^\star-t_0)^2}{2\beta} \leq \max_{\overline{\Omega}\times[0,T]} u - u(x_0,t_0),$$

from which we deduce that

$$\begin{aligned} x^\star &\to x_0 \quad \text{as } \alpha \downarrow 0,\ \forall \beta, C > 0 \\ t^\star &\to t_0 \quad \text{as } \beta \downarrow 0,\ \forall \alpha, C > 0. \end{aligned}$$

From the upper semicontinuity of u, it follows that

$$u(x^\star,t^\star) \to u(x_0,t_0) \quad \text{as } (\alpha,\beta) \to (0,0),$$

and all the convergences are uniform with respect to $C > 0$.
Now, since $u(x_0,t_0) > 0$ by assumption, we have $u(x^\star,t^\star) > 0$ for α, β sufficiently small, so that, even if $x^\star$ belongs to $\partial\Omega$, we can apply the definition of viscosity subsolution with test function

$$\varphi(x,t) = \frac{|x-x_0|^2}{2\alpha} + \frac{(t-t_0)^2}{2\beta} + Cd(x),$$

to

$$\frac{t^\star-t_0}{2\beta} + H\left(x^\star, t^\star, \frac{x^\star-x_0}{\alpha} + CDd(x^\star)\right) \leq 0.$$

It then follows that

$$H\left(x^\star, t^\star, \frac{x^\star-x_0}{\alpha} + CDd(x^\star)\right) \leq \frac{T}{2\beta}$$

and, by assumption (H1), there exists a constant $\Gamma_\beta > 0$ such that

$$\left|\frac{x^\star-x_0}{\alpha} + CDd(x^\star)\right| \leq \Gamma_\beta.$$

Choosing the constant C such that

$$C = C(\alpha,\beta) > \Gamma_\beta + \frac{\text{diam}(\Omega)}{\alpha}$$

we get the desired contradiction. □

Before giving the comparison result for generalized semisolution of problem (3), we present, as an intermediate result, the comparison principle for subsolutions and supersolutions assuming pointwise the boundary data.

In the sequel, we denote by $\partial_p Q$ the parabolic boundary of the cylinder $Q = \Omega \times (0,T)$, that is

$$\partial_p Q = \overline{\Omega} \times \{0\} \cup \partial\Omega \times [0,T].$$

Theorem 1. *Assume that $H \in C(\overline{Q} \times \mathbb{R}^N)$ satisfy* (H1) *and let u and v respectively be a subsolution and a supersolution of*

$$w_t + H(x,t,Dw) = 0 \qquad \text{in } \Omega \times (0,T),$$

with u u.s.c. in $\overline{\Omega} \times [0,T)$ bounded from above, and v l.s.c. in $\overline{\Omega} \times [0,T)$ bounded from below.

If u is monotone non-decreasing with respect to t, then

$$u \le v \qquad \text{on } \partial_p\Omega$$

implies

$$u \le v \qquad \text{in } \overline{\Omega} \times [0,T).$$

Proof. We extend, as usual, u and v for $t = T$ by setting

$$u(x,T) = \limsup_{\substack{y \to x \\ t \to T \\ t<T}} u(y,t), \qquad v(x,T) = \liminf_{\substack{y \to x \\ t \to T \\ t<T}} v(y,t)$$

and observe that, as in Proposition 1, u and v are respectively a viscosity subsolution and supersolution also in $\Omega \times \{T\}$.

Let us argue by contradiction assuming that $u - v$ has a positive maximum in $\overline{Q}$; for small enough fixed $\alpha > 0$, it then follows that

$$M = \max_{\overline{\Omega} \times [0,T]} [u(x,t) - v(x,t) - \alpha t] = u(x_0,t_0) - v(x_0,t_0) - \alpha t_0 > 0.$$

For every fixed $\varepsilon > 0$, we introduce the auxiliary function

$$\Phi_\varepsilon(x,y,t,s) = u(x,t) - v(y,s) - \alpha s - \frac{|x-y|^2}{2\varepsilon} - \frac{(t-s)^2}{2\varepsilon}$$

which attains by upper semicontinuity its maximum at some point $(x_\varepsilon, y_\varepsilon, t_\varepsilon, s_\varepsilon) \in \overline{\Omega}^2 \times [0,T]^2$ (we drop the dependence on α which is fixed); let us notice that

$$M_\varepsilon = \Phi_\varepsilon(x_\varepsilon, y_\varepsilon, t_\varepsilon, s_\varepsilon) \ge M > 0.$$

This implies

$$\frac{|x_\varepsilon - y_\varepsilon|^2}{2\varepsilon} + \frac{(t_\varepsilon - s_\varepsilon)^2}{2\varepsilon} < u(x_\varepsilon, t_\varepsilon) - v(y_\varepsilon, s_\varepsilon) \le \max_{\overline{\Omega} \times [0,T]} u - \min_{\overline{\Omega} \times [0,T]} v$$

and so, by compactness, there exists $(\bar{x},\bar{t}) \in \overline{Q}$ such that

$$(x_\varepsilon, y_\varepsilon, t_\varepsilon, s_\varepsilon) \to (\bar{x},\bar{x},\bar{t},\bar{t}) \qquad \text{as } \varepsilon \downarrow 0\,.$$

Moreover, since

$$M \le M_\varepsilon \le u(x_\varepsilon,t_\varepsilon) - v(y_\varepsilon,s_\varepsilon) - \alpha s_\varepsilon\,,$$

it follows that

$$\begin{aligned} M &\le \liminf_{\varepsilon\downarrow 0} M_\varepsilon \le \limsup_{\varepsilon\downarrow 0} M_\varepsilon \\ &\le \limsup_{\varepsilon\downarrow 0}[u(x_\varepsilon,t_\varepsilon) - v(y_\varepsilon,s_\varepsilon) - \alpha s_\varepsilon] \\ &\le u(\bar{x},\bar{t}) - v(\bar{x},\bar{t}) - \alpha\bar{t} \le M\,, \end{aligned}$$

and, consequently,

$$\begin{aligned} \lim_{\varepsilon\downarrow 0} u(x_\varepsilon,t_\varepsilon) &= u(\bar{x},\bar{t}),\\ \lim_{\varepsilon\downarrow 0} v(y_\varepsilon,s_\varepsilon) &= v(\bar{x},\bar{t}),\\ \lim_{\varepsilon\downarrow 0} M_\varepsilon &= M\,. \end{aligned}$$

Since $u(\bar{x},\bar{t}) - v(\bar{x},\bar{t}) = M + \alpha\bar{t} > 0$, we deduce that $(\bar{x},\bar{t})$ belongs to $\Omega\times(0,T]$, as well as $(x_\varepsilon,t_\varepsilon)$ and $(y_\varepsilon,s_\varepsilon)$ do for ε sufficiently small. Moreover, since for all (x,t), $(y,s)\in\overline{Q}$

$$(4) \qquad u(x_\varepsilon,t_\varepsilon) - \frac{|x_\varepsilon - y_\varepsilon|^2}{2\varepsilon} - \frac{(t_\varepsilon - s_\varepsilon)^2}{2\varepsilon} \ge u(x,t) - \frac{|x - y_\varepsilon|^2}{2\varepsilon} - \frac{(t - s_\varepsilon)^2}{2\varepsilon}$$

and

$$v(y_\varepsilon,s_\varepsilon) + \alpha s_\varepsilon + \frac{|x_\varepsilon - y_\varepsilon|^2}{2\varepsilon} + \frac{(t_\varepsilon - s_\varepsilon)^2}{2\varepsilon} \le v(y,s) + \alpha s + \frac{|x_\varepsilon - y|^2}{2\varepsilon} + \frac{(t_\varepsilon - s)^2}{2\varepsilon}\,,$$

by the definitions of a viscosity subsolution and supersolution we get

$$(5) \qquad \frac{t_\varepsilon - s_\varepsilon}{\varepsilon} + H\left(x_\varepsilon, t_\varepsilon, \frac{x_\varepsilon - y_\varepsilon}{\varepsilon}\right) \le 0,$$

$$(6) \qquad -\alpha + \frac{t_\varepsilon - s_\varepsilon}{\varepsilon} + H\left(y_\varepsilon, s_\varepsilon, \frac{x_\varepsilon - y_\varepsilon}{\varepsilon}\right) \ge 0\,.$$

Furthermore, using the fact that $t \mapsto u(x_\varepsilon,t)$ is nondecreasing, if $t_\varepsilon < T$ from (4) with $x = x_\varepsilon$ we deduce, for all $t > t_\varepsilon$,

$$\frac{1}{t - t_\varepsilon}\left[\frac{(t - s_\varepsilon)^2}{2\varepsilon} - \frac{(t_\varepsilon - s_\varepsilon)^2}{2\varepsilon}\right] \ge \frac{u(x_\varepsilon,t) - u(x_\varepsilon,t_\varepsilon)}{t - t_\varepsilon} \ge 0.$$

Sending $t \to t_\varepsilon^+$, the left term of the above inequality tends to $\frac{1}{\varepsilon(t_\varepsilon - s_\varepsilon)}$. Hence,

$$(7) \qquad \frac{t_\varepsilon - s_\varepsilon}{\varepsilon} \ge 0\,,$$

which obviously holds also if $t_\varepsilon = T$.

Combining (5) with (7), we obtain

$$H\left(x_\varepsilon, t_\varepsilon, \frac{x_\varepsilon - y_\varepsilon}{\varepsilon}\right) \leq 0$$

which, thanks to (H1), implies

$$\left|\frac{x_\varepsilon - y_\varepsilon}{\varepsilon}\right| \leq C$$

for some $C > 0$ independent of ε. We can then assume that there exists $\xi \in \mathbb{R}^N$ such that

$$\frac{x_\varepsilon - y_\varepsilon}{\varepsilon} \to \xi \qquad \text{as } \varepsilon \downarrow 0;$$

subtracting (6) from (5) and passing to the limit as $\varepsilon \downarrow 0$, we then get the contradiction $\alpha \leq 0$. □

Theorem 2. *Assume that* $H \in C(\overline{Q} \times \mathbb{R}^N)$ *satisfies* (H1); *let* u *be a nondecreasing with respect to* t *subsolution of problem* (3) *and let* v *be a supersolution such that*

$$u(x,t) = \limsup_{\substack{(y,s)\to(x,t)\\ y\in\Omega}} u(y,s)$$

for every $(x,t) \in \partial\Omega \times (0,T]$ *such that* $v(x,t) < 0$. *Then*

$$u \leq v \qquad \text{in } \overline{Q}\,.$$

Proof. For $\alpha > 0$, we set

$$v_\alpha(x,t) = v(x,t) + \alpha t\,.$$

It is immediate to check that v_α satisfies in the viscosity sense

$$\text{(8)} \qquad \begin{cases} -\alpha + (v_\alpha)_t + H(x,t,Dv_\alpha) \geq 0 & \text{in } \Omega \times (0,T], \\ v_\alpha(x,0) \geq u_0(x) & \text{in } \overline{\Omega}, \\ \max\{v_\alpha, -\alpha + (v_\alpha)_t + H(x,t,Dv_\alpha)\} \geq 0 & \text{on } \partial\Omega \times [0,T], \end{cases}$$

and we will get the statement from the inequality

$$\text{(9)} \qquad u \leq v_\alpha \qquad \text{in } \overline{Q}$$

in the limit as $\alpha \downarrow 0$.

If $u \leq v_\alpha$ on $\partial_p Q$ for all $\alpha > 0$, then, by Theorem 1, inequality (9) is true and we conclude. Thus we assume, by contradiction, that there exists $\alpha_0 > 0$ such that

$$m := \max_{\partial_p Q}(u - v_\alpha) > 0\,, \qquad \forall \alpha \leq \alpha_0\,.$$

Moreover, we observe that

$$M = \max_{\overline{Q}}(u - v_\alpha) = m\,, \qquad \forall \alpha \leq \alpha_0\,;$$

indeed, $v_\alpha + m$ is a supersolution of (3) satisfying

$$u \le v_\alpha + m \qquad \text{on } \partial_p Q$$

and so, again by Theorem 1,

$$u \le v_\alpha + m \qquad \text{in } \overline{Q}\,.$$

Let $(x_0, t_0) \in \partial_p Q$ such that

$$M = u(x_0, t_0) - v_\alpha(x_0, t_0) > 0\,.$$

Since, by Proposition 1, $u(x,0) \le u_0(x) \le v(x,0) = v_\alpha(x,0)$ in $\overline{\Omega}$, we have

$$(x_0, t_0) \in \partial\Omega \times (0, T]\,. \tag{10}$$

Moreover, from Proposition 2 it follows that $u(x_0, t_0) \le 0$ and thus

$$v_\alpha(x_0, t_0) < 0\,. \tag{11}$$

By assumption, there exists a sequence $\{(x_k, t_k)\}$ such that $x_k \in \Omega$ and

$$(x_k, t_k) \to (x_0, t_0), \quad u(x_k, t_k) \to u(x_0, t_0) \qquad \text{as } k \to \infty\,.$$

We set

$$\varepsilon_k = |x_k - x_0|, \quad \delta_k = |t_k - t_0|, \quad \lambda_k = d(x_k) = \operatorname{dist}(x_k, \partial\Omega)$$

and we define the function

$$\begin{aligned}\Phi_k(x,y,t,s) = u(x,t) - v_\alpha(y,s) - \frac{|x-y|^2}{\varepsilon_k} - \frac{(t-s)^2}{\delta_k}\\ - \left[\left(\frac{d(x)-d(y)}{\lambda_k} - 1\right)^-\right]^2 - |x-x_0|^2 - (t-t_0)^2\end{aligned}$$

which attains its maximum in $\overline{\Omega}^2 \times [0,T]^2$ at some point $(\bar{x}_k, \bar{y}_k, \bar{t}_k, \bar{s}_k)$. Setting then

$$M_k = \max_{\overline{\Omega}^2 \times [0,T]^2} \Phi_k(x,y,t,s) = \Phi_k(\bar{x}_k, \bar{y}_k, \bar{t}_k, \bar{s}_k)$$

we obtain

$$\begin{aligned}\Phi_k(x_k, x_0, t_k, t_0) &\le M_k \le u(\bar{x}_k, \bar{t}_k) - v_\alpha(\bar{y}_k, \bar{s}_k)\\ &\le \max_{\overline{\Omega}\times[0,T]} u - \min_{\overline{\Omega}\times[0,T]} v_\alpha = c_1\end{aligned} \tag{12}$$

and

$$\begin{aligned}\Phi_k(x_k, x_0, t_k, t_0) &= u(x_k, t_k) - v_\alpha(x_0, t_0) - \varepsilon_k - \delta_k - \varepsilon_k^2 - \delta_k^2\\ &\to u(x_0, t_0) - v_\alpha(x_0, t_0)\end{aligned} \tag{13}$$

as $k \to \infty$. Hence,

$$\frac{|\bar{x}_k - \bar{y}_k|^2}{\varepsilon_k} + \frac{(\bar{t}_k - \bar{s}_k)^2}{\delta_k} \le c_1 - \Phi_k(x_k, x_0, t_k, t_0) \le \text{constant.}$$

Therefore,

$$|\bar{x}_k - \bar{y}_k|\,,\ |\bar{t}_k - \bar{s}_k| \to 0 \quad \text{as } k \to \infty\,.$$

By compactness and by the upper semicontinuity of $u - v_\alpha$ it then follows

$$\limsup_{k\to\infty}\,[u(\bar{x}_k,\bar{t}_k) - v_\alpha(\bar{y}_k,\bar{s}_k)] \le M\,,$$

which, together with (12) and (13), implies

$$\lim_{k\to\infty} \Phi_k(\bar{x}_k,\bar{y}_k,\bar{t}_k,\bar{s}_k) = \lim_{k\to\infty} M_k = \lim_{k\to\infty}\,[u(\bar{x}_k,\bar{t}_k) - v_\alpha(\bar{y}_k,\bar{s}_k)] = M\,. \tag{14}$$

This in turn implies

$$\lim_{k\to\infty}\left\{\frac{|\bar{x}_k - \bar{y}_k|^2}{\varepsilon_k} + \frac{(\bar{t}_k - \bar{s}_k)^2}{\delta_k} + \left[\left(\frac{d(\bar{x}_k) - d(\bar{y}_k)}{\lambda_k} - 1\right)^-\right]^2 + |\bar{x}_k - x_0|^2 + (\bar{t}_k - t_0)^2\right\} = 0\,. \tag{15}$$

In particular, we find that

$$(\bar{x}_k,\bar{t}_k)\,,\ (\bar{y}_k,\bar{s}_k) \to (x_0,t_0) \quad \text{as } k\to\infty,$$

and, by (14) and the semicontinuity of u and v_α,

$$u(\bar{x}_k,\bar{t}_k) \to u(x_0,t_0) \quad v_\alpha(\bar{y}_k,\bar{s}_k) \to v_\alpha(x_0,t_0) \qquad \text{as } k\to\infty\,;$$

from (11) it then follows, for k sufficiently large, that

$$v_\alpha(\bar{y}_k,\bar{s}_k) < 0\,. \tag{16}$$

Furthermore, being

$$\left(\frac{d(\bar{x}_k) - d(\bar{y}_k)}{\lambda_k} - 1\right)^- \to 0 \text{ as } k\to\infty,$$

we have

$$\frac{d(\bar{x}_k) - d(\bar{y}_k)}{\lambda_k} - 1 \ge o_k(1)\,.$$

Hence, for k big enough,

$$d(\bar{x}_k) \ge d(\bar{y}_k) + \lambda_k(1 + o_k(1)) > 0\,,$$

which implies

$$\bar{x}_k \in \Omega\,. \tag{17}$$

By the maximality of $(\bar{x}_k,\bar{y}_k,\bar{t}_k,\bar{s}_k)$ and thanks to (16), (17) and Proposition 1, we can apply the definition of subsolution of problem (3) to u with test function

$$\varphi(x,t) = \frac{|x - \bar{y}_k|^2}{\varepsilon_k} + \frac{(t - \bar{s}_k)^2}{\delta_k} + \left[\left(\frac{d(x) - d(\bar{y}_k)}{\lambda_k} - 1\right)^-\right]^2 + |x - x_0|^2 + (t - t_0)^2$$

at the point $(\bar{x}_k, \bar{t}_k)$, as well as the definition of supersolution of (8) to v_α with test function

$$\psi(x,t) = -\left\{ \frac{|\bar{x}_k - y|^2}{\varepsilon_k} + \frac{(\bar{t}_k - s)^2}{\delta_k} + \left[\left(\frac{d(\bar{x}_k) - d(y)}{\lambda_k} - 1 \right)^- \right]^2 \right\}$$

at the point $(\bar{y}_k, \bar{s}_k)$. We obtain then the following inequalities:

$$\frac{2(\bar{t}_k - \bar{s}_k)}{\delta_k} + 2(\bar{t}_k - t_0) + H(\bar{x}_k, \bar{t}_k, p_k + q_k - \sigma_k Dd(\bar{x}_k)) \le 0 \tag{18}$$

and

$$-\alpha + \frac{2(\bar{t}_k - \bar{s}_k)}{\delta_k} + (\bar{t}_k - t_0) + H(\bar{y}_k, \bar{s}_k, p_k - \sigma_k Dd(\bar{y}_k)) \ge 0\,, \tag{19}$$

where we have set

$$p_k = \frac{2(\bar{x}_k - \bar{y}_k)}{\varepsilon_k}, \quad q_k = 2(\bar{x}_k - x_0), \quad \sigma_k = \frac{2}{\lambda_k}\left(\frac{d(\bar{x}_k) - d(\bar{y}_k)}{\lambda_k} - 1 \right)^- .$$

Arguing as in the proof of Theorem 1, from inequality (18) and the monotonicity of u with respect to t we obtain

$$H(\bar{x}_k, \bar{t}_k, p_k + q_k - \sigma_k Dd(\bar{x}_k)) \le 0$$

which implies, by (H1),

$$|p_k + q_k - \sigma_k Dd(\bar{x}_k)| \le C$$

for some positive C independent on k. From such an estimate, since $|Dd(\bar{x}_k)| = 1$, it follows that

$$\sigma_k \le C + |p_k + q_k| \le C_1(1 + |p_k|)\,,$$

and therefore, by the definition of p_k and by (15),

$$\sigma_k\, |\bar{x}_k - \bar{y}_k| \to 0 \qquad \text{as } k \to \infty\,.$$

Moreover, we have

$$|(p_k + q_k - \sigma_k Dd(\bar{x}_k)) - (p_k - \sigma_k Dd(\bar{y}_k))| \le |q_k| + L\sigma_k |\bar{x}_k - \bar{y}_k| \to 0$$

as $k \to \infty$, where L is the Lipschitz constant of $Dd(x)$ in a neighborhood of $\partial\Omega$. Therefore, having subtracted (19) from (18), we can pass to the limit for $k \to \infty$, getting the contradiction $\alpha \le 0$. □

Remark 1. In Theorems 1 and 2 the monotonicity assumption on the subsolution u can be relaxed to the following one: there exists a nonnegative constant c such that $t \in [0,T] \mapsto u(x,t) + ct$ is nondecreasing for every fixed $x \in \overline{\Omega}$.

Indeed, this implies that for every C^1 function $\varphi(t)$ such that $u(\cdot, x) - \varphi$ has a local maximum at $t_0 \in (0,T]$

$$\varphi'(t_0) \ge -c\,.$$

Moreover, if we replace assumption (H1) with

$$\lim_{|\xi|\to\infty} H(x,t,\xi) = -\infty \quad \text{uniformly w.r.t. } (x,t) \in \overline{\Omega} \times [0,T],$$

then we obtain analogous results for subsolutions and monotone nonincreasing supersolutions. □

The coercivity assumption (H1) has important consequences also for the regularity of the monotone in t subsolutions. Adapting the proof of Lemma 2.5, Section 2.4.2 of [2] we indeed have the following result.

Proposition 3. *Let $H \in C(\overline{Q} \times \mathbb{R}^N)$ satisfy* (H1) *and let u_0 be a Lipschitz continuous function in $\overline{\Omega}$ vanishing on $\partial\Omega$; if $u(x,t)$ is a bounded, u.s.c. in $\overline{Q}$ subsolution of problem* (3) *monotone nondecreasing with respect to t, then u is Lipschitz continuous in $\overline{Q}$.*

Proof. Let $x \in \Omega$ and $t \in [0,T]$ be fixed and, for $K > 0$ to be appropriately chosen, let $(\bar{y},\bar{s}) \in \overline{Q}$ be such that

$$\max_{\overline{Q}} \left\{ u(y,s) - K\sqrt{|y-x|^2 + (s-t)^2} \right\} = u(\bar{y},\bar{s}) - K\sqrt{|\bar{y}-x|^2 + (\bar{s}-t)^2}\,.$$

For $s = 0$ and $y \in \overline{\Omega}$, applying Proposition 1 and using the monotonicity of u, if $K \geq K_0$, where K_0 is the Lipschitz constant of u_0, it follows that

$$u(y,0) - K\sqrt{|y-x|^2 + t^2} \leq u_0(y) - K|y-x|$$
$$\leq u_0(x) - (K - K_0)|y-x| \leq u_0(x) \leq u(x,t)\,.$$

Hence $\bar{s} > 0$. Similarly, for $y \in \partial\Omega$ and $s \in [0,T]$, by Proposition 2 and choosing

$$K \geq \frac{\max_{\overline{Q}} |u|}{\operatorname{dist}(x,\partial\Omega)},$$

we obtain

$$u(y,s) - K\sqrt{|y-x|^2 + (s-t)^2} \leq -K|y-x|$$
$$\leq -K\operatorname{dist}(x,\partial\Omega) \leq -\max_{\overline{Q}} |u| \leq u(x,t)\,,$$

which implies $\bar{y} \in \Omega$.

Thus, if $(\bar{y},\bar{s}) \neq (x,t)$, we can use $\varphi(y,s) = K\sqrt{|y-x|^2 + (s-t)^2}$ as test function and $(\bar{y},\bar{s})$ as a maximum point of $u - \varphi$ in the definition of subsolution of (3) to obtain

$$K\frac{\bar{s}-t}{\sqrt{|\bar{y}-x|^2 + (\bar{s}-t)^2}} + H\left(\bar{y},\bar{s},K\frac{\bar{y}-x}{\sqrt{|\bar{y}-x|^2 + (\bar{s}-t)^2}}\right) \leq 0\,. \tag{20}$$

From the monotonicity of u with respect to t, arguing as in the proof of Theorem 1, it follows that

$$K\frac{\bar{s}-t}{\sqrt{|\bar{y}-x|^2 + (\bar{s}-t)^2}} \geq 0$$

and, consequently,

$$H\left(\bar{y},\bar{s},K\frac{\bar{y}-x}{\sqrt{|\bar{y}-x|^2+(\bar{s}-t)^2}}\right)\le 0\,.$$

The assumption (H1) then implies that there exists a positive constant C_1 independent of K such that

$$K\frac{|\bar{y}-x|}{\sqrt{|\bar{y}-x|^2+(\bar{s}-t)^2}}\le C_2\,,$$

and this in turn implies, using again (20) and the continuity of H, the existence of $C_2>0$ such that

$$K\frac{\bar{s}-t}{\sqrt{|\bar{y}-x|^2+(\bar{s}-t)^2}}\le C_2\,.$$

Combining these two inequalities we get

$$K\le K\frac{|\bar{y}-x|+(\bar{s}-t)}{\sqrt{|\bar{y}-x|^2+(\bar{s}-t)^2}}\le C_1+C_2\,;$$

if we choose $K>C_1+C_2$ we get then a contradiction from the above inequality. Hence $(\bar{y},\bar{s})=(x,t)$, so that

$$u(y,s)-u(x,t)\le K\sqrt{|y-x|^2+(s-t)^2}\qquad\forall\,(y,s)\in\overline{Q}\,.$$

It then follows, since K depends continuously on (x,t), that u is locally Lipschitz continuous in $\Omega\times[0,T]$; by the Rademacher theorem, u is almost everywhere differentiable and (see [2, 1]) satisfies

$$u_t+H(x,t,Du(x,t))\le 0\qquad\text{a.e. in } Q\,. \tag{21}$$

Moreover, since u is nondecreasing in t, it follows that $u_t\ge 0$ a.e. in Q. This implies

$$H(x,t,Du(x,t))\le 0\qquad\text{a.e. in } Q\,,$$

which in turn implies, by (H1), that $|Du(x,t)|$ is essentially bounded. Then, again from (21), also $u_t(x,t)$ is essentially bounded and we get the conclusion. □

3. On the parabolic regularization. This section is devoted to prove the existence, uniform estimates in the sup norm and monotonicity of the viscosity solution of a second order parabolic problem approximating (1).

We assume that conditions (H2) and (H3) are satisfied.

For every $\varepsilon>0$, let $u_0^\varepsilon\in C_0^\infty(\Omega)$ be a smooth function vanishing on $\partial\Omega$ satisfying

$$\left.\begin{array}{ll} u_0^\varepsilon\to u_0 & \text{uniformly in } \overline{\Omega},\\ Du_0^\varepsilon\to Du_0 & \text{locally uniformly in } \Omega,\\ \varepsilon\,\Delta u_0^\varepsilon\to 0 & \text{locally uniformly in } \Omega \end{array}\right\}\quad\text{as } \varepsilon\downarrow 0\,.$$

Set

$$h(x) = \begin{cases} H(x,0,Du_0(x)) & \text{if } x \in \Omega, \\ 0 & \text{if } x \notin \Omega, \end{cases}$$

and let $h_\varepsilon(x)$ be a standard mollified version of h. Then, h_ε is a smooth function with compact support in $\mathbb{R}^N$, locally uniformly convergent to $H(x,0,Du_0(x))$ in Ω and satisfying, thanks to (H3),

$$h_\varepsilon(x) \le 0 \qquad \text{in } \Omega\,. \tag{22}$$

Furthermore, let $\hat{H}_\varepsilon(x,t,\xi)$ be a bounded Lipschitz continuous function of class $C^1(\overline{Q}\times\mathbb{R}^N)$ nonincreasing with respect to t and locally uniformly convergent to H in $\overline{Q}\times\mathbb{R}^N$ and set

$$H_\varepsilon(x,t,\xi) = \hat{H}_\varepsilon(x,t,\xi) + \varepsilon\,\Delta u_0^\varepsilon(x) - \hat{H}_\varepsilon(x,0,Du_0^\varepsilon(x)) + h_\varepsilon(x). \tag{23}$$

Thus, H_ε is a bounded Lipschitz continuous function nonincreasing with respect to t and locally uniformly convergent to H in $\overline{Q}\times\mathbb{R}^N$.
We look for a viscosity solution of

$$\begin{cases} u_t^\varepsilon - \varepsilon\,\Delta u^\varepsilon + H_\varepsilon(x,t,Du^\varepsilon) = 0 & \text{in } Q, \\ u^\varepsilon(x,0) = u_0^\varepsilon(x) & \text{in } \Omega, \\ u^\varepsilon(x,t) = 0 & \text{on } \partial\Omega\times(0,T)\,. \end{cases} \tag{24}$$

Here by a viscosity solution we mean a continuous function u^ε in $\overline{\Omega}\times[0,T)$ satisfying the equation inside the domain in the viscosity sense, and pointwise equal to the boundary data (see [4]).

Theorem 3. *For any bounded Lipschitz continuous and C^1 function $G : \overline{Q}\times\mathbb{R}^N \to \mathbb{R}$ and for any smooth function $v_0 \in C_0^\infty(\Omega)$ there exists a unique classical solution v of*

$$\begin{cases} v_t - \Delta v + G(x,t,Dv) = 0 & \textit{in } Q, \\ v(x,0) = v_0(x) & \textit{in } \Omega, \\ v(x,t) = 0 & \textit{on } \partial\Omega\times(0,T)\,. \end{cases} \tag{25}$$

Moreover, there exists a constant $M > 0$ depending only on $\|G(x,t,0)\|_{L^\infty(Q)}$, $\|v_0\|_{L^\infty(\Omega)}$ and T such that

$$\|v\|_{L^\infty(Q)} \le M\,. \tag{26}$$

Proof. The uniqueness and the uniform bound (26) follow from the maximum principle for smooth solutions of (25) (see e.g. [12]) and from the fact that the functions $\pm(C_1 + C_2 t)$ are respectively a supersolution and a subsolution, provided that the constants C_1 and C_2 satisfy

$$C_1 \ge \|G(x,t,0)\|_{L^\infty(Q)}\,, \quad C_2 \ge \|v_0\|_{L^\infty(\Omega)}\,.$$

Concerning the existence of solutions of (25) we can first apply the general theory of pseudo-monotone operators (see [8, 9]) in order to obtain a weak solution v, that is a function v belonging to $L^2(0,T;H_0^1(\Omega))$ with $v_t \in L^2(0,T;H^{-1}(\Omega))$ satisfying

$$-\int_0^T < w_t, v > + \int_\Omega v(T)\,w(T) + \int_Q Dv\,Dw + \int_Q G(x,t,Dv)\,w = \int_\Omega v_0 w(0),$$

for any $w \in L^2(0,T;H_0^1(\Omega))$ with $w_t \in L^2(0,T;H^{-1}(\Omega))$.
Next we observe that v satisfies (in the previous weak sense)

$$v_t - \Delta v \in L^\infty(Q)$$

so that, from the smoothness of the boundary data and the parabolic regularity theory (see [7]), we obtain

$$v,\ v_t,\ v_{x_i},\ v_{x_i x_j} \in L^q(Q), \qquad \forall q > 1,\ \forall 1 \le i,\ j \le N.$$

In particular, v belongs to $C(\overline{Q})$ and, for every fixed $t \in [0,T]$, $v(\cdot,t)$ belongs to $C^1(\overline{\Omega})$.

Furthermore, deriving with respect to x_i the equation in (25) and using the same regularity result in its local form we obtain

$$v_{x_i t},\ v_{x_i x_j x_k} \in L^q_{\mathrm{loc}}(Q), \qquad \forall q > 1,\ \forall 1 \le i,\ j,\ k \le N,$$

so that v is in $C^1(Q)$, $v(\cdot,t)$ is in $C^2(\Omega)$ for every fixed $t \in [0,T]$ and v is a classical solution of (25). □

Applying the previous Theorem to problem (24) we get, for every $\varepsilon > 0$, the existence of a unique classical (and therefore viscosity) solution u^ε which, moreover, satisfies the uniform bound

$$\|u^\varepsilon\|_{L^\infty(Q)} \le M \qquad \forall \varepsilon > 0, \tag{27}$$

with M depending on $\|H(x,t,0)\|_{L^\infty(Q)}$, $\|u_0\|_{L^\infty(\Omega)}$ and T but being independent of ε.

Proposition 4. *If assumptions* (H2) *and* (H3) *hold then* $u^\varepsilon(x,t)$ *is monotone nondecreasing in* t *for each fixed* $x \in \overline{\Omega}$.

Proof. By the definition (23) of H_ε, by its monotonicity in t and by inequality (22) it follows that u_0^ε is a stationary classical subsolution of (24). The comparison principle applied to problem (24) yields then

$$u^\varepsilon(x,t) \ge u_0^\varepsilon(x), \qquad \forall (x,t) \in \overline{Q}. \tag{28}$$

Let $t_0 \in [0,T)$ be fixed; then (H2) and (28) imply that the function $v^\varepsilon(x,t) = u^\varepsilon(x,t+t_0)$ is a viscosity supersolution of (24) in $\Omega \times (0,T-t_0)$ so that, again by the comparison principle, it follows that

$$v^\varepsilon(x,t) = u^\varepsilon(x,t+t_0) \ge u^\varepsilon(x,t), \qquad \forall (x,t) \in \Omega \times (0,T-t_0)$$

as claimed. □

4. Proof of Theorem 0. Thanks to the estimate (27), the weak limits

$$\overline{u}(x,t) = \limsup_{\delta \downarrow 0} \{u^\varepsilon(y,s) : |y-x| \le \delta,\ |s-t| \le \delta,\ \varepsilon \le \delta\}$$

and

$$\underline{u}(x,t) = \liminf_{\delta \downarrow 0} \{u^\varepsilon(y,s) : |y-x| \le \delta,\ |s-t| \le \delta,\ \varepsilon \le \delta\}$$

are well-defined. It is not hard to check, using Proposition 4, that $\overline{u}$, $\underline{u}$ are monotone nondecreasing in t.

By a standard technique in the theory of viscosity solutions (see e.g. [1, 2]) one also proves that $\overline{u}(x,t)$ is a subsolution and $\underline{u}(x,t)$ is a supersolution of problem (1).

Moreover, we have $\overline{u}(x,0) \geq u_0(x)$ and $\overline{u}(x,t) \geq 0$ in $\partial\Omega \times (0,T)$ by definition, and $\overline{u}(x,0) \leq u_0(x)$, $\overline{u}(x,t) \leq 0$ in $\partial\Omega \times (0,T)$ by Proposition 1 and 2, respectively. Hence

$$\begin{aligned} \overline{u}(x,0) &= u_0(x) \quad && \text{in } \overline{\Omega}, \\ \overline{u}(x,t) &= 0 && \text{on } \partial\Omega \times (0,T). \end{aligned}$$

Analogously, by Proposition 1 and by definition, it follows that

$$\begin{aligned} \underline{u}(x,0) &= u_0(x) \quad && \text{in } \overline{\Omega}, \\ \underline{u}(x,t) &\leq 0 && \text{on } \partial\Omega \times (0,T). \end{aligned}$$

Following [3], we redefine $\overline{u}$ on $\partial\Omega \times (0,T)$ in order to apply Theorem 2; setting

$$\overline{u}_*(x,t) = \begin{cases} \overline{u}(x,t) & \text{if } (x,t) \in \Omega \times [0,T], \\ \limsup\limits_{\substack{(y,s)\to(x,t) \\ y\in\Omega}} \overline{u}(y,s) & \text{if } (x,t) \in \partial\Omega \times [0,T], \end{cases}$$

we obtain that $\overline{u}_*$ is still a subsolution of (1) and monotone nondecreasing with respect to t, with $\overline{u}_*(x,t) \leq \overline{u}(x,t)$. Moreover, $\overline{u}_*$ and $\underline{u}$ satisfy the assumptions of Theorem 2, so that

$$\overline{u}_* \leq \underline{u} \qquad \text{in } \overline{Q}.$$

Being $\overline{u}_* \geq \underline{u}$ by definition, one concludes that

$$\overline{u}_* = \underline{u} \qquad \text{in } \overline{Q},$$

and that the sequence $\{u^\epsilon\}$ locally uniformly converges to the solution $u = \overline{u}_* = \underline{u}$.

We further observe that the upper semicontinuous function $\overline{u}(x,t)$, which coincides with u inside Q, is a viscosity solution in Q and it satisfies in the classical sense the boundary conditions. This fact expresses the incompatibility between the equation inside the domain Q and the boundary conditions: if the boundary data are pointwise assumed, the continuity of the solution must not be expected.

Finally, the uniqueness of the solution is an immediate consequence of the Comparison Theorem 2 while the Lipschitz continuity follows from Proposition 3 and assumptions (H1) and (H3) (which in particular imply that u_0 is Lipschitz continuous in $\overline{\Omega}$). □

References

[1] M. Bardi, I. Capuzzo Dolcetta: *Optimal Control and Viscosity Solutions of Hamilton-Jacobi-Bellman Equations*, Birkhäuser, Boston 1997

[2] G. Barles: *Solutions de viscosité des équations de Hamilton-Jacobi*, Math. Appl. **17**, Springer, Berlin 1994

[3] G. Barles, B. Perthame: *Comparison principle for Dirichlet-type Hamilton-Jacobi equations and singular perturbations of degenerated elliptic equations*, Appl. Math. Opt. **21** (1990), 21-44

[4] M. G. Crandall, H. Ishii, P. L. Lions: *User's guide to viscosity solutions of second order partial differential equations*, Bull. Amer. Math. Soc. **27** (1992), 1-67

[5] H. Ishii: *A boundary value problem of the Dirichlet type for Hamilton-Jacobi equations*, Ann. Scuola Norm. Pisa **16** (1989), 105-135

[6] H. Ishii: *Hamilton-Jacobi equations with discontinuous hamiltonians on arbitrary open sets*, Bull. Fac. Sci. Engin. Chuo Univ. **28** (1985), 33-77

[7] O. Ladyzhenskaya, V. Solonnikov, N. Uraltseva: *Linear and Quasilinear Equations of Parabolic Type* [in Russian], Nauka, Moscow 1967; Engl. transl.: Amer. Math. Soc., Providence RI 1968

[8] J. L. Lions: *Quelques méthodes de résolution des problèmes aux limites non linéaires*, Gauthier Villars, Paris 1969

[9] J. L. Lions: *Sur certaines équations paraboliques non linéaires*, Bull. Soc. Math. France **93** (1965), 155-175

[10] P. L. Lions: *Generalized solutions of Hamilton-Jacobi equations*, Research Notes Math. **69**, Pitman, Boston 1982

[11] P. L. Lions, B. Perthame: *Remarks on Hamilton-Jacobi equations with measurable time dependent hamiltonians*, Nonlin. Anal. TMA **11, 5** (1987), 613-622

[12] M. H. Protter, H. F. Weinberger: *Maximum principles in differential equations*, Prentice Hall, New York 1967

Italo Capuzzo Dolcetta, Dipartimento di Matematica, Università di Roma "La Sapienza", Piazzale A. Moro 2, I-00185 Roma, Italy; `capuzzo@mat.uniroma1.it`

Fabiana Leoni, Dipartimento di Matematica, Università di Roma "La Sapienza", Piazzale A. Moro 2, I-00185 Roma, Italy; `leoni@mat.uniroma1.it`

Progress in Nonlinear Differential Equations
and Their Applications, Vol. 40

Some Remarks on a Nonlinear Model of Competitive Equilibrium

ANTONIO CARBONE, PETR P. ZABREJKO

Ad Alfonso con affetto e stima

Summary: We provide an alternative approach to a problem proposed by Scarf and Hansen to characterize a competitive equilibrium in a nonlinear complementary problem.

Keywords: Complementary problem, competitive equilibrium, fixed point principle.

Classification: 47N10, 47H10.

In this note we are concerned with the following problem which was proposed by Scarf and Hansen [6] (see also [3]) to characterize a competitive equilibrium: given vectors $b \in \mathbb{R}^m$, $c \in \mathbb{R}^n$, a matrix $A \in \mathbb{R}^{m \times n}$, and a function $d : \mathbb{R}^m \to \mathbb{R}^m$, find vectors $p_* \in \mathbb{R}^m$, $y_* \in \mathbb{R}^n$ with nonnegative components such that the following conditions are satisfied:

(1) $$p_* \geq 0, \qquad c - A^t p_* \geq 0,$$

(2) $$y_* \geq 0, \qquad b + Ay_* - d(p_*) \geq 0,$$

(3) $$(c - A^t p_*)^t y_* = 0,$$

and

(4) $${p_*}^t(b + Ay_* - d(p_*)) = 0.$$

(Here and in what follows the inequality $x \leq y$ is meant coordinate-wise, and A^t denotes the transposed matrix of A.) All the terms occurring in this problem, as well as the set of conditions (1)-(4), have a precise meaning in this economical model. Thus, b is the vector of endowments, c is the vector of unit costs of operating the activities, A is the so-called "technology matrix" containing (positive) output and (negative) input entries. The unknown vectors p and y describe the prices and activity levels, respectively. Finally, $p \mapsto d(p)$ is the market demand function; since we do not require d to be linear, the system (1)-(4) is a *nonlinear equilibrium problem*.

The significance of the conditions (1)-(4) is the following: (1) means that no activity earns a positive profit, (2) means that no commodity is in excess demand, (3) means that an activity earning a deficit is not used, while an operated activity has no loss, and (4) means that a commodity in excess supply has zero price, while a positive price implies market clearance.

The equilibrium problem (1)-(4) is a typical example of a (finite-dimensional) *complementarity problem.* In fact, if we write $z = (y, p)$ and define a function $F : \mathbb{R}^{n+m} \to \mathbb{R}^{n+m}$ by

$$(5) \qquad F(z) = F(y,p) = (c - A^t p_*, b + Ay_* - d(p_*)),$$

the above equilibrium problem may be stated equivalently as follows: *find a vector $z_* = (y_*, p_*)$ in the nonnegative cone $\mathbb{R}_+^{n+m}$ of $\mathbb{R}^{n+m}$ such that*

$$(6) \qquad F(z_*) \in \mathbb{R}_+^{n+m}, \qquad \langle z_*, F(z_*)\rangle = 0,$$

where $\langle \cdot, \cdot \rangle$ denotes the usual scalar product in $\mathbb{R}^{n+m}$.

Usually, such a complementary problem is reduced to *a fixed point problem*; for the problem (6), this may be done quite easily as follows. Denote by P the *metric projection* of $\mathbb{R}^{n+m}$ onto the cone $\mathbb{R}_+^{n+m}$, i.e. $P(x)$ is, for any $x \in \mathbb{R}^{n+m}$, the unique point in $\mathbb{R}_+^{n+m}$ such that

$$||x - P(x)|| = \inf \{||x - z|| : z \in \mathbb{R}_+^{n+m}\}.$$

Defining then an operator $T : \mathbb{R}_+^{n+m} \to \mathbb{R}^{n+m}$ by $T(x) = P(x - F(x))$, *every solution z_* of the complementary problem* (6) *is a fixed point of the operator T, and vice versa.* This is an obvious consequence of the (geometrically evident) fact that the point $P(x)$ of best approximation to x may be characterized by the orthogonality condition $\langle P(x), x - P(x)\rangle = 0$.

Many interesting results and examples illustrating the interconnections between complementary problems and fixed point theorems can be found in the book [2]. However, when applying these connections in practice, one may encounter serious difficulties. Suppose, for example, that we want to apply the classical *Brouwer fixed point principle* to the operator T given above. This requires not only a continuity condition on the operator T (which may be achieved by imposing an appropriate continuity condition on the market demand function d), but also the existence of an invariant set $C \subseteq \mathbb{R}_+^{n+m}$ for T which is nonempty, convex and compact. Now, the natural candidate for this set, namely $C = \mathbb{R}_+^{n+m}$ is nonempty, convex, and closed, but *not compact*! Thus one has to impose additional *apriori estimates* to the problem which often are either unnatural or very hard to verify. For example, the following condition proposed by Karamardian [4] is simple in theory but difficult to check in practice: *Suppose that one can find a compact convex set $C \subseteq \mathbb{R}_+^{n+m}$ such that, for every $x \in \mathbb{R}_+^{n+m} \setminus C$, there exists an $x \in C$ with $\langle z, F(x)\rangle > \langle x, F(x)\rangle$. Then the complementary problem for F has a solution.*

Let us describe another approach to the original problem on the existence of an equilibrium (1)-(4) and the corresponding complementary problem for F. Consider the following pair of dual linear optimization problems with parameter q:

(A) Maximize the linear function

$$(7) \qquad \pi(q;p) = \langle d(q) - b, p\rangle \qquad (q \in \mathcal{P})$$

on the set $\mathcal{P}$ which is defined by the formula

$$(8) \qquad \mathcal{P} = \{p : p \geq 0, \ A^* p \leq c\}.$$

(B) Minimize the linear function

$$\eta(y) = \langle y, c\rangle \tag{9}$$

on the set $\mathcal{Q}(q)$ ($q \in \mathcal{P}$), which is defined by the formula

$$\mathcal{Q}(q) = \{y : y \geq 0,\ Ay \geq d(q) - b\}. \tag{10}$$

Observe that in the first problem the maximizing function depends on the parameter $q \in \mathcal{P}$, but the set does not depend on this parameter. In the second problem we have the opposite situation: the minimizing function does not depend on the parameter, but the set does!

Using the classical duality theory of linear optimization problems, under standard assumptions on the solvability of both problems we have

$$\max_{p\in\mathcal{P}} \pi(q;p) = \min_{y\in\mathcal{Q}(q)} \eta(y) \tag{11}$$

for all values of the parameter $q \in \mathcal{P}$.

Let

$$T(q) = \left\{p \in \mathcal{P} : \langle d(q) - b, p\rangle = \max_{p\in\mathcal{P}} \langle d(q) - b, p\rangle\right\}. \tag{12}$$

As is well-known, for each $q \in \mathcal{P}$, the set $T(q)$ is a nonempty convex and closed subset of the set $\mathcal{P}$. This means that (12) defines a multivalued operator T defined on $\mathcal{P}$ and leaving $\mathcal{P}$ invariant.

Let p_* be a fixed point of this operator T, i.e. $p_* \in T(p_*)$. By definition this means that

$$p_* \in \left\{\tilde{p} \in \mathcal{P} : \langle d(p_*) - b, \tilde{p}\rangle = \max_{p\in\mathcal{P}} \langle d(p_*) - b, p\rangle\right\}, \tag{13}$$

and, due to (11),

$$\pi(p_*, p_*) = \max_{p\in\mathcal{P}} \pi(p_*, p) = \min_{y\in\mathcal{Q}(p_*)} \eta(y).$$

In particular, if y_* is an arbitrary solution of the second problem on the set $\mathcal{Q}(p_*)$ then

$$\langle d(p_*) - b, p_*\rangle = \langle y_*, c\rangle. \tag{14}$$

Theorem 1. *Let p_* be a fixed point of the operator T, and let $y_* \in \mathcal{Q}(p_*)$ satisfy the condition* (14). *Then (y_*, p_*) is an equilibrium for the problem* (1)-(4) *or, equivalently, a solution of the corresponding complementarity problem for F.*

Proof. It is sufficient to prove the equalities (3) and (4). But, by definition, we have

$$\langle y_*, c - A^t p_*\rangle \geq 0, \tag{15}$$

and

$$\langle b + Ay_* - d(p_*), p_*\rangle \geq 0. \tag{16}$$

Adding both inequalities we get

$$\langle y_*, c - A^t p_* \rangle + \langle b + Ay_* - d(p_*), p_* \rangle = \langle y_*, c \rangle + \langle b - d(p_*), p_* \rangle = 0.$$

Thus, the sum of two nonnegative terms is equal to zero, and so both these numbers are zero. □

Theorem 1 reduces the original problem to the investigation fixed points of a multivalued operator T acting in a closed and convex set $\mathcal{P}$.

Let us assume that the demand function $d(p)$ is continuous. In this case the function $\pi(q;p)$ is continuous with respect to the parameter $q \in \mathcal{P}$. By virtue of well-known results on the continuity of the set of minimum points in linear optimization problems one can see that the operator T is lower semicontinuous. This means that we can use the classical Kakutani fixed point principle.

Theorem 2. *Assume that the set $\mathcal{P}$ is bounded and the demand function $d(p)$ is continuous. Then the equilibrium problem* (1)-(4) *or, equivalently, the corresponding complementarity problem for F is solvable.*

Proof. Under the hypotheses of the theorem, the multivalued operator T leaves the compact convex set $\mathcal{P}$ invariant and is lower semicontinuous. Due to the Kakutani fixed point theorem it has a fixed point $x_* \in \mathcal{P}$. Furthermore, the problem (7) for $q = p_*$ is solvable; due to the classical duality theorem, the problem (8) for $q = p_*$ is also solvable. If $y_* \in \mathcal{Q}(p_*)$ is a solution of problem (8), then the pair (y_*, p_*) is a solution of the original equilibrium problem (1)-(4) by Theorem 1. □

We omit the proof the following statement which repeats the proof of Theorem 2.

Theorem 3. *Assume that there exists a compact convex set $\mathcal{P}_0 \subseteq \mathcal{P}$ such that $T(\mathcal{P}_0) \cap \mathcal{P}_0 \neq \emptyset$ and the demand function $d(p)$ is continuous. Then the equilibrium problem* (1)-(4) *or, equivalently, the corresponding complementarity problem for F is solvable.*

It seems that the first condition of Theorem 3 is artificial and cumbersome. However, if the range $d(\mathcal{P})$ of the set $\mathcal{P}$ for the demand function $d(p)$ is sufficiently small (which actually means that $d(p)$ is a slowly changing function) the existence of such an invariant set is natural; this fact can be obtained on the base of standard results concerning properties of extremal sets for linear optimization problems. In particular, if the image $d(\mathcal{P})$ is a compact set one can construct a compact and invariant for T set $\mathcal{P}_0$ as the convex hull of a finite number of points of the set $\mathcal{P}$.

The following result is also a corollary of Theorem 3.

Theorem 4. *Assume that, for each $q \in \mathcal{P}$, the inequalities*

$$\sup_{p \in \mathcal{P}} \langle d(q) - b, p \rangle < \infty \qquad (q \in \mathcal{P}) \tag{17}$$

hold and the demand function $d(p)$ is continuous. Then the equilibrium problem (1)-(4) *or, equivalently, the corresponding complementarity problem for F is solvable.*

Proof. It is well-known (see e.g. [1,5]) that the set $\mathcal{P}$ can be represented in the form

$$\mathcal{P} = \mathcal{P}_0 + W,$$

where $\mathcal{P}_0$ is a compact convex set and W a wedge. Both the set $\mathcal{P}_0$ and the wedge W are uniquely defined; the first of them is called the *essential set* and the second is the *asymptotic wedge* of $\mathcal{P}$. The linear optimization problem (7) for some $q \in \mathcal{P}$ is solvable if and only if

$$d(q) - b \in -W^\circ, \tag{18}$$

where W° is the polar to W; moreover, this solvability condition is equivalent to the inequality (17) with the same $q \in \mathcal{P}$. Furthermore, in the case of solvability of the problem (7) for $q \in \mathcal{P}$, the equality

$$\max_{p\in\mathcal{P}} \pi(q,p) = \max_{p\in\mathcal{P}_0} \pi(q,p)$$

holds. This means that for each $q \in \mathcal{P}$, in the case of solvability the problem (7) with this q, we have

$$T(\mathcal{P}_0) \cap \mathcal{P}_0 \neq \emptyset.$$

Thus, the compact convex set $\mathcal{P}_0$ satisfies the basic condition of Theorem 3; applying this theorem we obtain the statement of Theorem 4. □

The hypotheses of Theorem 4 are effective: using standard methods of optimization theory one can find the essential set $\mathcal{P}_0$ and the asymptotic wedge W in explicit form and check the condition (17) or the equivalent condition (18) without difficulties.

Let us make some final remarks. First, the statement of Theorem 4 really means that the original nonlinear problem (1)-(4) or the equivalent complementarity problem for F is always solvable in the case of solvability of the family of linear optimization problems (7). We omit the equivalent statement in terms of solvability of the family of dual problems (8); the conditions of solvability of these problems are more complicated than the conditions of Theorem 4.

Second, using a more delicate reasoning one can generalize Theorem 4 to the case when the optimization problems (7) and (8) are solvable only for some values of the parameter $q \in \mathcal{P}$ (actually it is sufficient to consider only values q from the set $\mathcal{P}_0$).

Finally, Theorems 1-4 can be generalized to the case when the demand function $d(p)$ is semicontinuous in a suitable sense.

References

[1] J.-P. AUBIN: *Mathematical Methods of Game and Economic Theory*, North-Holland Publ. Comp., Amsterdam 1979

[2] K. C. BORDER: *Fixed Point Theorems and Applications to Economics and Game Theory*, Cambridge Univ. Press, Cambridge 1985

[3] G. ISAC: *Complementarity Problems*, Lect. Notes Math. **1528**, Springer, Berlin 1992

[4] S. KARAMARDIAN: *A generalized complementarity problem*, J. Optim. Theory Appl. **8** (1971), 161-168

[5] R. Rockafellar: *Convex Analysis*, Princeton Univ. Press, Princeton 1970

[6] H. E. Scarf, T. Hansen: *Computation of Economic Equilibria*, Yale Univ. Press, New Haven 1983

Antonio Carbone, Università della Calabria, Dipartimento di Matematica, I-87036 Arcavacata di Rende (CS), Italy; carbonea@unical.it

Petr P. Zabrejko, Belgosuniversitet, Matematicheskij Fakul'tet, pr. Skariny 4, BY-220050 Minsk, Belorussia; zabreiko@mmf.bsu.unibel.by

Progress in Nonlinear Differential Equations
and Their Applications, Vol. 40

Almost Discrete Convergence

EMANUELE CASINI, PIER LUIGI PAPINI

*Dedicated with great sympathy to Alfonso Vignoli
on the occasion of his 60th birthday*

Summary: We define two notions of convergence which seem to be of some interest, also for their meaning in probability theory, and which, apparently, have not yet been considered explicitly in the literature. We study these notions and compare them with the most familiar notions of convergence.

Keywords: Measurable function, discrete convergence, almost discrete convergence, discrete convergence in measure, Lusin theorem, Severini-Egorov theorem.

Classification: 28A20.

Acknowledgement: The authors wish to thank the referee who pointed out some unclear and misleading statements in a previous version of the paper.

1. Introduction. Let X be a nonempty set, and let f, f_n ($n \in \mathbb{N} = \{1, 2, \ldots\}$) be real valued functions defined on X. We may speak of convergence of the sequence $\{f_n\}$ to f in several ways. We list a few notions among those commonly considered, and we indicate the abbreviations we shall use in the following: convergence everywhere (E), almost everywhere (AE), uniform (U), almost uniform (AU), in measure (M), in L_p ($1 \leq p \leq \infty$). Note that convergence in L_∞ means uniform convergence outside a set of measure 0, and so it implies, but is different from, (AU) convergence. The implications existing among the classical types of convergence are indicated in many books: see e.g. [12, p. 207] or the nice picture on p. 75 of [2], or also [11] for these and several other kinds of convergence. All the indicated notions, apart from (E) and (U) convergence, make sense only if $X = (X, S, \mu)$ is a measure space and f, f_n are measurable functions. We denote the null function by Θ; also, we use the usual abbreviation a.e. for almost everywhere.

2. Discrete convergence. In [6] the following notion was introduced:

Definition 1. A function f is said to be the *discrete limit* of a sequence $\{f_n\}$ if for every $x \in X$ there exists $n_0 = n_0(x)$ such that $f_n(x) = f(x)$ for $n \geq n_0$. In this case we say that $\{f_n\}$ converges *discretely* to f and write $f_n \xrightarrow{D} f$.

It is clear that this notion of convergence can behave nicely, but it is somehow too restrictive to be used; in particular, it implies convergence everywhere. Note that, for example, any discrete function can be considered as discrete limit of functions assuming only a finite number of values.

We may weaken the notion of discrete convergence in the following way.

Definition 2. We say that $\{f_n\}$ converges to f *discretely almost everywhere* if there exists a set M such that $\mu(M) = 0$ and $f_n \xrightarrow{D} f$ in $X \setminus M$. In this case we shall write: $f_n \xrightarrow{D\,a.e.} f$.

The above terminology is motivated by the fact that $f_n \xrightarrow{D} f$ (or also $f_n \xrightarrow{D\,a.e.} f$) exactly when, for every $x \in X$, $f_n(x) \to f(x)$ (respectively, $f_n \xrightarrow{a.e.} f$) in the discrete topology of $\mathbb{R}$.

It is well known that the classical properties of the space of measurable functions $f : X \to Y$ like completeness or Severini-Egorov's theorem also hold if the image space Y is a complete (generalized) pseudo-metric space (see e. g. [19]). We remark that, for this extension, the notion of measurable function heavily depends on the topology of Y. In particular, in case $Y = \mathbb{R}$ with the discrete topology, measurability of f implies that the values $f(x)$ are countable for almost all $x \in X$. Here we consider instead "measurable functions" as functions which are measurable, according to the measure space X, when $Y = \mathbb{R}$ is endowed with the Euclidean topology.

The notion considered in Definition 2 is still very strong; for example, it implies (AE) convergence (thus (AU) convergence if $\mu(X) < \infty$), while the converse is trivially false. Yet, it is of some interest for applications. For example if we consider phenomena which can be described in terms of Markov chains with absorbing states, then the sequence of random variables $\{X_n\}$ describing the system at time n may converge under some natural assumptions to the limit distribution in this sense (see also [15]).

The notion of discrete convergence was considered in [16], where also the notion of discrete a.e. convergence was implicitly used; the convergence $f_n \xrightarrow{D} f$ was indicated there with the phrase "$\{f_n\}$ merges with f".

Let us introduce some notation. Given a measure space X and two real functions f, g on X, set

$$E_{f,g} = \{x \in X : f(x) = g(x)\}$$

and

$$E'_{f,g} = \{x \in X : f(x) \neq g(x)\}.$$

Also, for $f, g : X \to \mathbb{R}$ and $\alpha > 0$, let

$$E^{\alpha}_{f,g} = \{x \in X : \ |f(x) - g(x)| > \alpha\},$$

and observe that

(1)
$$E'_{f,g} = \bigcup_{\alpha>0} E^{\alpha}_{f,g}\,.$$

Given a sequence $\{f_n\}$, for $n \in \mathbb{N}$ set

$$E_n = \{x \in X : \ f_{n+k}(x) = f_n(x) \ \text{ for all } \ k \in \mathbb{N}\} = \bigcap_{k=1}^{\infty} E_{f_n, f_{n+k}}\,.$$

Thus,

$$X \setminus E_n = \bigcup_{k=1}^{\infty} E'_{f_n,f_{n+k}} = \bigcup_{i,j=n}^{\infty} E'_{f_i,f_j}. \tag{2}$$

Of course, $\{E_n\}$ is a nondecreasing sequence of subsets of X; moreover, $f_n \xrightarrow{D} f$ for some f is equivalent to $X = \bigcup_{n=1}^{\infty} E_n$. Also, if $f_n \xrightarrow{D} f$, then

$$X \setminus E_n = \{x \in X : \, x \in E'_{f_{n+k},f} \text{ for some } k \geq 0\} = \bigcup_{k=n}^{\infty} E'_{f_k,f}. \tag{3}$$

Note that

$$E'_{f_n,f} \subseteq X \setminus E_n \tag{4}$$

for every $n \in \mathbb{N}$.

Remark 1. If $f_n \xrightarrow{D} f$ on a sequence $\{X_j\}$ of subsets of X, then $f_n \xrightarrow{D} f$ also on $\bigcup_{j=1}^{\infty} X_j$. The same is true for the union of any family of sets. Moreover, given $\{f_n\}$, by defining f on $Y = \bigcup_{n=1}^{\infty} E_n$ in the obvious way, we see that $f_n \xrightarrow{D} f$ in Y.

We put into evidence a simple fact.

Proposition 1. *For a sequence $\{f_n\}$ of measurable real functions in X consider the following conditions:*

(i) $\lim_{n\to\infty} \mu(X \setminus E_n) = 0$;

(ii) $f_n \xrightarrow{D\,a.e.} f$ *for some measurable function* f;

(ii') *for any $\varepsilon > 0$ there exists a set A_ε such that $\mu(A_\varepsilon) < \varepsilon$ and $f_n \xrightarrow{D} f$ on $X \setminus A_\varepsilon$ for some measurable function f.*

Then (i) $\Rightarrow$ (ii) $\Leftrightarrow$ (ii'). *Moreover, if $\mu(X) < \infty$, then the three conditions are equivalent.*

Proof. (i) $\Rightarrow$ (ii): Given $\{f_n\}$, define f on $\bigcup_{n=1}^{\infty} E_n$ in the obvious way; we have $f_n \xrightarrow{D} f$ on $\bigcup_{n=1}^{\infty} E_n$. If (i) holds, then $\mu(X \setminus E_n)$ is finite for n large enough and $\{X \setminus E_n\}$ is a nonincreasing sequence, so

$$\mu(X \setminus \bigcup_{n=1}^{\infty} E_n) = \mu\left(\bigcap_{n=1}^{\infty} (X \setminus E_n)\right) = \lim_{n\to\infty} \mu(X \setminus E_n) = 0.$$

Consequently, the equivalence class of f is well defined in X, and we have (ii).

(ii) $\Rightarrow$ (ii') is obvious.

(ii') $\Rightarrow$ (ii): Given $n \in \mathbb{N}$, let A_n such that $\mu(X \setminus A_n) < 1/n$ and $f_n \xrightarrow{D} f$ in A_n. Then $f_n \xrightarrow{D} f$ on $\bigcup_{n=1}^{\infty} A_n$ (the limit function f is defined in the obvious way). Moreover, $\mu(X \setminus \bigcup_{n=1}^{\infty} A_n) < 1/n$ for every n, thus $\mu(X \setminus \bigcup_{n=1}^{\infty} A_n) = 0$: so we have (ii).

Now let $\mu(X) < \infty$. Assume (ii), so $f_n \xrightarrow{D\,a.e.} f$; then

$$0 = \mu(X \setminus \bigcup_{n=1}^{\infty} E_n) = \mu\left(\bigcap_{n=1}^{\infty} (X \setminus E_n)\right) = \lim_{n\to\infty} \mu(X \setminus E_n),$$

so (i) holds. This concludes the proof. □

Remark 2. The next example shows that the implication (ii) ⇒ (i) is not true in general if $\mu(X) = \infty$. The proof of the last part of Proposition 1 shows that it is true e.g. if some "finiteness restriction" of this type (weaker that (i)) is satisfied (cf. [3, p. 629]), namely:

(f) there exists a natural number n such that $\mu(X \setminus E_n) < \infty$.

Example 1. The sequence $f_n(x) = \chi_{[n,+\infty)}(x)$, in the usual Lebesgue space $(\mathbb{R}, S, \mu)$, satisfies (ii) but not (i) (we have $f_n \xrightarrow{D} \Theta$, the null function over $\mathbb{R}$). Also, $\{f_n\}$ is not (AU) convergent (nor is any of its subsequences) to Θ; the same holds for (M) convergence.

Note that (i) implies (see (ii)) that $f_n \xrightarrow{D\,a.e.} f$ for some f, so (see (3)):

(j) $\lim_{n\to\infty} \mu\left(\bigcup_{k=n}^{\infty} E'_{f_k,f}\right) = 0.$

Conversely (see again (3)), condition (j) for some f implies (i); then also (see (4)):

(jj) $\lim_{n\to\infty} \mu(E'_{f_n,f}) = 0$.

3. Almost discrete convergence and discrete convergence in measure. Now we shall define two notions which are in general stronger and weaker, respectively, than discrete a.e. convergence. Again, we shall assume X to be a measure space.

Definition 3. We say that a sequence of measurable functions $\{f_n\}$ converges *almost discretely* to a measurable function f, and we write $f_n \xrightarrow{AD} f$, if condition (j) holds; i.e., equivalently:

(j') for every $\varepsilon > 0$ there exists $n_\varepsilon \in \mathbb{N}$ such that $\mu(\{x \in X : f_k(x) \neq f(x) \text{ for some } k \geq n\}) < \varepsilon$ for $n > n_\varepsilon$.

Definition 4. We say that a sequence of measurable functions $\{f_n\}$ converges *discretely in measure* to a measurable function f, and we write $f_n \xrightarrow{DM} f$, if condition (jj) holds; i.e., equivalently:

(jj') for every $\varepsilon > 0$ there exists $n_\varepsilon \in \mathbb{N}$ such that $\mu(E'_{f_n,f}) < \varepsilon$ for $n > n_\varepsilon$.

We obtain (DM) convergence from (D) convergence, by a process similar to that giving (AU) from (U) convergence or, similarly, to that giving (M) from (AE) convergence.

The idea of considering (AD) convergence is suggested by the so called Lusin theorem (see e.g. [12, pp. 110-111]); we recall that this same result had already been proved by Vitali, see [20]. This theorem implies that the class of Lebesgue measurable functions defined on $[a, b]$ is exactly the class of (AD) limits of sequences of continuous functions

([9], see also [17] and [1]). The importance of this fact was pointed out by Tonelli, who did not show the equivalence between the two classes, but indicated a different approach to Lebesgue integral for bounded functions in terms of such limits ([18], see also [16]).

It is clear that (DM) convergence implies (M) convergence. If $\mu(X) = \infty$, then (D) convergence does not imply (M) convergence, so it does not imply (jj) either, see Example 1; in particular, (ii) $\not\Rightarrow$ (jj).

The following example shows that, also when $\mu(X) < \infty$, (DM) convergence does not imply a.e. convergence (i.e., (jj) $\not\Rightarrow$ (ii), and so in particular (jj) $\not\Rightarrow$ (j)).

Example 2. Let $X = [0,1]$ (with the usual Lebesgue measure). Consider a sequence $\{I_n\}$ of subintervals of $[0,1]$ satisfying the following conditions: each point of $[0,1]$ belongs to infinitely many I_n and $\mu(I_n) \to 0$ as $n \to \infty$. For example, we could take $I_1 = [0,1], I_2 = [0,1/2], I_3 =]1/2,1], I_4 = [0,1/3], I_5 =]1/3,2/3], I_6 =]2/3,1], I_7 = [0,1/4]$, and so on.

If we set $f_n = \chi_{I_n}$ $(n \in \mathbb{N})$, then clearly $\{f_n\}$ is (DM) convergent to Θ; but f_n does not converge (AE) (so neither discretely a.e.) to Θ.

Note that $\frac{1}{n}(f_1 + f_2 + \cdots + f_n) \overset{DM}{\not\longrightarrow} \Theta$ in this example.

Remark 3. It is easy to see that (AU) convergence does not imply (U) convergence a.e.; also, we have seen that (DM) convergence does not imply a.e. discrete convergence. Proposition 1 shows that instead (AD) convergence implies discrete a.e. convergence; this is clear also from Remark 1.

The following statement, which reminds Severini-Egorov's theorem, follows immediately from Proposition 1 and the comments at the end of Section 2.

Proposition 2. *Let* $\mu(X) < \infty$. *Then* (ii) $\Leftrightarrow$ (i) $\Leftrightarrow$ (j), *and so* (ii) *implies* (j). *Consequently, in this case discrete a.e. convergence implies, and hence is equivalent to,* (AD) *convergence.*

Consider now the following "vanishing restriction", which is clearly weaker than (i) (see [3, p.629]):

(v) for all $\alpha > 0$ we have $\lim_{n\to\infty} \mu\left(\bigcup_{i,j=n}^{\infty} E^{\alpha}_{f_i,f_j}\right) = 0.$

This condition is equivalent to (AU) convergence of $\{f_n\}$ to some function f; it can be expressed also in the following way (see again [3] or [11, p. 28]):

(v') there exists f such that for all $\alpha > 0$ we have $\lim_{n\to\infty} \mu\left(\bigcup_{i=n}^{\infty} E^{\alpha}_{f_i,f}\right) = 0.$

Clearly, (AD) convergence implies (AU) convergence, since (j) $\Rightarrow$ (v').

4. Convergence and metrics. Now we will show that it is possible to describe (DM) convergence in terms of metrics.

Theorem 1. *Let X be a measure space and $f, g : X \to \mathbb{R}$ measurable. Set*

$$d(f,g) = \mu(E'_{f,g}). \tag{5}$$

Then d is a generalized pseudo-distance on the space of measurable real functions. If $\mu(X) < \infty$, then d is a pseudo-distance.

Proof. The proof is almost trivial: it is enough to observe that for all measurable functions $f, g, h : X \to \mathbb{R}$ we have $E'_{f,g} \subseteq E'_{f,h} \cup E'_{h,g}$. □

Let $f, g : X \to \mathbb{R}$ be measurable functions. By setting $f \sim g$ when $\mu(E'_{f,g}) = 0$ we obtain in a natural way a generalized metric d', and even a metric in case $\mu(X) < \infty$, on the space $M(X)$ of all equivalence classes with respect to $\sim$.

Clearly, $\lim_{n\to\infty} d(f_n, f) = 0$ means that $f_n \xrightarrow{DM} f$. Moreover, the function d defined by (5) has the following properties:

(a) $d(fh, gh) \leq d(f, g)$ for every triplet f, g, h;

(b) $d(f + h, g + h) = d(f, g)$ for every triplet f, g, h;

(c) $d(\alpha f, \Theta) = d(f, \Theta)$ for $\alpha \neq 0$ and any f;

and so also

(b') $d(f + g, \Theta) \leq d(f, \Theta) + d(g, \Theta)$ for every pair f, g;

(c') $d(\alpha f, \alpha g) = d(f, g)$ for every f, g and $\alpha \neq 0$.

The properties (a), (b), (c) of d imply the following

Proposition 3. *If $f_n \xrightarrow{DM} f$, $g_n \xrightarrow{DM} g$, and $\lambda \in \mathbb{R}$, then the sequences $f_n g_n$; $f_n + g_n$; λf_n* (DM) *converge to fg; $f + g$; λf respectively. Similar results are true for the sequences $f_n \vee g_n$; $f_n \wedge g_n$; $|f_n|$; f_n^+; f_n^-; and also for $1/f_n$ if f_n and f are different from 0 everywhere.*

Analogous results hold for (AD), discrete a.e., or discrete convergence. Recall, however, that for (M) convergence these facts in general are not true (see [12, pp. 201-205]).

Now take a decreasing sequence $\varepsilon_n \to 0$. Since $E'_{f,g} = \bigcup_{n=1}^{\infty} E_{f,g}^{\varepsilon_n}$ we have

$$d(f, g) = \lim_{n\to\infty} \mu(E_{f,g}^{\varepsilon_n});$$

also, we have

$$d(f, g) = \sup\{\mu(E_{f,g}^{\sigma}) : \sigma > 0\}.$$

We recall that convergence in measure can be obtained by using a metric, for example, by putting

$$\delta(f, g) = \inf\{\sigma + \mu(E_{f,g}^{\sigma}) : \sigma > 0\}.$$

As usual, f and g denote real functions on a measure space. It is clear that, for any $\sigma > 0$, we have $\delta(f, g) \leq \sigma + \mu(E_{f,g}^{\sigma}) \leq \sigma + d(f, g)$; therefore

$$\delta(f, g) \leq d(f, g).$$

In particular, this shows again that $f_n \xrightarrow{DM} f$ implies $f_n \xrightarrow{M} f$.

There are some analogies among results concerning (DM) and (M) convergence. We are going to prove that the metric space induced by d is complete, similarly to that induced by δ. Other comparisons will be done after the proof of completeness.

Theorem 2. *The space $M(X)$ is complete with respect to the metric d'.*

Proof. Let $\{f_n\}$ be a Cauchy sequence with respect to d. We want to show that $f_{n_k} \xrightarrow{DM} f$ for some subsequence $\{f_{n_k}\}$ of $\{f_n\}$. For any $k \in \mathbb{N}$ there exists an index n_k such that $d(f_n, f_m) = \mu(E'_{f_n,f_m}) < 1/2^k$ for $n, m \geq n_k$. According to that, we choose in this way an increasing sequence $\{n_k\}$. Now set

$$E'_k = \bigcup_{j=k}^{\infty} E'_{f_{n_j},f_{n_{j+1}}};$$

we have then

$$\mu(E'_k) < \sum_{i=k}^{\infty} \frac{1}{2^i} = \frac{1}{2^{k-1}}.$$

Consider the sequence $\{f_{n_k}\}$: for $x \in X \setminus E'_k$ we have $f_{n_k}(x) = f_{n_{k+i}}(x)$ for all $i \in \mathbb{N}$. Now define f in X so that for $x \in X \setminus E'_k$, its value agrees with that of the sequence $\{f_{n_k}(x)\}$: We can do this for arbitrarily large k; then, given $\varepsilon > 0$, $\mu(\{x \in X : f_{n_k}(x) \neq f(x)\}) < \varepsilon$ for k large enough, that is $d(f_{n_k}, f) \to 0$ as $k \to \infty$. This implies $f_{n_k} \xrightarrow{DM} f$, which completes the proof. □

In proving Theorem 2, we have constructed a sequence $\{f_{n_k}\}$ satisfying (j) (thus also (ii)): in fact, given $\varepsilon > 0$, for $k \in \mathbb{N}$ such that $2^{1-k} < \varepsilon$ we have $f_{n_k} \xrightarrow{D} f$ on $X \setminus E'_k$, and $\mu(E'_k) < 1/2^{k-1} < \varepsilon$. We may therefore state the following proposition, which reminds one of the results relating (M) convergence to (AU) or (AE) convergence.

Proposition 4. *Let $f_n \xrightarrow{DM} f$; then there exists a subsequence $\{f_{n_k}\}$ of $\{f_n\}$ satisfying* (j); *so, in particular, $f_{n_k} \xrightarrow{D\,a.e.} f$.*

Moreover, if $\mu(X) < \infty$ and $f_n \xrightarrow{D\,a.e.} f$, then there is a subsequence which (AD) *converges to f.*

(DM) convergence (similarly to (M) convergence) in general is not compatible with existence of a norm, as the last part of Example 2 shows; it is compatible with a norm only when the measure space X is of a very special type (see [11, Theorem 3.1.4]). But the following example shows that (DM) convergence is not compatible with a quasi-norm either. (Recall that a quasi-norm is obtained by replacing, in the properties defining a norm, the homogeneity condition by a weaker one).

Indeed, let $f(x) \equiv 1$ on $[0, 1]$; if we take a sequence $\varepsilon_n \to 0$, we do not have $\varepsilon_n f \xrightarrow{DM} \Theta$.

Theorem 3. (AD) *convergence is not compatible with a metric.*

Proof. Assume that (AD) is obtained by using a distance d_1. Take a sequence $\{f_n\}$ such that $f_n \xrightarrow{DM} f$, but $f_n \not\xrightarrow{AD} f$. This means that for some $\alpha > 0$, there exists a subsequence $\{f_{n_k}\}$ of $\{f_n\}$ such that

(o) $d_1(f_{n_k}, f) > \alpha$ for all k.

Since $f_n \xrightarrow{DM} f$, according to Proposition 4 we can find a subsequence of $\{f_{n_k}\}$ which converges (AD) to f, contradicting (o). This proves the assertion. □

Remark 4. The reasoning done to prove the previous proposition shows that (AD) convergence is not generated by a first countable topology; a similar result holds for discrete as well as for discrete a.e. convergence (see also [4]).

5. A few more examples and remarks. We have seen that $f_n \xrightarrow{DM} f$ implies $f_n \xrightarrow{M} f$, $f_n \xrightarrow{D\,a.e.} f$ implies $f_n \xrightarrow{AE} f$, and $f_n \xrightarrow{AD} f$ implies $f_n \xrightarrow{AU} f$. We show now by means of another three examples that other implications involving these kinds of convergence do not hold. In these examples, we take $X = [0,1]$ with μ being the Lebesgue measure. We denote by $\{f_{n_k}\}$ an arbitrary subsequence of $\{f_n\}$.

Example 3. Let

$$f_n(x) = \begin{cases} e^n & \text{if } 0 < x < 1/n, \\ 0 & \text{otherwise.} \end{cases}$$

Then $f_n \xrightarrow{D} \Theta$, but $f_{n_k} \stackrel{L_p}{\not\to} \Theta$ for $0 < p \leq \infty$.

Example 4. Take the (bounded) sequence

$$f_n(x) = \begin{cases} 1 & \text{if } 0 < x < 1/n, \\ 0 & \text{otherwise.} \end{cases}$$

Then $f_n \xrightarrow{D} \Theta$, but $f_{n_k} \stackrel{U}{\not\to} \Theta$.

Example 5. Let $f_n(x) \equiv 1/n$. Then $f_n \xrightarrow{U} \Theta$, but $f_{n_k} \stackrel{DM}{\not\to} \Theta$ and $f_{n_k} \stackrel{D\,a.e}{\not\to} \Theta$.

We summarize part of the discussion with the following two schemes concerning all types of convergence considered above. The first scheme holds for any space, while the second scheme holds for spaces of finite measure.

$$\begin{array}{ccc} \text{(D a.e.)} & \Rightarrow & \text{(AE)} \\ \Uparrow & & \Uparrow \\ \text{(AD)} & \Rightarrow & \text{(AU)} \\ \Downarrow & & \Downarrow \\ \text{(DM)} & \Rightarrow & \text{(M)} \end{array} \qquad\qquad \begin{array}{ccc} \text{(D a.e.)} & \Rightarrow & \text{(AE)} \\ \Updownarrow & & \Updownarrow \\ \text{(AD)} & \Rightarrow & \text{(AU)} \\ \Downarrow & & \Downarrow \\ \text{(DM)} & \Rightarrow & \text{(M)} \end{array}$$

To conclude, we recall that "Baire classes" for discrete convergence (as well as for some other types of convergence) have been studied in [5-8,12-14]; see, in particular, Section 3 of [7].

References

[1] G. Androulakis: *A counterexample to a question of R. Haydon, E. Odell and H. Rosenthal*, Proc. Amer. Math. Soc. **126** (1998), 1425-1428

[2] R. G. Bartle: *The Elements of Integration*, J. Wiley & Sons, New York 1966

[3] R. G. Bartle: *An extension of Egorov's theorem*, Amer. Math. Monthly **87** (1980), 628-633

[4] Z. Bukovská, T. Šalát: *Some types of convergence of sequences of real valued functions*, Acta Math. Univ. Comenian. **58/59** (1991), 215-220

[5] Á. Császár: *Extensions of discrete and equal Baire functions*, Acta Math. Hungar. **56** (1990), 93-99

[6] Á. Császár, M. Laczkovich: *Discrete and equal convergence*, Studia Sci. Math. Hungar. **10** (1975), 463-472

[7] Á. Császár, M. Laczkovich: *Some remarks on discrete Baire classes*, Acta Math. Acad. Sci. Hungar. **33** (1979), 51-70

[8] Á. Császár, M. Laczkovich: *Discrete and equal Baire classes*, Acta Math. Hungar. **55** (1990), 165-178

[9] R. E. Dressler, K. R. Stromberg: *The Tonelli integral*, Amer. Math. Monthly **81** (1974), 67-68

[10] J. Gerlits: *Remarks on discrete convergence*, Studia Sci. Math. Hungar. **11** (1976), 145-149

[11] E. Lukacs: *Stochastic Convergence*, D. C. Heath & Co., Lexington MA 1968

[12] M. E. Munroe: *Measure and Integration*, Addison-Wesley Publ. Co., Reading MA 1971

[13] V. Prokaj: *A note on equal convergence*, Acta Math. Hungar. **73** (1996), 155-158

[14] V. Prokaj: *A note on pointwise convergence*, Acta Math. Hungar. **75** (1997), 161-164

[15] J. S. Rosenthal: *Faithful coupling of Markov chains: now equals forever*, Advances Appl. Math. **18** (1997), 372-381

[16] D. N. Sarkhel: *Lebesgue versus Denjoy integral*, Indian J. Math. **37** (1995), 1-4

[17] A. H. Stone: *Lusin's theorem*, Atti Sem. Mat. Fis. Univ. Modena **44** (1996), 351-357

[18] L. Tonelli: *Fondamenti di calcolo delle variazioni I*, Zanichelli, Bologna 1921

[19] M. Väth: *Volterra and Integral Equations of Vector Functions*, M. Dekker, New York 1999

[20] G. Vitali: *Una proprietà delle funzioni misurabili*, Rend. Ist. Lomb. Sci. Lett. **38** (1905), 599-603

Emanuele Casini, Università di Como-Varese, Dipartimento di Scienze Chimiche, Fisiche e Matematiche, Via Lucini 3, I-22100 Como, Italy; casini@fis.unico.it

Pier Luigi Papini, Università di Bologna, Dipartimento di Matematica, Piazza di Porta S. Donato 5, I-40126 Bologna, Italy; papini@dm.unibo.it

Progress in Nonlinear Differential Equations
and Their Applications, Vol. 40

Nonlinear Stability of Eigenvalues of Compact Self-Adjoint Operators

RAFFAELE CHIAPPINELLI

Dedicated to Alfonso Vignoli on the occasion of his 60th birthday

Summary: Variational methods are used to study the effect of suitably restricted nonlinear perturbations upon the eigenvalues of a compact selfadjoint operator.

Keywords: Gradient operator, critical point, minimax principle.

Classification: 35J65, 35P30, 47H12, 58E05.

1. Introduction. Let H be a real, infinite-dimensional Hilbert space and let $\mathcal{K}(H)$ denote the vector space of all linear compact selfadjoint operators in H, equipped with the usual operator norm. If $T \in \mathcal{K}(H)$ and $\lambda_0 \neq 0$ is an eigenvalue of T, then given any $B \in \mathcal{K}(H)$ with $\|B\| < |\lambda_0|$, $T + B$ has an eigenvalue λ within $\|B\|$ of λ_0, i.e. such that $|\lambda - \lambda_0| \leq \|B\|$. This can be easily demonstrated on using the minimax characterization of the eigenvalues for members of $\mathcal{K}(H)$, see e.g. [7,21]. Indeed, for any integer $n \geq 0$ let

$$\mathcal{U}_n = \{V \subset H : V \text{ subspace of dimension } \leq n\} \tag{1.1}$$

and for $n \geq 1$ set

$$c_n = \inf_{\mathcal{U}_{n-1}} \sup_{S \cap V^\perp} (Tu, u) \tag{1.2}$$

where $S = \{u \in H : ||u|| = 1\}$ and $V^\perp$ is the subspace orthogonal to V; (u, v) denotes the given scalar product of $u, v \in H$ and $||u||$ the corresponding norm. Then $c_1 \geq c_2 \geq \ldots \geq c_n \geq \ldots \geq 0$, and if $c_n > 0$ T has n positive eigenvalues: precisely, $c_i = \lambda_i^+$ for $i = 1, \ldots, n$ where (λ_i^+) denotes the (possibly finite) sequence of *all* positive eigenvalues of T, arranged in decreasing order and counting multiplicities. A similar description holds for the negative eigenvalues of T.

Thus let λ_0 be a fixed eigenvalue of T, say $\lambda_0 = \lambda_k^+$. Given $B \in \mathcal{K}(H)$, we have

$$(Tu, u) - ||B|| \leq ((T + B)u, u) \leq (Tu, u) + ||B||$$

for all $u \in S$. Thus for any subspace V of H,

$$\sup_{u \in S \cap V^\perp} (Tu, u) - ||B|| \leq \sup_{u \in S \cap V^\perp} ((T + B)u, u) \leq \sup_{u \in S \cap V^\perp} (Tu, u) + ||B||$$

and taking the infimum over $\mathcal{U}_{k-1}$ we have

$$\lambda_0 - ||B|| \leq c'_k := \inf_{\mathcal{U}_{k-1}} \sup_{u \in S \cap V^\perp} ((T+B)u, u) \leq \lambda_0 + ||B||.$$

Assuming that $||B|| < \lambda_0$, this implies by the aforementioned criterion that c'_k is an eigenvalue of $T+B$ and thus proves the claim. (Indeed, this shows that $T+B$ has k positive eigenvalues $\mu_i^+ = c'_i \quad (i = 1, \ldots, k)$ with $|\mu_i^+ - \lambda_i^+| \leq ||B||$ for each i.)

It is natural to ask whether this *stability property* of the eigenvalues of T can be extended to *nonlinear* perturbations of T. Obviously one should first say what is an *eigenvalue* for a nonlinear operator, and we do this in the obvious way by transporting the definition from the linear theory:

Definition 1.1. Let E be a real Banach space and let $F : E \to E$ with $F(0) = 0$. A number $\lambda_0 \in \mathbb{R}$ is said to be an *eigenvalue* of F if there exists $u_0 \in E$, $u_0 \neq 0$ such that $F(u_0) = \lambda_0 u_0$; in this case u_0 is said to be an *eigenvector* associated with λ_0.

Needless to say, this definition is not satisfactory in general. For instance, (λ_0, u_0) might be an isolated solution of $F(u) = \lambda u$ in $\mathbb{R} \times E$, i.e. $F(u) - \lambda u \neq 0$ for (λ, u) near (λ_0, u_0) with $(\lambda, u) \neq (\lambda_0, u_0)$, a situation which hardly matches with our idea of eigenvalue. On the other hand, Definition 1.1 is certainly sensible for the class of *positively homogeneous* operators, i.e. those $F : E \to E$ such that $F(tu) = tF(u)$ for $t > 0$ and all $u \in E$; indeed for such an F, if u_0 is an eigenvector corresponding to λ_0, then so does every point on the ray $\{tu_0 : t > 0\}$.

Motivated by this remark, we restrict our attention to such operators, and shall also assume that they are *completely continuous* (i.e. continuous and compact), which enables us to set

$$||F|| = \sup_{u \neq 0} \frac{||F(u)||}{||u||} = \sup_{||u||=1} ||F(u)|| \tag{1.3}$$

where the last equality follows by homogeneity. We stay for simplicity within the context of a Hilbert space H, and impose on F the further condition of being a *gradient* operator, i.e. $F = \nabla f$ for some C^1 functional f defined on H (see Section 2). These requirements are mere substitutes for the compactness and selfadjointness of a linear map, see e.g. Berger [2]. Let us finally denote by $\mathcal{F}(H)$ the class of all completely continuous and positively homogeneous gradient operators $F : H \to H$ with $F(0) = 0$.

We can now formulate a "nonlinear" stability result for mappings of the form $F = T + N$, with $T \in \mathcal{K}(H)$ and $N \in \mathcal{F}(H)$, in terms of the *isolation distance* [13]

$$d := d(\lambda_0) = dist(\lambda_0, \sigma(T) \setminus \{\lambda_0\})$$

of a nonzero eigenvalue λ_0 of T in the spectrum $\sigma(T)$ of T.

Theorem 1. *Let $T \in \mathcal{K}(H)$ and let λ_0 be a nonzero eigenvalue of T with isolation distance $d = d(\lambda_0)$. If $N \in \mathcal{F}(H)$ has norm $||N|| < d/2$, then $T + N$ has an eigenvalue λ with $|\lambda - \lambda_0| < ||N||$.*

Theorem 1 will be proved in Section 3 as a specific application of a Critical Point Theorem for C^1 functionals which we discuss in Section 2 after the presentation of some preliminary material. When applied to the context of an $F \in \mathcal{F}(H)$ this will provide the existence of eigenvalues λ of F characterized as minimax (or maximin) values

$$\lambda = \sup_{\mathcal{A}} \inf_{A} (F(u), u) \tag{1.4}$$

of the "quadratic form" $(F(u), u)$ of F, as in the linear case. Comparing (1.4) with the formula (1.2) for the $c_n = \lambda_n^+$, or rather with the dual formula (see e.g. [7,21])

$$c_n = \sup_{\mathcal{U}_n} \inf_{S \cap V} (Tu, u) \tag{1.5}$$

we note that the class $\mathcal{A}$ on which the quadratic form has to be considered is not merely the intersection of S with finite-dimensional subspaces of H but has to be replaced by a different and more involved topological class of compact subsets of S, which will be shortly described in Section 2.

Remark 1.1. Theorem 1 could be regarded in the framework of nonlinear spectral theory. Spectra of nonlinear operators were first studied in papers by Kachurovskii [12] and Neuberger [17], and then considered in the book by Martin [16, p.69]. New strength to the theory came by the work of A. Vignoli and his co-workers (see e.g. [10,11]), to whom I owe much of the interest in the subject. For a complete and up to date survey on the matter, we recommend the paper by Appell and Dörfner [1]. Very recently, a new definition of spectrum for nonlinear operators has been proposed by Feng [9]. In particular, she proves several results about the existence of eigenvalues of positively homogeneous operators, but does not consider gradient mappings in Hilbert space.

2. Preliminaries and a Critical Point Theorem. Recall that a mapping $F : H \to H$ is said to be a *gradient operator* if there exists a function $f : H \to \mathbb{R}$ (the *potential* of F) such that

$$(F(u), v) = f'(u)v \qquad \forall u, v \in H$$

where $f'(u)$ denotes the derivative of f at the point u, which is a bounded linear form on H; in this case we write $F = \nabla f$. Assuming F continuous, i.e. f of class C^1, F and f are related via the formula (see e.g. [2, Section II.2.5])

$$f(u) = \int_0^1 (F(tu), u)\, dt. \tag{2.0}$$

If $F = \nabla f$, then we can relate the normed eigenvectors of F with the critical points of f on the spheres in H. Indeed, recall that $u_0 \in H$ is a *(free) critical point* of f if $f'(u_0) = 0$, while given a "hypersurface" $M = \{u \in H : g(u) = \alpha\}$, i.e. the level set of another C^1 functional $g : H \to \mathbb{R}$, we say that $u_0 \in M$ is a *constrained critical point* of f on M if $f'(u_0) = \lambda g'(u_0)$ for some "Lagrange multiplier" $\lambda \in \mathbb{R}$. Thus if $F = \nabla f$ and $G = \nabla g$, then the solutions u of the problem

$$F(u) = \lambda G(u), \qquad g(u) = R$$

are precisely the constrained critical points of f on the surface

$$M_R = \{u \in H : g(u) = R\}.$$

We look for normalized eigenvectors of F, i.e. solutions of $F(u) = \lambda u$ belonging to $S_R = \{u \in H : \|u\| = R\}$ for a given $R > 0$. Now if F is positively homogeneous, then any R will do the same job and moreover by (2.0), the potential f of F is merely reduced to $\frac{1}{2}(F(u), u)$ - as if F were linear selfadjoint. Just in the same way, we can thus formalize the above-mentioned relation in terms of the "quadratic form" of F:

$$Q(u) = (F(u), u). \tag{2.1}$$

Proposition 2.1. *Any critical point u_0 of Q on $S = \{u \in H : \|u\| = 1\}$ is a norm-one eigenvector of F, corresponding to the eigenvalue $\lambda_0 = Q(u_0)$. Conversely, if $F(u_0) = \lambda_0 u_0$ with $\|u_0\| = 1$, then $\lambda_0 = Q(u_0)$ and u_0 is a critical point of Q on S.*

Proof. S is a level set of the functional $g(u) = \|u\|^2$. At a critical point u_0 of Q on S we have

$$Q'(u_0)v = \lambda_0 g'(u_0)v \qquad \forall v \in H$$

for some λ_0, i.e. (since $Q(u) = (F(u), u) = 2f(u)$)

$$2(F(u_0), v) = 2\lambda_0(u_0, v) \qquad \forall v \in H$$

whence $F(u_0) = \lambda_0 u_0$. Taking the scalar product with u_0 we obtain $\lambda_0 = (F(u_0), u_0) = Q(u_0)$. Thus,

eigenvalues of F

= values of Q at its critical points on S

= *critical values* of Q on S. □

Now if we look at a general C^1 functional f, its simplest critical values on S are the absolute minimum and maximum on S (of course, one has to make sure that these are attained, which requires some further assumption - compensating for the lack of compactness of S - such as the Palais-Smale condition explained below). Definitely more difficult is the search of additional, i.e. "intermediate" critical values, and a useful tool for their discovery is the *Lusternik-Schnirelmann theory* introduced in [15] for functionals acting on compact manifolds, and later extended to Hilbert or Banach manifolds by e.g. Krasnoselskii [14], Browder [3], Palais [18], Schwartz [22], Rabinowitz [19,20]. We recommend e.g. the survey [18] for a beautiful illustration of the theory, and bound ourselves to recall that critical values of f are found by *minimax* or *maximin*: for instance,

$$c = \sup_{A \in \mathcal{A}} \inf_{u \in A} f(u)$$

where $\mathcal{A}$ is a (suitably chosen) class of subsets of S.

In [4] and [5] we have employed these ideas and methods to obtain a result on the existence of "saddle points" of f on S. In order to state it, one further definition is needed:

Definition 2.1. f is said to satisfy the *Palais-Smale (PS) condition on S* if any sequence $(u_n) \subset S$ such that $f(u_n)$ is bounded and $f'_S(u_n) \to 0$ contains a convergent subsequence.

Here $f'_S(u)$ is defined by the equality $f'_S(u)v = (F(u),v) - (F(u),u)(u,v)$ for $u \in S$ and $v \in H$, where $F = \nabla f$; that is to say, $f'_S(u)$ is the linear form associated with the tangential component to S of $\nabla f(u)$,

$$\nabla_S f(u) := F(u) - (F(u),u)u. \tag{2.2}$$

Theorem 2 (Constrained Saddle Point Theorem). *Let f be a C^1 functional on H and let V, W be supplementary closed subspaces of H (i.e., $H = V \oplus W$) with $\dim V < \infty$. Assume that there exist constants $\alpha < \beta$ so that*

$$\begin{cases} f(u) \geq \beta, & u \in S \cap V, \\ f(u) \leq \alpha, & u \in S \cap W. \end{cases}$$

Then if f satisfies (PS) on S, f has a critical value $c \geq \beta$. Furthermore, if V_0 is a nontrivial subspace of V and $f(u) \leq \gamma$ on $S \cap (V_0 \oplus W)$, then

$$\beta \leq c \leq \gamma. \tag{2.3}$$

Sketch of the proof. Set

$$c = \sup_{A \in \mathcal{A}} \inf_{u \in A} f(u) \tag{2.4}$$

where

$$\mathcal{A} = \{A \subset S \setminus W : A \text{ compact, noncontractible in } S \setminus W\}. \tag{2.5}$$

Recall that a subset A of a topological space X is said to be *contractible* (in X) if it can be deformed continuously (in X) to a point, i.e. if there exists a continuous map $U : [0,1] \times A \to X$ such that $U(0,x) = x$ and $U(1,x) = x_0$ for all $x \in A$ and some $x_0 \in X$. Thus, $\mathcal{A}$ is the family of all compact subsets of $S \setminus W$ having *Lusternik-Schnirelmann category* [18,22] greater than one in $S \setminus W$.

We note that $\mathcal{A}$ is nonempty as it contains $S \cap V$, which is not contractible into itself (and therefore in $S \setminus W$ neither, see e.g. [14, Lemma VI.2.6] since $\dim V < \infty$. Thus by (2.4),

$$c \geq \inf_{u \in S \cap V} f(u) \geq \beta. \tag{2.6}$$

One further property of the class $\mathcal{A}$ is that its elements have to be "big enough", in the sense that they must meet $V_0 \oplus W$ with V_0 an arbitrary (nontrivial) subspace of V (see e.g.[14, Lemma VI.2.7]). As a consequence, we can immediately prove the last statement in Theorem 2: indeed, given any $A \in \mathcal{A}$,

$$\inf_{u \in A} f(u) \leq \sup_{u \in S \cap (V_0 \oplus W)} f(u)$$

and thus it follows by the definition (2.4) of c that $c \leq \gamma$ if γ is as in Theorem 2. Therefore, the estimate (2.3) follows directly from the definition and properties of $\mathcal{A}$. The main statement of Theorem 2, i.e. that c is a critical value of f, is a consequence of the assumption $\alpha < \beta$ and of standard deformation techniques from critical point theory (see e.g. [19,20]); we refer to [4] or [5] for details. □

Remark 2.1. To draw the conclusions of Theorem 2, it is enough (see [20, Remark A.17]) that f satisfy the "local" (PS) condition

$$(PS)_c \qquad (f(u_n) \to c, \quad f'_S(u_n) \to 0) \implies (u_n) \text{ has a convergent subsequence.}$$

Remark 2.2. Theorem 2 can be viewed as a constrained version of P. Rabinowitz' Saddle Point Theorem [20]. It refines an earlier version due to Krasnoselskii [14], used in bifurcation for gradient mappings.

3. Nonlinear eigenvalue problems. Let us now concentrate on the eigenvalue problem

$$F(u) \equiv Tu + N(u) = \lambda u \tag{3.0}$$

with $T \in \mathcal{K}(H)$ and $N \in \mathcal{F}(H)$ (see the Introduction). Let p and q be defined by

$$p = \inf_{\|u\|=1} (N(u), u), \qquad q = \sup_{\|u\|=1} (N(u), u). \tag{3.1}$$

Theorem 3. *Let $\lambda_0 > 0$ be an eigenvalue of T and let $d(\lambda_0)$ be its isolation distance in $\sigma(T)$. Assume that*

$$(N1) \qquad q - p < d(\lambda_0)$$

$$(N2) \qquad 0 < \lambda_0 + p.$$

Then $T + N$ has an eigenvalue $\lambda > 0$ with

$$\lambda_0 + p \leq \lambda \leq \lambda_0 + q. \tag{3.2}$$

A similar statement holds if $\lambda_0 < 0$, replacing (N2) *with $\lambda_0 + q < 0$.*

Sketch of the proof. We apply Theorem 2 with

(i) $f = Q$, the quadratic form of F:

$$Q(u) = (Tu, u) + (N(u), u) \equiv Q_0(u) + (N(u), u); \tag{3.3}$$

(ii) V the (orthogonal) sum of the eigenspaces corresponding to all eigenvalues λ of T with $\lambda \geq \lambda_0$;

(iii) V_0 the eigenspace corresponding to λ_0;

(iv) $W = V^\perp$, the orthogonal complement to V in H.

By the variational characterization of the eigenvalues of T, we have $\lambda_0 = \min\{(Tu,u) : u \in S \cap V\}$. Also letting $\nu = \sup\{(Tu,u) : u \in S \cap W\}$, then $\lambda_0 > \nu \geq 0$ and ν is the nearest point to the left of λ_0 in $\sigma(T)$. Precisely, if $\nu > 0$ then ν is attained and is the next neighbour to λ_0 in the decreasing sequence of all positive eigenvalues of T, while if $\nu = 0$ then $\nu \in \sigma(T)$ anyway because T is compact. Thus $d(\lambda_0) \leq \lambda_0 - \nu \leq \lambda_0$ and we have

$$(3.4) \qquad \begin{cases} Q_0(u) \geq \lambda_0, & u \in S \cap V, \\ Q_0(u) \leq \nu, & u \in S \cap W, \end{cases}$$

whereas by the definition (3.1) of p and q we have

$$(3.5) \qquad \begin{cases} Q(u) \geq \lambda_0 + p \equiv \beta, & u \in S \cap V, \\ Q(u) \leq \nu + q \equiv \alpha, & u \in S \cap W. \end{cases}$$

Since $\lambda_0 + p > \nu + q$ by virtue of the assumption (N1), we see that the condition $\beta > \alpha$ of Theorem 2 is satisfied. Define c as in (2.4).Then by (2.3),

$$\lambda_0 + p \leq c \leq \lambda_0 + q$$

because $Q(u) \leq \lambda_0 + q \equiv \gamma$ on $S \cap (V_0 \oplus W)$. Now we use (N2) to prove that Q satisfies (PS) at level c, see Remark 2.1. Indeed, let $(u_n) \subset S$ be such that $Q(u_n) \to c$ and $Q'_S(u_n) \to 0$ or equivalently (see (2.2))

$$(3.6) \qquad F(u_n) - Q(u_n)u_n \to 0.$$

By compactness of F we can assume, passing if necessary to a subsequence, that $F(u_n)$ converges to some $z \in H$. Using (3.6), we then first see that $Q(u_n)u_n$ converges (to z) and then, since $Q(u_n) \to c \geq \lambda_0 + p > 0$, that (u_n) itself converges, as required.

Thus by Theorem 2, c is a critical value of Q on S. Finally by Proposition 2.1, c is an eigenvalue of F and the claim is proved. □

Remark 3.1. Theorem 1 in the Introduction follows easily from Theorem 3: just note that $p \geq -||N||$, $q \leq ||N||$ and so, if $2||N|| < d(\lambda_0)$, then

$$(N1) \qquad q - p \leq 2||N|| < d(\lambda_0)$$

and

$$(N2) \qquad 0 < d(\lambda_0) - ||N|| \leq \lambda_0 + p.$$

Remark 3.2. More general forms of the above results can be found in [4]. In particular, it is not essential that N be positively homogeneous in order to prove that $T + N$ possesses eigenvectors u_r *of arbitrary norm* $r > 0$ with eigenvalues λ_r near λ_0. In the more general case however, λ_r will depend upon $r > 0$. Furthermore, if $N(u) = o(||u||)$ as $||u|| \to 0$, then $\lambda_r \to \lambda_0$ as $r \to 0$; that is to say, λ_0 is a bifurcation point of $T + N$. This well-known fact (see e.g. [19, Section 4]) can thus be regarded in a more general context.

4. Applications to semilinear elliptic problems. Theorem 3 can be used to discuss the existence and position of eigenvalues for semilinear elliptic problems of the form

$$\begin{cases} Lu = \mu(u + f(x,u)) & \text{in } \Omega, \\ u = 0 & \text{on } \partial\Omega, \end{cases} \tag{4.1}$$

where Ω is a bounded open set in $\mathbb{R}^N (N \geq 1)$ with boundary $\partial\Omega$, and L is a uniformly elliptic formally selfadjoint operator,

$$Lu := -\sum_{i,j=1}^{N} \frac{\partial}{\partial x_j}(a_{ij}(x)\frac{\partial u}{\partial x_i}) + a_0(x)u \tag{4.2}$$

with L^∞ coefficients $a_{ij} = a_{ji} \quad (i,j = 1,\dots,N)$ while $a_0 \in L^{N/2}(\Omega)$ with $a_0 \geq 0$ a.e. in Ω. The nonlinearity is given by the "piecewise linear" function

$$f(x,s) = \alpha(x)s^+ - \beta(x)s^- \tag{4.3}$$

where $s^+ = \max\{s, 0\}, s = s^+ - s^-$ and $\alpha, \beta \in L^\infty(\Omega)$ with $\alpha, \beta \geq 0$ a.e. in Ω. Then also the functions c, d defined a.e. in Ω by the rules

$$c(x) = \min\{\alpha(x), \beta(x)\}, \qquad d(x) = \max\{\alpha(x), \beta(x)\} \tag{4.4}$$

are measurable, bounded and nonnegative and we set

$$C = \operatorname{ess\,inf}\{c(x) : x \in \Omega\}, \qquad D = \operatorname{ess\,sup}\{d(x) : x \in \Omega\}. \tag{4.5}$$

We put (4.1) in relation with the linear problem

$$\begin{cases} Lu = \mu u & \text{in } \Omega, \\ u = 0 & \text{on } \partial\Omega \end{cases} \tag{4.6}$$

which has an infinite sequence $0 < \mu_1^0 < \mu_2^0 \leq \dots$ of eigenvalues.

Theorem 4. *Let C, D be as in (4.5) and let $\mu_n^0 < \mu_{n+1}^0$ be two consecutive eigenvalues of (4.6). If*

$$D - C < \frac{\mu_{n+1}^0 - \mu_n^0}{\mu_{n+1}^0} \tag{4.7}$$

then (4.1) has an eigenvalue μ_n with

$$\frac{\mu_n^0}{1+D} \leq \mu_n \leq \frac{\mu_n^0}{1+C}. \tag{4.8}$$

Remark 4.1 If $\alpha = \beta$ (i.e., the problem is *linear*), then the estimate (4.8) is a straightforward consequence of the Courant minimax principle [6]. In this case it holds for all $n \in \mathbb{N}$, and there is no restriction like (4.7)!

Sketch of the proof. Let $H = H_0^1(\Omega)$ be the first Sobolev space on Ω equipped with the scalar product

$$(4.9) \qquad (u,v) = \sum_{i,j=1}^{N} \int_\Omega a_{ij}(x) \frac{\partial u}{\partial x_i}\frac{\partial v}{\partial x_j}\, dx + \int_\Omega a_0(x) uv\, dx$$

and consider the weak form of (4.1), i.e. look for $u \in H$, $u \neq 0$ and $\mu \in \mathbb{R}$ such that

$$(4.10) \qquad (u,v) = \mu \{ \int_\Omega uv\, dx + \int_\Omega (\alpha(x)u^+ v - \beta(x) u^- v)\, dx \}$$

for all $v \in H$. We next define operators T and N in H by the rules

$$(Tu,v) = \int_\Omega uv\, dx, \qquad (N(u),v) = \int_\Omega (\alpha(x)u^+ v - \beta(x)u^- v)\, dx$$

for $u, v \in H$, so that (4.10) becomes

$$(4.11) \qquad Tu + N(u) = \lambda u$$

where $\lambda = \mu^{-1}$ (note that $\mu = 0$ is not an eigenvalue of (4.1)).
It is straightforward to check (see e.g. [20, Appendix B], [7, Chapter 1], or [8, Chapters 2 and 3]) that $T \in \mathcal{K}(H)$ and $N \in \mathcal{F}(H)$. In particular, N is positively homogeneous and its quadratic form is

$$(4.12) \qquad (N(u),u) = \int_\Omega [\alpha(x) u^{+2} + \beta(x) u^{-2}]\, dx$$

so that

$$(4.13) \qquad C \int_\Omega u^2(x)\, dx \leq (N(u),u) \leq D \int_\Omega u^2(x)\, dx$$

by the definition (4.5) of C and D. Therefore, letting $Q_0(u) = (Tu,u)$ we have

$$(4.14) \qquad C \leq \tilde{p} := \inf_{Q_0(u)=1} (N(u),u), \qquad \tilde{q} := \sup_{Q_0(u)=1} (N(u),u) \leq D.$$

This suggests to introduce a technical modification in the theory discussed before. Namely we assume - in addition to the hypotheses on T and N already stated - that T is positive and N is nonnegative (i.e. $(Tu,u) > 0$ for $u \neq 0$, $(N(u),u) \geq 0$ for all $u \in H$), and consider the quantities $\tilde{p}, \tilde{q}$ rather than p, q as in (3.1). Thus, for all $u \in H$,

$$\tilde{p} Q_0(u) \leq (N(u),u) \leq \tilde{q} Q_0(u)$$

whence

$$(1+\tilde{p}) Q_0(u) \leq Q(u) \leq (1+\tilde{q}) Q_0(u).$$

Now given an eigenvalue λ_0 of T, let ν, V, V_0, W have the same meaning as in the proof of Theorem 3. We then have

$$\begin{cases} Q(u) \geq (1+\tilde{p})\lambda_0, & u \in S \cap V, \\ Q(u) \leq (1+\tilde{q})\nu, & u \in S \cap W, \end{cases}$$

and thus in order to apply the Constrained Saddle Point Theorem, we need require that $(1+\tilde{q})\nu < (1+\tilde{p})\lambda_0$. However $\tilde{q}\nu \leq \tilde{q}\lambda_0$ and so the previous inequality is satisfied if $\nu + \tilde{q}\lambda_0 < \lambda_0 + \tilde{p}\lambda_0$, i. e.

$$(N3) \qquad \tilde{q} - \tilde{p} < 1 - \frac{\nu}{\lambda_0}.$$

Assuming this and defining c as in (2.4), we obtain (rather than (3.2))

$$(1+\tilde{p})\lambda_0 \leq c \leq (1+\tilde{q})\lambda_0.$$

One may check that (N3) also ensures the (PS) condition at c, so that c is indeed a critical value of Q, and thus an eigenvalue of $T+N$.

We now apply these remarks to our elliptic eigenvalue problem (4.1), on taking $\lambda_0 = 1/\mu_n^0, \nu = 1/\mu_{n+1}^0$. Thus $1 - \nu/\lambda_0 = (\mu_{n+1}^0 - \mu_n^0)/\mu_{n+1}^0$ and (N3) follows from (4.7) since $\tilde{q} \leq D, \tilde{p} \geq C$ in the present situation. Therefore we obtain an eigenvalue c of $T+N$ satisfying the estimate

$$(1+C)\lambda_0 \leq c \leq (1+D)\lambda_0$$

i.e. (setting $c = 1/\mu$) an eigenvalue μ of (4.1) satisfying the desired bound (4.8). This completes the proof of Theorem 4. □

References

[1] J. APPELL, M. DÖRFNER: *Some spectral theory for nonlinear operators*, Nonlinear Anal. TMA **28** (1997), 1955-1976

[2] M. S. BERGER: *Nonlinearity and Functional Analysis*, Academic Press, New York 1977

[3] F. E. BROWDER: *Existence theorems for nonlinear partial differential equations*, in: Proc. Symp. Pure Math. **16**, Amer. Math. Soc., Providence 1970, p. 1-60

[4] R. CHIAPPINELLI: *Bounds on eigenvalues of nonlinearly perturbed compact selfadjoint operators*, Panamerican Math. J. **8, 2** (1998), 1-29

[5] R. CHIAPPINELLI: *Constrained critical points and eigenvalue approximation for semilinear elliptic operators*, Forum Math. **11** (1999), 459-481

[6] R. COURANT, D. HILBERT: *Methods of Mathematical Physics I*, Wiley, New York 1953

[7] D. G. DE FIGUEIREDO: *Positive solutions of semilinear elliptic problems*, in: Differential Equations, Proceedings of the 1st Latin American School [Eds.: D. G. DE FIGUEIREDO and C. S. HÖNIG], Lect. Notes Math. **957**, Springer, Berlin 1982, p. 34-87

[8] D. G. DE FIGUEIREDO: *Lectures on the Ekeland Variational Principle with Applications and Detours*, Tata Institute of Fundamental Research, Bombay 1989

[9] W. FENG: *A new spectral theory for nonlinear operators and its applications*, Abstract Appl. Anal. **2** (1997), 163-183

[10] M. FURI, M. MARTELLI, A. VIGNOLI: *Contributions to the spectral theory for nonlinear operators in Banach spaces*, Annali Mat. Pura Appl. **118** (1978), 229-294

[11] M. FURI, A. VIGNOLI: *A nonlinear spectral approach to surjectivity in Banach spaces*, J. Funct. Anal. **20** (1975), 304-318

[12] R. I. KACHUROVSKII: *Regular points, spectrum and eigenfunctions of nonlinear operators* [in Russian], Dokl. Akad. Nauk SSSR **188** (1969), 274-277; Engl. transl.: Soviet Math. Doklady **10** (1969), 1001-1005

[13] T. KATO: *Perturbation Theory for Linear Operators*, Springer, Berlin 1976

[14] M. A. KRASNOSELSKII: *Topological Methods in the Theory of Nonlinear Integral Equations* [in Russian], Gostekhizdat, Moscow 1956; Engl. transl.: Macmillan, New York 1964

[15] L. LUSTERNIK, L. SCHNIRELMANN: *Méthodes topologiques dans les problèmes variationnels*, Hermann, Paris 1934

[16] R. H. MARTIN: *Nonlinear Operators and Differential Equations in Banach Spaces*, J. Wiley & Sons, New York 1976

[17] J. W. NEUBERGER: *Existence of a spectrum for nonlinear tranformations*, Pacific J. Math. **31** (1969), 157-159

[18] R. S. PALAIS: *Critical point theory and the minimax principle*, in: Proc. Symp. Pure Math. **15**, Amer. Math. Soc., Providence 1970, p. 185-212

[19] P. RABINOWITZ: *Variational methods for nonlinear eigenvalue problems*, in: Eigenvalues of Nonlinear Problems [Ed.: G. PRODI], Cremonese, Roma 1974, p. 141-195

[20] P. RABINOWITZ: *Minimax Methods in Critical Point Theory with Applications to Differential Equations*, CBMS Regional Conference Series Math. **65**, Amer. Math. Soc., Providence RI 1986

[21] F. RIESZ, B. SZ.-NAGY: *Leçons d'analyse fonctionelle*, Gauthier-Villars, Paris 1972

[22] J. T. SCHWARTZ: *Generalizing the Lusternik-Schnirelmann theory of critical points*, Comm. Pure Appl. Math. **17** (1964), 307-315

RAFFAELE CHIAPPINELLI, Dipartimento di Matematica, Università di Siena, Via del Capitano 15, I-53100 Siena, Italy; `chiappinelli@unisi.it`

Progress in Nonlinear Differential Equations
and Their Applications, Vol. 40

A Bifurcation Theorem for Lagrangian Intersections

ELEONORA CIRIZA, JACOBO PEJSACHOWICZ

A Alfonso cariñosamente

Summary: Our main result is as follows. Let N be a closed manifold and let $L = \{L_t\}$ be an exact, compactly supported family of Lagrangian submanifolds of the symplectic manifold $M = T^*(N)$ such that L_0 admits a generating family quadratic at infinity. Let $p : [0,1] \to M$ be a path of intersection points of L_t with N. Assume that L_t is transversal to N at $p(t)$ for $t = 0, 1$, and that the Maslov intersection index $\mu(L, N, p)$ is different from zero. Then arbitrarily close to the branch p there are intersection points of L_t such that N does not belong to p.

Keywords: Bifurcation theory, fixed point, symplectomorphism, Lagrangian submanifold, Lagrangian intersection.

Classification: 58E07, 58C30, 58F05.

1. Introduction. Using a general result about bifurcation of periodic orbits of Hamiltonian vector fields, in [8] was proved a bifurcation theorem for fixed points of symplectomorphisms. This theorem can be viewed as a kind of (weak) analogue in bifurcation theory of the classical Arnold conjecture about the number of fixed points of a symplectomorphism. Roughly speaking, while Arnold's conjecture states that symplectomorphisms should have more fixed points than the ones prescribed by the Lefschetz-Hopf fixed point index the result proved in [8] implies that one parameter families of symplectomorphisms bifurcate more often than prescribed by the standard bifurcation theory.

Arnold's conjecture is one of the starting points of symplectic topology and because of this it is present in the literature under many different forms and degrees of generality. In this paper we want to push further the parallelism described above to a natural generalization of the classical Arnold conjecture which had deserved considerable attention. Namely the estimate for the number of intersection points of two Lagrangian submanifolds of a symplectic manifold.

The cause that forces a Hamiltonian deformation $L_1 = \phi(L)$ of a compact Lagrangian submanifold L of M to have a huge intersection with L can be explained as follows: by a well known theorem of Weinstein the submanifold L has a neighborhood symplectomorphic to a neighborhood of the zero section in the cotangent bundle T^*L. If L is simply connected and if L_1 is a Lagrangian submanifold that is C^1 close to L then L_1 is given by the image of the differential $dS : L \to T^*L$ of a smooth function $S : L \to \mathbb{R}$ and therefore will have as many intersection points with L as critical points has a function on L. The latter is bounded from below by Lusternik-Schnirelmann inequalities and Morse inequalities, if the critical points are non-degenerate. Of course $L_1 = \phi(L)$ need not be C^1-close to L. But when $M = T^*(L)$ using an Hamiltonian isotopy ϕ_t with $\phi_1 = \phi$ one can still produce a family of functions $S : L \times \mathbb{R}^k \to \mathbb{R}$ with k big enough such that critical points of S correspond to intersections of L with L_1. This is a theorem of Sikorav [16]. Using this theorem one can still get estimates on the number of intersection points but weaker that in the previous case. Functions S as before are usually called generating families, Morse families or generating phase and were first introduced by Hörmander in [11] for rather different purpose.

We will show that intersections of one parameter families of Lagrangian submanifolds with a given one have stronger bifurcation properties than intersections of general submanifolds of right codimension essentially for the same reasons as above. For families L_t close enough in the C^1 topology to a given Lagrangian submanifold L_0 bifurcation of intersection points of L_t with L_0 reduces, by the above described process, to bifurcation of critical points of one parameter families of smooth functions. In this setting bifurcation arises whenever the spectral flow, or what is the same, the difference between the Morse indices of the end points of the trivial branch is non-zero. This gives a stronger invariant than the usual bifurcation index obtained by comparing the sign of the determinant of the Jacobian matrix of the gradient at the end points of the trivial branch. Via generating families we will show that the assumption of being C^1 close can be substituted with a more general one without modifying the conclusions. It seems that yet more general results can be obtained through the analysis of Lagrangian boundary value problems along the lines of [9,10]. However, this approach applied to Lagrangian intersections faces much harder technical problems than in the fixed point case and this is the main reason why we have chosen to use generating functions here. This later is conceptually simpler and has the advantage of being purely finite dimensional. We are grateful to J. C. Sikorav, who suggested this approach to us and provided the necessary references.

The paper has two sections. In the first section we present our main results while the second is devoted to proofs. We tried to preserve the expository nature of the paper by avoiding technical arguments at cost of reducing the generality. Even if we made some effort in order to make the paper self contained, not every result we use will be proved here. For topics such as generating families and symplectic reduction a good general reference are Weinstein's lecture notes [19].

2. The main results. Let $M = T^*(N)$ be the cotangent bundle of a closed manifold N endowed with the standard symplectic structure. We will consider bifurcations of intersections of $N \equiv 0_N$, identified with the zero section O_N of the bundle $T^*(N)$, with an exact one-parameter family of Lagrangian submanifolds $L = \{L_t\}_{t\in[0,1]}$ such that L_t coincides with L_0 outside of a compact subset of M. To say it more precisely we will consider families $L_t = i_t(L_0)$ where $i_t\colon L_0 \to M$ is a smooth family of Lagrangian embeddings with $i_t \equiv i_0$ outside of a compact subset of L_0. Such a family is said to be *compactly supported.* Moreover L is called *exact* if the one-form $i^*\omega(\frac{\partial}{\partial t}, -)$ is exact on $L_0 \times [0,1]$. We will not discus the natural topology in the space of all Lagrangian submanifolds of a given manifold (see [20]). Let us remark only that a family i_t as above induces a continuous path in the space $C^\infty(L_0, M)$ with respect to the fine C^1 topology. Therefore L_t is C^1 close to L_s whenever t is close enough to s.

Let $I = [0,1]$ and let $p: I \to M$ be a smooth path such that $p(t) \in L_t \cap N$. A point $p(t_*) \in L_{t_*} \cap N$ is called *bifurcation point* from the given path p of intersection points if any neighborhood of $(t_*, p(t_*))$ in $[0,1] \times M$ contains points (t,q) with $q \in L_t \cap N$, $q \neq p(t)$. It follows easily from the implicit function theorem that a necessary condition for $p(t_*)$ to be a bifurcation point of intersection is that the manifold L_t^* fails to be transversal to N at $p(t_*)$. Denoting by T_pM the tangent space at p to a manifold M this means.that for $p_* = p(t_*)$ one has that $T_{p_*}L_{t_*} + T_{p_*}N$ is a proper subset of $T_{p_*}M$. Since $\dim T_{p_*}L_{t_*} = \dim T_{p_*}N = \frac{1}{2}\dim M$ this turns out to be equivalent to

$$T_{p_*}L_{t_*} \cap T_{p_*}N \neq \{0\}. \tag{1}$$

Of course this condition is not sufficient as can be easily shown with examples. In what follows we shall assume that the manifolds L_0 and L_1 are transversal to N, that is for $t = 0,1$, condition (1) holds, and we will give sufficient conditions for the existence of at least one bifurcation point $p(t_*)$ with $0 < t_* < 1$. For this we need to define the Maslov intersection index of two families of Lagrangian submanifolds L_t and N_t of a symplectic manifold M along a given path $p: I \to M$ of intersection points.

Since the interval I is contractible, the pullback $p^*(TM)$ by p of the tangent bundle of M is a trivial bundle whose fiber over t is the tangent space $T_{p(t)}M$. Let us take any symplectic trivialization $\psi: p^*(TM) \to I \times \mathbb{R}^{2n}$ of this bundle. In a standard way the images under the trivialization maps $\psi_t: T_{p(t)}M \to \mathbb{R}^{2n}$ of the tangent spaces $T_{p_*}L_{t_*}$ and $T_{p_*}N_{t_*}$ determine two paths $l(t)$ and $n(t)$ in the space Λ_n of all Lagrangian subspaces of $\mathbb{R}^{2n}$. Assuming that (1) holds, the paths l,n have transversal intersection at the end points and hence their relative Maslov index $\mu(l,n)$ is well defined (see the next section and [14]). This index is an integer which counts with appropriate multiplicities the points in $(0,1)$ where $l(t) \cap n(t) \neq \{0\}$. From the invariance of the Maslov index under the action of the symplectic group it follows that $\mu(l,n)$ is independent of the choice of trivialization. We will call it (once more!) the *Maslov intersection index* of the family $L = \{L_t\}$ with $N = \{N_t\}$ along p, and we will denote it by $\mu(L,N,p)$. In what follows we will show that under some extra assumption the nonvanishing of $\mu(L,N,p)$ is a sufficient condition for the existence of at least one bifurcation point.

In order to state our main theorem we need also to recall the definition of generating family. Let E be a finite dimensional vector space. Let $S: N \times E \to \mathbb{R}$ be a smooth function such that the differential dS is transversal to the submanifold $N^0 = T^*(N) \times E \times \{0\}$ of $T^*(N \times E) \equiv T^*(N) \times E \times V$. If this holds, by the implicit function theorem, the set $C = \{(n,v) : dS_n(v) = 0\}$ of vertical critical points of S is a submanifold of $N \times E$ of the same dimension as N. Here, as in the rest of the paper, the variable appearing as a subindex does not have the usual meaning of partial derivative: S_n denotes the function $S_n: E \to \mathbb{R}$ defined by $S_n(v) = S(n,v)$ and S_v denotes the function $S_v: N \to \mathbb{R}$ defined in the same way.

Let $e: C \to T^*(N)$ defined by $e(n,v) = dS_v(n)$. It is not difficult to show that the map e is a Lagrangian immersion (but generally not an embedding) of the manifold C into T^*N.

Given a Lagrangian submanifold L of M we will say that S is a *generating family* for L if there is a diffeomorphism h from C onto L such that $e = ih$. Here $i: L \to M$ is the inclusion. Notice that in this case e is an embedding, but the definition generalizes to immersed manifolds as well. The generating family S is said to be *quadratic at infinity* if there is a non-degenerate quadratic form Q on E such that $S(n, v) = Q(v)$ for $\|v\|$ big enough.

Theorem 2.1. *Let N be a closed manifold and let $L = \{L_t\}$ be an exact, compactly supported family of Lagrangian submanifolds of $M = T^*(N)$ such that L_0 admits a generating family quadratic at infinity. Let $p : [0,1] \to M$ be a path of intersection points of L_t with N.*

Assume L_t is transversal to N at $p(t)$ for $t = 0, 1$ and that the Maslov intersection index $\mu(L, N, p) \neq 0$. Then there exist a $t_* \in (0,1)$ such that $p(t_*)$ is a point of bifurcation for intersection points of L_t with N from the trivial branch p.

Remark 2.2. If $L_0 = N$ then the first assumption of Theorem 2.1 holds by taking $S = 0$.

Remark 2.3. The assumptions about L_t do not represent a serious restriction and can be easily overcome since in any way we are interested on what happens on a neighborhood of N only. The true limitation of the approach based on generating functions is due to the compactness of N. Eventually this also can be relaxed to some extent following [6].

We wish to discuss now a local form of Theorem 2.1 which provides sufficient conditions in order that a point $p_* = p(t_*)$ of non-transversal intersection of L_{t_*} with N be a bifurcation point. As before, given any symplectic trivialization $p^*(TM) \simeq [0,1] \times \mathbb{R}^{2n}$ the Lagrangian subbundles $T_{p(t)}L_t$, $T_{p(t)}N$ determine paths $l(t)$ and $n(t)$ in Λ_n. Choosing a supplementary subspace W to $l(t_*)$ in $\mathbb{R}^{2n}$ for t close enough to t_* one can define a linear map $B : l(t_*) \to W$ such that $l(t) = \{v + B(t)v : v \in l(t_*)\}$, i.e., $l(t)$ is the "graph of $B(t)$" . It can be shown (see [14]) that the quadratic form Q^l defined on $l(t_*)$ by $Q^l(v) = \frac{d}{dt}|_0 \omega(v, B(t)v)$ is independent of the choice of W. Similarly one defines a form Q^n on $n(t_*)$.

Following [14,15] we define the *crossing form* of l with n at t^* by

$$Q(t^*) = Q^n - Q^l \text{ restricted to } n(t_*) \cap l(t_*). \tag{2}$$

Corollary 2.4. *Assume that the crossing form $Q(t^*)$ is non-degenerate. If the signature* $\operatorname{sign} Q \neq 0$ *then $p(t_*)$ is a (a fortiori isolated) bifurcation point for intersections of the family $L = \{L_t\}$ with N.*

The graph of a symplectomorphism of a symplectic manifold V is a Lagrangian submanifold of $V \times V^-$, where V^- is the same manifold but endowed with the symplectic form $-\omega$. Fixed points corresponds to intersections of the graph with the diagonal Δ. Let $Symp_0(V)$ be the connected component of the identity in the group of all symplectomorphisms of V. For a given path of symplectomorphisms ϕ_t in $Symp_0(V)$ and a path of fixed points $p(t)$ of ϕ_t whose ends are non-degenerate fixed points (i.e. such that $T_{p(t)}\phi_t$ is nonsingular for $t = 0, 1$) it is easy to show that the Maslov intersection index $\mu(\mathrm{Graph}\phi, \Delta, p \times p)$ defined above coincides with the relative Conley-Zehnder index constructed in [8].

Corollary 2.5 [8]. *Let (V,ω) be a closed symplectic manifold with trivial first De Rham cohomology group. Let $\phi : [0,1] \to Symp_0(V)$ be a path of symplectomorphisms and $p : [0,1] \to U$ be a smooth path such that $p(t)$ is a fixed point of ϕ_t.*
If the Conley-Zehnder index $CZ(\phi,p)$ of ϕ along p is defined and $CZ(\phi,p) \neq 0$ there is a bifurcation of fixed points of ϕ from the trivial branch p.

Boundary value problems for Hamiltonian systems are of variational type whenever the boundary values belongs to Lagrangian submanifolds. As we mentioned before Theorem 2.1 can be proved by looking at the corresponding action functional as a kind of generalized generating family with infinite dimensional phase. Going backwards, under some extra assumptions, bifurcation of solutions of Lagrangian boundary value problems can be reduced to intersection problems of the type considered in Theorem 2.1

A *canonical relation* is a Lagrangian submanifold R of $M = V \times V^-$. A canonical relation and C^∞-function $H : V \times \mathbb{R} \to \mathbb{R}$ define a Lagrangian boundary value problem as follows: let X_t be the symplectic gradient of H, i.e. X_t is the time dependent vector field defined by $\omega(X_t, v) = dH_t(v)$. We will consider the following boundary value problem defined on smooth paths $u : [0,1] \to V$

$$\begin{cases} \dot{u}(t) = X_t u(t), \\ (u(0), u(1)) \in R. \end{cases} \tag{3}$$

For example if $R = \Delta$, solutions of (3) correspond to 1-periodic orbits of the Hamiltonian vector field X. If $R = L_1 \times L_2$, where L_1, L_2 are the Lagrangian submanifolds of V then (3) is a generalization for systems of the classical Sturm-Liouville boundary conditions. This type of boundary value problems arise typically in calculus of variations. There $R \subset T^*(N)$ is the co-normal bundle of a submanifold of N and therefore non-compact. Here instead we will assume that R is compact. We assume also that the vector field X_t is complete and hence we have for each $s, t \in \mathbb{R}$ a symplectomorphism $\phi_{t,s}$ defined by $\phi_{t,s}(X) = u(t)$ where $u(\tau)$ is the unique solution of the initial value problem

$$\begin{cases} \dot{u}(\tau) = X_\tau(u(\tau)), \\ u(s) = x. \end{cases} \tag{4}$$

The symplectomorphism $\phi_{1,0}$ will be denoted by ϕ^1 and we will call it the *time 1-map*. The smooth map $ev : C^\infty([0,1], V) \to V \times V$ defined by $ev(u) = (u(0), u(1))$ induces a one to one correspondence between the solutions of (3) and intersections of the Graph ϕ^1 with R. This allows us to use Theorem 2.1 to study bifurcation of solutions of boundary value problems.

Let R be a compact canonical relation and let $H : [a,b] \times V \times R \to \mathbb{R}$ be a smooth family of Hamiltonian functions on $V \times R$. We will assume:

(H_1) For each $\lambda \in [a,b]$ the time dependent gradient $X_{\lambda,t}$ of H_λ is complete, and let $\phi\colon [a,b] \times V \to V$ be defined by $\phi(\lambda, v) = \phi^1_\lambda(v)$ where ϕ^1_λ is the time 1-map of X_λ.

(H_2) There is a smooth path $\tau\colon [a,b] \to C^\infty([0,1], V)$ where $\tau(\lambda) \equiv \tau_\lambda \in C^\infty([0,1], V)$ is a solution of

$$\begin{cases} \dot{\tau}_\lambda(t) = X_{\lambda,t}\tau_\lambda(t), \\ (\tau_\lambda(0), \tau_\lambda(1)) \in \mathbb{R} \end{cases} \tag{5}$$

We will call the family τ_λ the trivial branch of solutions of (5).
(H_3) If $p(\lambda) = (\tau_\lambda(0), \tau_\lambda(1))$, we have R is transversal to Graph of ϕ^1_λ at $p(\lambda)$ for $\lambda = a, b$.

Theorem 2.6. *If H_1, H_2, H_3 hold and $\mu(\mathrm{Graph}\phi, R, p) \neq 0$ then there exist non trivial solutions of (5) bifurcating from the trivial branch. More precisely, there exist a point $(\lambda_*, \tau(\lambda_*)) \in [0,1] \times C^\infty([0,1] \times V$ such that any neighborhood of the point in that space contains points (λ, u) where u is a solution of (5) different from $\tau(\lambda)$.*

Finally let us explain why from Theorem 2.1 it follows that intersection points of families of Lagrangian manifolds bifurcate more often than what is prescribed by the standard bifurcation theory.
Suppose that we have the previous situation, $L_t, N,\ p(t) \in L_t \cap N$, but now we have only that L_t and N are submanifolds of M of complementary dimension. Then the only homotopy invariant of the path p detecting bifurcation is an integer mod 2 that can be defined as follows: choose continuously a basis $(e_1(t), ..., e_n(t))$ of $T_{p(t)}L_t$ and a basis $(f_1(t), ..., f_n(t))$ of $T_{p(t)}N$ and define $\sigma \in \mathbb{Z}_2 = \{-1, 1\}$ by

$$\sigma = \begin{cases} 1 & \text{if } (e_1(t), ..., e_n(t), f_1(t), ..., f_n(t) \text{ have the same orientation at } t = 0, 1, \\ -1 & \text{if } (e_1(t), ..., e_n(t), f_1(t), ..., f_n(t) \text{ have opposite orientation at } t = 0, 1. \end{cases}$$

Here as with any intersection theory we are assuming that M is oriented. We leave as exercise to the reader to show that bifurcation of intersection points arise whenever $\sigma = -1$. It follows from the work of Ize [12] that σ is the only homotopy invariant detecting bifurcation in this case. When L_t and N are Lagrangian it is easy to see that $\sigma = (-1)^{\mu(L,N,p)}$ and therefore whenever $\mu(L, N, p)$ is even the bifurcation predicted does not follow from the general theory.

3. Proofs. We will reduce the proof of Theorem 2.1 to a well known result about bifurcation of critical points of one parameter families of functionals. The reduction goes through a result of Sikorav about generating families together with the isotopy extension lemma of Chaperon [4].
Sikorav's result (see Proposition 1. 2 and Remark 1.7 in [16]) deals with deformations of a Lagrangian submanifold under Hamiltonian isotopies (i.e. isotopies given by the flow of a time-dependent Hamiltonian vector field) and can be formulated as follows:
If ϕ_t is a Hamiltonian isotopy of T^*N and if $L_0 \subset T^*N$ is generated by a family quadratic at infinity then there exists a smooth family of functions $S_t : N \times \mathbb{R}^k \to \mathbb{R}$ quadratic at infinity such that $\phi_t(L_0)$ is generated by the family S_t.
On the other hand Chaperon [4,5] proved that any one parameter exact compactly supported family of Lagrangian embeddings $L_t = i_t(L_0)$ can be extended to an Hamiltonian isotopy of the ambient manifold.
Putting both results together we have

Proposition 3.1. *For any smooth one parameter family L_t of Lagrangian submanifolds of $T^*(N)$ verifying the hypothesis of Theorem 2.1 there exists a smooth $S : [0,1] \times N \times \mathbb{R}^k \to \mathbb{R}$, C^∞ quadratic at infinity such that S_t generates L_t.*

The second result we will need deals with bifurcation of critical points of smooth functions. Let N be any manifold and let $S : I \times N \to \mathbb{R}$ be a one parameter family of smooth functions on N. Let $\tau: I \to N$ be a path such that $\tau(t)$ is a critical point of S_t (*the trivial branch of critical points*). At any critical point $\tau(t)$ there is a well defined symmetric bilinear form $H(S_t, \tau(t))$ on $T_{p(t)}N$ called the *Hessian* of S_t at $\tau(t)$ (see [13]). We will assume that τ_0 and τ_1 are non-degenerate critical points (i.e., the Hessian at τ_0 and τ_1 are non-degenerate). Let us recall that the *Morse index* $\mu(S, x)$ of a function S at a non-degenerate critical point x is defined as the dimension of the negative eigenspace of $H(S, x)$.

The following proposition is well known in bifurcation theory. It is usually proved using either homotopy invariance of the Conley index [2] or the invariance under small perturbations of the local critical group of a critical point. For the reader's convenience we will include a proof that follows almost at verbatim an argument due to Berger [3], who proved a special case.

Proposition 3.2. *Let $S : I \times N \to \mathbb{R}$ and $\tau: I \to N$ be as above. If $\mu(S_1, \tau(1)) \neq \mu(S_0, \tau(0))$ then there exist $t_* \in (0,1)$ and a sequence $(t_i, x_i) \in I \times N$ such that x_i is a critical point of S_{t_i} different from $\tau(t_i)$ and $t_i \to t_*$, $x_i \to \tau(t_*)$.*

Proof. By contractibility of I, the tubular neighborhood of the embedded path $t \mapsto (t, \tau(t))$ in $I \times N$ is diffeomorphic to $\mathbb{R}^n$. Using this, the general situation can be easily reduced to the case $S : I \times \mathbb{R}^n \to \mathbb{R}$, $\tau \equiv 0$ with 0 being non-degenerate as critical point of S_0 and S_1. Assume that there are no non-trivial critical points of the family S_t close to 0. Then we can consider the kth-local critical group, $C_k(S_t, 0)$, associated to the isolated critical point 0 of S_t. This group is defined by

$$C_k(S_t, 0) \equiv H_k(S_t^0 \cap \mathcal{N}, S_t^0 \cap \mathcal{N} \setminus \{0\}),$$

where $H_k(\cdot,\cdot)$ is the kth relative singular homology group with coefficients in $\mathbb{Z}$, $S_t^0 \equiv \{x : S_t(x) \leq 0\}$ and $\mathcal{N}$ is a small enough neighborhood of 0. It follows from Morse theory that in absence of bifurcation points each critical group $C_k(S_t, 0)$ is independent of t. By the Morse lemma, whenever 0 is a non-degenerate critical point of S_t the local groups are given by

$$C_k(S_t, 0) = \begin{cases} \mathbb{Z} & \text{if } k = \mu(S_t, 0), \\ 0 & \text{otherwise.} \end{cases} \tag{6}$$

But this contradicts $\mu(S_1, \tau(1)) \neq \mu(S_0, \tau(0))$ and Proposition 3.2 is proved. □

We will need one more proposition relating the change in Morse indices with the Maslov intersection index. Before we state it let us shortly review the construction of the relative Maslov index.

The *Maslov index* of a closed path in Λ_n is defined in [1] as follows: for any Lagrangian $L \in \Lambda_n$ there exists a unitary operator T of $\mathbb{C}^n \simeq \mathbb{R}^{2n}$ sending $\mathbb{R}^n_0 = \mathbb{R}^n \times \{0\}$ into L. It follows from this that $U(n)$ acts transitively on Λ_n, the isotropy subgroup of $\mathbb{R}^n_0$ being $O(n)$. Therefore $\Lambda_n \simeq U(n)/O(n)$ and the map ρ sending L into $\det^2 T$ is well defined on Λ_n. Given any closed path $\gamma: S^1 \to \Lambda_n$ the Maslov index $m(\gamma)$ is the winding number of $\rho\gamma$. Given two paths $l, n: I \to \Lambda_n$ with transversal end points, the path $l \times n$ has end points transversal to the diagonal $\Delta \in \Lambda(\mathbb{R}^{2n} \times \mathbb{R}^{2n-})$. Since the set of all

Lagrangian subspaces transversal to a given one is contractible, if we take any path δ joining in this set the endpoints of $l \times n$, the Maslov index of the closed path made by $l \times n$ followed by δ will be independent of the choice of δ. The index of this closed path is by definition the relative Maslov index $\mu(l,n)$ introduced in the previous section. A similar construction can be found in [18] (see also [14] for a different approach).

It follows from Theorem 2.3 in [14] that $\mu(l,n)$ verifies all the axioms characterizing the Maslov index. In particular, if $t_1, ..., t_k$ are the only points of non-transversal intersection and if the crossing form (2) at each point t_i is non-degenerate we have

$$\mu(l,n) = \sum_{i=0}^{k} \operatorname{sign} Q(t_i). \tag{7}$$

With this understood let us consider again the intersection index $\mu(L,N,p)$ defined as above.

Proposition 3.3. *Let $S : I \times N \to \mathbb{R}$, $\tau : I \to N$ be as above. Let L_t be the image of $dS_t : N \to T^*(N)$. Identifying N with the 0-section of $T^*(N)$ we have that L_t is transversal to N for $t = 0, 1$; moreover*

$$\mu(L,N,\tau) = \frac{1}{2}[\mu(S_1, \tau(1)) - \mu(S_0, \tau(0))].$$

Proof. As before we can assume $N = \mathbb{R}^n$ and $\tau(t) \equiv 0$. Using the inner product we can identify $T^*\mathbb{R}^n$ with $T\mathbb{R}^n$. Under this identification dS_t goes into the section of $T\mathbb{R}^n = \mathbb{R}^n \times \mathbb{R}^n$ given by the gradient $x \mapsto (x, \nabla S_t(x))$, and N is the zero section. The family L_t is nothing but the the graph of the gradient ∇S_t. The tangent space to L_t at $\tau(t) = (0,0)$ is the graph of the symmetric operator A_t defined as the derivative $D\nabla S_t(0)$ at 0. Therefore the Hessian $H_t(S_t, 0)$ is given by $H_t(0)(u,v) = \langle A_t u, v\rangle$ and since for $t = 0,1$ we have Ker $A_t = 0$, it follows that L_0, L_1 are transversal to $N = \mathbb{R}^n \times \{0\}$. By definition of the Maslov intersection index $\mu(L,N,\tau) = \mu(l,n)$ where $l: [0,1] \to \Lambda_n$ is given by $l(t) =$ graph A_t (which is Lagrangian since A_t is symmetric) and $n(t) = \mathbb{R}^n \times \{0\}$. By the localization property of the Maslov index (Theorem 2.3 in [14])

$$\mu(l,n) = \frac{1}{2}[\operatorname{sign} H(S_1, 0) - \operatorname{sign} H(S_0, 0)]$$

which can be easily shown to be the same as

$$\frac{1}{2}[\mu(S_1, \tau(1)) - \mu(S_0, \tau(0))],$$

and the assertion follows. □

Now let us turn to the

Proof of Theorem 2.1. Let $S: [0,1] \times N \times \mathbb{R}^k \to \mathbb{R}$ be the path of generating families given by Proposition 3.1. Thus each $L_t = e_t(C_t)$ where $C_t = \{(n,v) : v$ is critical of $S_{t,n}: \mathbb{R}^k \to \mathbb{R}$ given by $S_{t,n}(v) = S_t(n,v)\}$. Since here each e_t is an embedding it is easy to see that e_t induces a bijection between critical points of $S_t: N \times \mathbb{R}^k \to \mathbb{R}$ and intersection points $L_t \cap N$. Therefore the path of intersection points p has a corresponding path $\tau: I \to N \times \mathbb{R}^k$ of critical points of S_t. We claim that since L_0, L_1 are transversal to N at $p(0), p(1)$ it follows that $\tau(0)$ and $\tau(1)$ are non-degenerate critical points. This is a direct consequence of the linear algebra of symplectic reductions. Indeed, let $N' = N \times \mathbb{R}^k$ and let us consider the symplectic manifold $V = T^*(N') = T^*(N) \times \mathbb{R}^{2k}$. The manifold $\{0\} \times \mathbb{R}^k$ is an isotropic submanifold of V and the symplectic reduction of $V = T^*(N')$ modulo this submanifold is $T^*(N)$. On the other hand N' and dS_t are Lagrangian submanifolds of V whose symplectic reductions are N and L_t respectively. By the standard properties of the symplectic reductions [9] the fact that L_t intersects transversely N at $p(t)$, for $t = 0, 1$, implies that dS_t intersects transversely N'. But this is equivalent to non-degeneracy of the critical point $\tau(t)$ for those values of t.

Moreover, by Proposition 3 of [18] (see also [9]), we have that $\mu(L, N, p)$ is invariant under symplectic reductions. Hence from the hypothesis of Theorem 2.1 using Proposition 3.3 we get

$$\mu(S_1, \tau(1)) - \mu(S_0, \tau(0)) = \mu(dS, N, \tau) = \mu(L; N; p) \neq 0.$$

By Proposition 3.2 we will find a sequence of critical points of S_t bifurcating from the trivial branch. Via the map e those critical points correspond to nontrivial intersections of L_t with N. □

Proof of Corollary 2.4. This follows from Theorem 2.1 and (7). □

Proof of Corollary 3.3. We will use some results in [8]. To simplify we will take $p(t) \equiv p$ constant. Since $H^1(V; \mathbb{R}) = 0$ all elements of $Symp_0(V)$ of the group of all symplectomorphisms can be realized as the time one map of a 1-periodic Hamiltonian vector field. Moreover it is proved in [8] that there exist a smooth family of time dependent Hamiltonian functions $H': I \times I \times V \to \mathbb{R}$ such that ϕ_t is the time-one map of the Hamiltonian time-dependent vectorfield $X'_t : I \times V \to TV$ (here the time will be s in order to keep the previous notations).

By Weinstein's theorem [20] any Lagrangian submanifold N of a symplectic manifold V has a neighborhood $U(N)$ that is symplectomorphic to a neighborhood of the zero section in the cotangent bundle T^*N of N. We will apply this theorem to the diagonal Δ in $V \times V^-$ and then modify the Hamiltonian and the flow $\phi_{t,s}$ outside of a neighborhood of p in order to be able to use Theorem 2.1 in the cotangent bundle $T^*\Delta$. For this take any metric on V and consider a ball $B = B(p, 2r)$ such that the closure $\overline{B \times B}$ is contained in $U = U(\Delta)$. Let $f: V \to \mathbb{R}$ be such that $0 \leq f \leq 1$, $f \equiv 1$ on $B(p, r)$ and $f \equiv 0$ outside B. Define $H: I \times I \times V \to \mathbb{R}$ by

$$H(t, s, v) = f(v) H'(t, s, v). \tag{8}$$

Let $\psi_{t,s}$ be the flow of $X_t = X(H_t)$ and let $\psi_t \equiv \psi_{t,1}$ be the corresponding time-1 map. By definition of H, $\psi_t(v) = v$ outside of B and therefore the manifolds $L_t = \text{Graph } \psi_t \cap U$ coincide with Δ outside $\overline{B \times B}$ and can be viewed as a one parameter family of Lagrangian submanifolds of $T^*\Delta$ with compact support. Using that $H^1(V, \mathbb{R}) = 0$ we get that $L \equiv L_t$ is exact. Moreover by Sikorav's theorem L_0 possesses a generating family being ψ_0 isotopic by Hamiltonian isotopy to the identity map of V. Thus we can apply Theorem 2.1 to L and Δ and find a sequence of (t_n, p_n) with $t_n \to t^*$, $p_n \to p$, $p_n \neq p$ and $\psi_{t_n}(p_n) = p_n$. We claim that for n big enough p_n is a fixed point of ϕ_n.

Indeed, by continuous dependence of initial conditions and parameters for solutions of differential equations and compactness of $I \times I$ we can find a ball $D = B(p, \delta)$ such that for all $v \in D$ and all $(t, s) \in I \times I$ $\psi_{t,s}(v) \in B(p, r)$ and therefore by definition of H for points $v \in D$ we have $\psi_{t,s}(v) = \phi_{t,s}(v)$. In particular on D we get $\phi_t = \psi_t$ and therefore p_n are fixed points of ϕ_t for n big enough. □

Theorem 2.6 can be proved along the same lines.

References

[1] V. I. Arnold: *A characteristic class entering in quantization conditions*, Funct. Anal. Appl. **1** (1967), 1-13

[2] P. H. Bartsch: *Topological Methods for Variational Problems with Symmetries,* Lect. Notes Math. **1560**, Springer, Berlin 1993

[3] M. S. Berger: *Bifurcation theory and the type numbers of M. Morse,* Proc. Nat. Acad. Sci. USA **69** (1972), 1737-1738

[4] M. Chaperon: *Questions de géométrie symplectique,* Astérisque **105-106** (1983), 231-249

[5] M. Chaperon: *On generating families*, in: The Floer memorial volume [Ed.: H. Hofer], Progress Math. **133**, Birkhäuser, Basel 1995, p. 283-296

[6] M. Chaperon: *Some remarks on generating families*, in: Proc. Moscow Geometry Symposium, to appear

[7] Yu. V. Chekanov: *Critical points of quasi-functions and generating families of Legendrian manifolds* [in Russian], Funkt. Anal. Prilozh. **30** (1996), 118-128

[8] E. Ciriza: *Bifurcation of periodic orbits of time dependent Hamiltonian systems on symplectic manifolds*, Rend. Sem. Mat. Torino, to appear

[9] E. Ciriza, J. Pejsachowicz: *The Maslov index of Fredholm Lagrangians and bifurcation for compact perturbations of unbounded self adjoint Fredholm operators,* preprint

[10] M. Fitzpatrick, J. Pejsachowicz, L. Recht: *Spectral flow and bifurcation of critical points of strongly-indefinite functionals, Part II: Bifurcation of periodic orbits of Hamiltonian systems,* J. Diff. Equ., to appear

[11] L. Hörmander: *Fourier integral operators I,* Acta Math. **127** (1971), 79-183

[12] J. Ize: *Necessary and sufficient conditions for multiparameter bifurcation,* Rocky Mountain J. Math. **18** (1988), 305-337

[13] J. Milnor: *Morse Theory,* Princeton Univ. Press, Princeton 1963

[14] J. ROBBIN, D. SALAMON: *The Maslov index for paths,* Topology **32** (1993), 827-844

[15] J. ROBBIN, D. SALAMON: *The spectral flow and the Maslov index,* Bull. London Math. Soc. **27** (1993), 1-33

[16] J. C. SIKORAV: *Problèmes d'intersections et de points fixes en géométrie hamiltonienne,* Comm. Math. Helv. **62** (1987), 61-72

[17] J. C. SIKORAV: *Sur les immersions lagrangiennes dans un fibré cotangent,* C. R. Acad. Sci. Paris **32** (1986), 119-122

[18] C. VITERBO: *Intersection de sous variétés Lagrangiennes, fonctionnelles d'action et indice des systèmes Hamiltoniens,* Bull. Soc. Math. France **115** (1987), 361-390

[19] A. WEINSTEIN: *Lectures on Symplectic Manifolds,* Reg. Conf. Ser. Math. **29**, Amer. Math. Soc., Providence RI 1977

[20] A. WEINSTEIN. *Lagrangian submanifolds and Hamiltonian systems,* Ann. Math. **98** (1998), 377-410

ELEONORA CIRIZA, Dipartimento di Matematica, Università di Roma "Tor Vergata", Via della Ricerca Scientifica, I-00133 Roma, Italy; `ciriza@mat.uniroma2.it`

JACOBO PEJSACHOWICZ, Dipartimento di Matematica, Politecnico di Torino, Corso Duca degli Abruzzi 24, I-10129 Torino, Italy; `pejsachowicz@polito.it`

Progress in Nonlinear Differential Equations
and Their Applications, Vol. 40

La valutazione di opzioni implicite nei mutui bancari

ANDREA CONSIGLIO, MASSIMO COSTABILE, CARLO MARI, IVAR MASSABÒ

Ad Alfonso per una accurata gestione delle sue cucuzze

Summary: We analyze the problem of pricing implicit options embedded in mortgages and provide a general framework which can be extended to other types of implicit contingent claims. In particular, we deal with the problem of pricing *prepayment* options and *cap* for floating rate mortgages.
A prepayment option allows the house-holder to prepay her mortgage, at any epoch before maturity, without incurring in any penalty. We show that such option is equivalent to an american call option where the underlying asset is the market value of the outstanding debt and the strike price is the outstanding debt given by the amortization plan.
By fixing the maximum rate which must be paid on the outstanding debt, the house-holder can protect herself from dramatic oscillation of the interest rate. Such option is equivalent to a set of caplet written on a function of the outstanding debt.
Empirical results are provided by fitting the Black, Derman and Toy model to the term structure of interest rate and volatility.

Keywords: Option pricing, term structure models, mortgages.

Classification: 92K99.

1. Introduzione. Un mutuo è una operazione finanziaria che consiste nello scambio di un importo monetario C, il capitale mutuato, disponibile al tempo t_0, con una successione di pagamenti, le rate del mutuo $\{R_1, R_2, \cdots, R_m\}$, non necessariamente di importo costante, esigibili ai tempi $\{t_1, t_2, \cdots, t_m\}$, con $t_0 \leq t_1 < t_2 < \cdots < t_m$. Faremo riferimento ad alcune tipologie di mutuo particolarmente diffuse, a tasso fisso o indicizzate, caratterizzate da rate posticipate ($t_0 < t_1$) e periodiche con $t_k = t_{k-1} + \tau, \quad k = 1, 2, \cdots, m$.

Il problema della valutazione di un mutuo può essere impostato a partire da alcuni principi di base. Se indichiamo infatti con $v(t, s)$ la funzione di sconto relativa alla struttura a termine dei tassi di interesse in vigore al tempo t (stimata ad esempio nel mercato dei titoli di Stato o quella stimata a partire dai tassi swap o dai tassi Libor etc. e corretta opportunamente per il rischio di credito) utilizzata dalla banca erogante si ha,

$$v(t, s) = \Big[1 + i(t, s) + \delta\Big]^{-(s-t)} \tag{1}$$

dove $i(t,s)$ rappresenta la struttura a termine dei tassi di interesse opportunamente stimata al tempo t e δ la maggiorazione per il rischio di credito.[1] Analizzeremo in dettaglio il caso dei mutui a tasso fisso e rate costanti e quello dei mutui indicizzati; l'estensione ad altre tipologie (per esempio, a tasso fisso con rate variabili) è immediata.

Nel caso di un mutuo a rate costanti, l'importo della rata R si determina sulla base dell'equivalenza finanziaria,

$$C = R\sum_{k=1}^{m} v(t_0,t_k), \tag{2}$$

da cui,

$$R = \frac{C}{\sum_{k=1}^{m} v(t_0,t_k)}. \tag{3}$$

Nella totalità dei mutui a tasso fisso, tra le altre caratteristiche contrattuali viene specificato il valore del tasso nominale (convertibile $\frac{1}{\tau}$ volte in un anno nel caso in esame), che si determina univocamente una volta calcolato il tasso interno di rendimento dell'operazione finanziaria considerata,

$$\sum_{k=1}^{m}(1+i)^{-k} = \sum_{k=1}^{m} v(t_0,t_k). \tag{4}$$

La (4), risolta rispetto ad i, ammette una ed una sola soluzione $i>0$ (dato che $m > \sum_{k=1}^{m} v(t_0,t_k)$) e fornisce il valore del tasso nominale,

$$j = \frac{i}{\tau} \tag{5}$$

cui si fa riferimento nella valutazione comparativa di mutui a tasso fisso. In molti casi il valore del tasso annuale di riferimento viene determinato sulla base della legge esponenziale secondo la formula,

$$j_e = (1+i)^{\frac{1}{\tau}}. \tag{6}$$

A partire dal tasso i è immediato determinare la struttura del piano di ammortamento. Le rate del mutuo, di importo costante R, risultano infatti decomponibili in una quota interessi I_k e in una quota capitale C_k $k=1,2,\cdots,m$, variabili nel tempo e determinabili in base alle relazioni ricorrenti,

$$I_k = iD_{k-1}, \qquad k=1,2,\cdots,m, \tag{7}$$

$$C_k = R - I_k \qquad k=1,2,\cdots,m. \tag{8}$$

dove D_{k-1} rappresenta il debito residuo all'epoca t_{k-1}(vale ovviamente la relazione $D_0 = C$) subito dopo il pagamento della $(k-1)$-esima rata R.

[1] In questo studio δ è assunto costante. Per una trattazione più generale che però esula dagli scopi del presente lavoro, occorre fare riferimento a modelli stocastici che incorporino in maniera non naive il rischio di credito (cfr. ad es. Jarrow, Lando e Turnbull [12], Nielsen e Ronn [15]).

I valori del debito residuo D_k, alle epoche t_k, $k = 1, 2, \cdots, m$, necessari per chiudere il set di equazioni (7) e (8) e per redigere il piano di ammortamento, si determinano in base alla relazione ricorrente,

$$(9) \qquad D_k = D_{k-1} - C_k, \qquad k = 1, 2, \cdots, m.$$

Poichè $D_m = 0$, dalle equazioni (9) segue immediatamente la condizione di chiusura elementare,

$$(10) \qquad \sum_{k=1}^{m} C_k = C,$$

che garantisce il rimborso completo del capitale C mediante il pagamento delle quote di capitale C_k, ai tempi t_k, $k = 1, 2, \cdots, m$.[2]

E' possibile che un mutuo stipulato a tasso fisso possa diventare particolarmente oneroso per il mutuatario conseguentemente ad un processo di riduzione dei tassi. La ricontrattazione o il rimborso anticipato di un mutuo è previsto dalla normativa vigente, infatti secondo il D. Lgs. n. 385 del 01/09/93 art. 40 "I debitori hanno facoltà di estinguere anticipatamente, in tutto o in parte, il proprio debito, corrispondendo alla banca un compenso, contrattualmente stabilito, correlato al capitale restituito anticipatamente". Il compenso è determinato in maniera forfettaria sulla base del valore del debito residuo e pertanto senza una corretta valutazione della rischiosità e del "prezzo" dell'*opzione* di rimborso incorporata per legge nei mutui bancari.

Uno degli scopi del lavoro è pertanto quello di formalizzare la modellizzazione dell'opzione di rimborso anticipato (OdRA) nel caso di un mutuo a tasso fisso, nel tentativo di fornire uno schema di valutazione completo per la determinazione del prezzo da imputare ai mutuatari per poter esercitare il diritto di rimborso anticipato del debito prima della sua naturale scadenza.

Nel caso di un mutuo indicizzato, l'importo della quota interessi viene determinato sulla base del valore del tasso di interesse in vigore all'inizio del periodo di riferimento della rata da corrispondere e relativo ad un orizzonte temporale coincidente con quello della rata stessa, secondo lo schema,

$$(11) \qquad I_k = D_{k-1}\Big[\frac{1}{v(t_{k-1}, t_k)} - 1\Big], \qquad k = 1, 2, \cdots, m.$$

La quantità,

$$(12) \qquad r(t_{k-1}, t_k) = \frac{1}{v(t_{k-1}, t_k)} - 1$$

è il tasso d'interesse al tempo t_{k-1} secondo la struttura a termine di valutazione e relativo all'orizzonte temporale $[t_{k-1}, t_k]$. A differenza di un mutuo a tasso fisso, gli importi delle quote interessi, I_k, sono quantità non note all'istante di stipula. Contrattualmente è nota la regola di indicizzazione, mentre gli importi I_k sono determinabili al tempo t_{k-1} sulla base del valore del tasso osservato a quell'epoca.

Il capitale mutuato può venire restituito durante la vita del contratto secondo un piano di rimborso qualsiasi, a patto che la condizione di chiusura (10) venga rispettata. Gli importi delle rate risultano allora non costanti e determinabili in base alla relazione,

$$(13) \qquad R_k = C_k + I_k, \qquad k = 1, 2, \cdots, m.$$

[2] Per una trattazione completa, si veda Moriconi [14], Peccati [17].

Dimostreremo, inoltre, che per effetto del meccanismo di indicizzazione, l'OdRA ha valore nullo nel caso di mutui a tasso variabile. Al contrario, è rilevante il problema della valutazione di mutui indicizzati in presenza di *cap*, cioè di limitazioni, fissati contrattualmente, per il valore del tasso di riferimento da utilizzare nella determinazione dell'importo delle quote interessi. Vedremo in particolare che il problema della valutazione del *cap* potrà essere ricondotto a quello della valutazione di un portafoglio di opzioni put di tipo europeo, scritte su titoli a cedola fissa ben definiti dalle caratteristiche tecniche del mutuo considerato.

Il lavoro si compone di altre due sezioni. Nella prima di esse discuteremo in dettaglio lo schema teorico generale sviluppato per la valutazione di alcune tipologie di opzioni implicite nei mutui bancari. In particolare, analizzeremo il caso dell'OdRA nei mutui a tasso fisso e nei mutui indicizzati e quello della valutazione del *cap* nei mutui indicizzati. La metodologia sviluppata può essere facilmente estesa a tipologie di opzioni più complesse.

La terza sezione contiene i risultati della sperimentazione numerica effettuata utilizzando il modello di Black, Derman e Toy. Sono state valutate OdRA e *cap* implicite in mutui di durata variabile dai 5 ai 30 anni. Le simulazioni sono state realizzate utilizzando diverse strutture di volatilità (costanti o rilevate nel mercato), nonché diversi livelli dei parametri che caratterizzano il modello.

2. Uno schema di riferimento. Le opzioni in finanza sono contratti che conferiscono all'acquirente, a fronte del pagamento di un premio, il diritto, ma non l'obbligo, di acquistare (opzioni *call*) o di vendere (opzioni *put*) ad un prezzo prefissato (prezzo d'esercizio) una attività (attività sottostante). Nel caso in cui sia possibile esercitare la facoltà di acquisto o di vendita dell'attività sottostante solo in corrispondenza di una specifica data, l'opzione è detta di tipo europeo mentre, se l'attività sottostante può essere acquistata o venduta in un istante di tempo qualsiasi tra la stipula del contratto e la scadenza, l'opzione è detta di tipo americano.

L'esigenza degli istituti finanziari di porre in essere strategie gestionali sempre più sofisticate ha condotto nel corso del tempo ad una crescente diffusione delle opzioni finanziarie, sia da un punto di vista quantitativo che in relazione all'articolazione dei prodotti offerti. In tale ambito un ruolo di primo piano è svolto dalle opzioni su tassi d'interesse. La tipologia delle opzioni su tassi d'interesse è molto ampia poiché ricadono in questa categoria tutte le opzioni il cui payoff risulti collegato alla dinamica dei rendimenti dei mercati di tipo obbligazionario. In molte situazioni le opzioni su tassi d'interesse non sono scambiate autonomamente ma vengono inserite all'interno di contratti complessi allo scopo di fornire copertura finanziaria nei confronti di movimenti non attesi della struttura dei rendimenti.

In molti casi l'attività finanziaria sottostante è rappresentata da titoli obbligazionari con o senza cedola, in presenza o meno di regole di indicizzazione. Le variazioni nel prezzo del sottostante sono determinate dai movimenti della curva dei rendimenti e, di conseguenza, il valore dell'opzione è collegato in modo univoco alla dinamica della struttura per scadenza dei tassi d'interesse.

Relativamente alle opzioni call di tipo europeo, l'acquirente troverà conveniente avvalersi della facoltà di esercizio esclusivamente nel caso in cui alla scadenza il valore

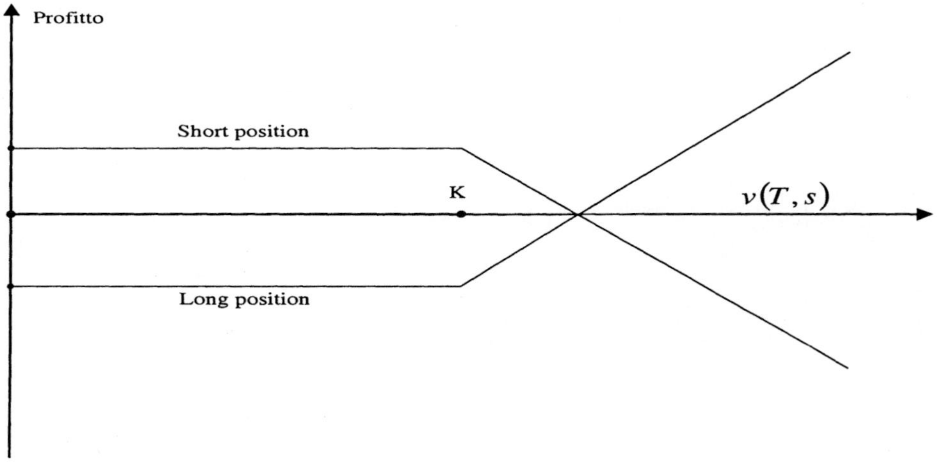

Figure 1: Profitto di opzioni call europee.

dell'attività sottostante risulti superiore al prezzo d'esercizio. In questo caso, infatti, sarà possibile acquistare l'attività sottostante ad un prezzo inferiore rispetto al suo valore di mercato e ottenere come payoff la differenza tra i due valori. Nel caso contrario l'opzione non verrà esercitata essendo conveniente acquistare l'attività sottostante direttamente nel mercato. Il payoff a scadenza di un'opzione call è quindi rappresentato dal maggiore tra due valori, il primo rappresentato dalla differenza tra il prezzo del titolo sottostante e il prezzo di esercizio e il secondo pari a zero.

E' possibile definire dal punto di vista formale il guadagno (payoff) a scadenza di un'opzione call europea scritta al tempo t al prezzo O_t su un titolo che rimborsa un valore unitario (zcb : *zero-coupon bond*) con scadenza in s, $(t < s)$. Se indichiamo, infatti, con T il tempo di esercizio $(t < T < s)$ e con K il prezzo d'esercizio, il payoff a scadenza è dato da,

$$O_T = \max\left[v(T,s) - K, 0\right]. \tag{14}$$

La Figura 1 rappresenta la funzioni del profitto, per l'acquirente (long position) nonchè per il venditore (short position), alla scadenza T di un'opzione call di tipo europea.

E' possibile notare come alla scadenza, per valori del titolo sottostante inferiori al prezzo d'esercizio, non essendo conveniente esercitare l'opzione, il detentore della posizione lunga subisca una perdita di entità pari al premio corrisposto per l'acquisto del titolo. Per valori del sottostante superiori al prezzo d'esercizio, l'esercizio dell'opzione diventa vantaggioso, ed il profitto cresce all'aumentare del valore del titolo sottostante. Analoghe considerazioni possono essere estese al caso della posizione corta.

Per quanto riguarda le opzioni put di tipo europeo il payoff a scadenza è,

$$P_T = \max\left[K - v(T,s), 0\right], \tag{15}$$

La Figura 2 riporta le funzioni del profitto relative ad un'opzione put di tipo europea. E' rilevante sottolineare che un'opzione di tipo americano ha valore non inferiore alla corrispondente opzione europea scritta sulla stessa attività sottostante e avente uguali

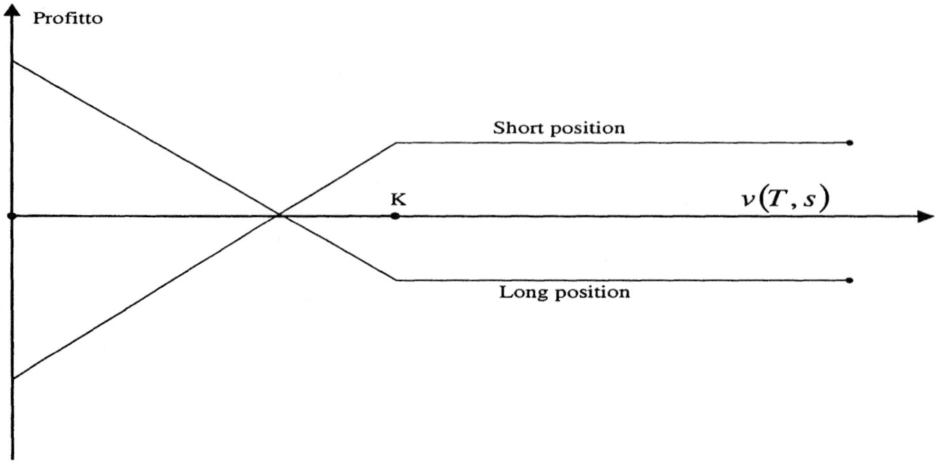

Figure 2: Profitto di opzioni put europee.

prezzo d'esercizio e vita a scadenza. Infatti l'opzione americana garantisce al detentore gli stessi diritti della corrispondente opzione europea ed in più fornisce la facoltà di esercizio in un istante di tempo qualsiasi tra la stipula e la scadenza del contratto.

Nel caso delle opzioni di tipo americano è significativo prendere in considerazione il payoff dell'opzione in istanti di tempo diversi dalla scadenza, in quanto la decisione di esercitare anticipatamente l'opzione si fonda sul confronto tra il *valore intrinseco* ed il *valore temporale* dell'opzione. In particolare, il valore intrinseco della call americana è rappresentato, all'istante di tempo τ, $t < \tau < T$, dalla quantità $v(\tau, s) - K$. Nel caso in cui tale valore risulti superiore al valore temporale dell'opzione all'istante τ, o_τ, cioè il valore dell'opzione se non esercitata, risulterà conveniente l'esercizio anticipato. Pertanto, il valore dell'opzione call di tipo americano risulta essere

$$O_\tau = \max\left[v(\tau, s) - K, o_\tau\right] \tag{16}$$

2.1. La valutazione dell'OdRA. In presenza dell'OdRA, il mutuatario ha la facoltà di rimborsare anticipatamente il debito, in un istante di tempo qualsiasi durante la vita del contratto, ad un prezzo, convenuto contrattualmente, pari al valore del debito residuo all'istante del rimborso. Il mutuo pertanto risulta caratterizzato da un'opzione *call* incorporata di tipo americano. L'attività sottostante l'opzione è costituita dalla successione delle rate non ancora pagate che possono venire "riacquistate" dal mutuatario al prezzo di esercizio determinato dal valore corrente del debito residuo D_t, che è noto per contratto ad ogni istante di tempo $t_0 \leq t \leq t_m$.

Poichè l'opzione di rimborso può essere esercitata in qualunque istante di tempo t ($t_0 \leq t \leq t_m$), al prezzo di esercizio D_t, il valore della *call* al tempo t è dato da,

$$O_t = \max\{V_t - D_t, o_t\}, \tag{17}$$

dove,

$$V_t = R \sum_{k: t_k > t} v(t, t_k), \tag{18}$$

è il valore al tempo t, secondo la struttura a termine utilizzata dalla banca, dell'indebitamento residuo del mutuatario;

$$D_t = D_k(1+j_e)^{t-t_k}, \tag{19}$$

è il valore del debito residuo al tempo t, con $t_k \leq t < t_{k+1}$; t_k rappresenta la data di scadenza dell'ultima rata pagata e o_t è il valore temporale dell'opzione. La condizione (17) descrive sinteticamente il fatto che il mutuatario ha la facoltà di riacquistare ad ogni istante di tempo t, al prezzo D_t, il proprio debito il cui valore "di mercato" è pari a V_t. Pertanto, se la differenza $V_t - D_t$ è maggiore di o_t, risulterà conveniente esercitare l'opzione al tempo t; nel caso contrario converrà considerare la possibilità di esercizio ad un istante di tempo successivo. È opportuno sottolineare che l'opzione call americana implicita, è caratterizzata da un prezzo di esercizio non costante ma che dipende in maniera deterministica dal tempo, secondo la relazione (9). Questo fatto non crea difficoltà particolari di carattere teorico per la valutazione del prezzo dell'opzione, occorre soltanto tenerne conto nella procedura di calcolo. Nelle nostre simulazioni numeriche abbiamo considerato l'ipotesi, peraltro realistica, di esercizio dell'opzione ai tempi $t_1, t_2, \cdots, t_m$, ai prezzi $D_1, D_2, \cdots, D_m$ rispettivamente, cioè immediatamente dopo il pagamento di una rata.

E' da notare che il valore dell'opzione di rimborso anticipato è indipendente dal tipo di trasformazione eventuale del mutuo. In altri termini, il prezzo dell'OdRA è lo stesso sia che si decida di rinegoziare il mutuo a tasso fisso (più basso ovviamente), che si decida di trasformarlo in indicizzato o che si decida di estinguerlo anticipatamente.

Il caso dei mutui a tasso variabile progettati secondo lo schema di riferimento specificato nell'Introduzione, si presta ad alcune considerazioni interessanti. In primo luogo il valore attuale dell'indebitamento residuo, calcolato con la struttura dei tassi della banca, coincide ad ogni istante di tempo di pagamento delle rate $\{t_1, t_2, \cdots, t_m\}$, con il valore del debito residuo (Mari [13]). Infatti l'importo delle rate,

$$R_k = C_k + D_{k-1}\left[\frac{1}{v(t_{k-1}, t_k)} - 1\right], \quad k = 1, 2, \cdots, m, \tag{20}$$

è decomponibile nella somma di due quantità, delle quali una, $C_k - D_{k-1}$, è deterministica ed esigibile al tempo t_k, mentre l'altra,

$$\frac{D_{k-1}}{v(t_{k-1}, t_k)} \tag{21}$$

è equivalente all'importo deterministico D_{k-1} ed esigibile al tempo t_{k-1}. Pertanto, il valore al tempo t_j, $t_j < t_k$, dell'importo $C_k - D_{k-1}$ risulta uguale a,

$$(C_k - D_{k-1})v(t_j, t_k)$$

mentre il valore al tempo t_j dell'importo aleatorio (21) è dato da $D_{k-1}v(t_j, t_{k-1})$. Il valore attuale al tempo t_j dell'indebitamento residuo risulta pertanto,

$$V_{t_j} = \sum_{k=j+1}^{m} \Big[D_{k-1}v(t_j, t_{k-1}) + (C_k - D_{k-1})v(t_j, t_k)\Big]. \tag{22}$$

Poichè $C_k = D_{k-1} - D_k$, la (22) può scriversi,

$$V_{t_j} = \sum_{k=j+1}^{m} \Big[D_{k-1} v(t_j, t_{k-1}) - D_k v(t_j, t_k) \Big]$$
$$= D_j v(t_j, t_j) - D_{j+1} v(t_j, t_{j+1}) + D_{j+1} v(t_j, t_{j+1})$$
$$-D_{j+2} v(t_j, t_{j+2}) + \cdots + D_{m-1} v(t_j, t_{m-1}) - D_m v(t_j, t_m).$$

Ricordando che $D_m = 0$ si ottiene,

$$V_{t_j} = D_j, \qquad j = 1, 2, \cdots, m. \tag{23}$$

La valutazione dell'OdRA è quindi immediata nell'ipotesi che possa essere esercitata soltanto ai tempi $t_1, t_2, \cdots, t_m$, subito dopo il pagamento della rata. Infatti sulla base della relazione (23), poichè ad ogni data di esercizio il valore di mercato del debito residuo V_{t_k}, coincide con il prezzo di esercizio D_k, $k = 1, 2, \cdots, m$, il prezzo dell'opzione è pari a 0. Ovviamente il valore dell'OdRA sarà strettamente maggiore di zero nell'ipotesi che l'opzione sia esercitabile ad ogni istante di tempo t $(t_0 \leq t \leq t_m)$. La base per la valutazione è ancora fornita dalla (17) a patto di definire correttamente il valore del debito residuo D_t per ogni $t_{k-1} < t < t_k$, $k = 1, 2, \cdots, m$. La dimostrazione formale di quanto sopra è contenuta nell' Appendice A.

2.2. La valutazione del ***cap*****.** Ben diversa è la situazione della valutazione di un mutuo indicizzato caratterizzato dalla presenza di un *cap*, cioè di una limitazione superiore al tasso d'interesse. In presenza del *cap*, la regola di indicizzazione può essere riscritta nel modo seguente,

$$R_k^* = C_k + D_{k-1} \min \Big\{ \frac{1}{v(t_{k-1}, t_k)} - 1, \tau j^* \Big\}, \quad k = 1, 2, \cdots, m, \tag{24}$$

dove j^* rappresenta il valore del *cap* (*cap rate*), cioè il valore del tasso massimo che la banca può richiedere al mutuatario per contratto.

La valutazione del prezzo della copertura dal rischio di variazioni verso l'alto della struttura dei tassi di interesse, può essere impostata considerando gli importi aleatori ai quali la banca deve rinunciare per effetto del *cap*,

$$X_k = R_k - R_k^* = D_{k-1} \max \Big\{ \frac{1}{v(t_{k-1}, t_k)} - (1 + \tau j^*), 0 \Big\} \tag{25}$$

$(k = 1, 2, \ldots, m)$, esigibili ai tempi $t_1, t_2, \cdots, t_m$ rispettivamente. L'importo X_k misurabile al tempo t_{k-1}, è equivalente all'importo aleatorio,

$$P_{k-1} = X_k v(t_{k-1}, t_k) = \max \{ D_{k-1} - D_{k-1}(1 + \tau j^*) v(t_{k-1}, t_k), 0 \}, \tag{26}$$

esigibile al tempo t_{k-1}. E' immediato constatare che la quantità P_{k-1} rappresenta la condizione a scadenza di una opzione put europea caratterizzata dal tempo di esercizio t_{k-1} e dal prezzo di esercizio D_{k-1}, scritta su uno *zcb* con valore di rimborso,

$$D_{k-1}(1 + \tau j^*) \tag{27}$$

al tempo t_k. Il prezzo della copertura (V_{cap})è dato pertanto dalla somma dei prezzi delle m opzioni put europee equivalenti,[3]

$$V_{cap} = \sum_{k=1}^{m} P_{0,k}, \tag{28}$$

dove $P_{0,k}$ rappresenta il prezzo al tempo $t_0 = 0$ della k-esima opzione.

3. Risultati empirici. L'incertezza nei mercati dei titoli a reddito fisso è un fattore determinante per la valutazione di strumenti finanziari di tipo *interest rate sensitive* pertanto, lo studio di opzioni sui tassi d'interesse non può prescindere dalla definizione di modelli stocastici di evoluzione della struttura a termine dei rendimenti.

Nei cosiddetti modelli "unifattoriali" il tasso a breve, r_t, è l'unica variabile di stato descrittiva del mercato (cfr. ad es. Vasicek [18], Cox, Ingersoll e Ross [4]).[4]

Più recentemente, Ho e Lee [7], e successivamente, Heath, Jarrow e Morton [5,6] propongono una classe di modelli della struttura a termine dei tassi d'interesse in cui l'evoluzione dei prezzi di zcb é consistente con strutture iniziali dei tassi arbitrarie.

Nel modello di Black, Derman e Toy (BDT) [2], che abbiamo utilizzato nelle simulazioni numeriche, il processo del tasso a breve è consistente con la struttura dei tassi e delle voltatilità osservabili nel mercato. Le caratteristiche principali del modello sono:

- la distribuzione limite del tasso a breve è log-normale, ciò preclude la possibilità di tassi futuri negativi;
- gli input del modello, struttura a termine dei rendimenti e struttura della volatilità, sono facilmente osservabili nel mercato;
- la dinamica del tasso a breve è rappresentata tramite un albero binomiale ricombinante.[5]

Sia $S = \{0 = s_0, s_1, \cdots s_N = N\Delta\}$ uno scadenzario discreto, dove $s_i = i\Delta$ e Δ è l'ampiezza dell'intervallo fra due periodi successivi dell'albero binomiale. Nell'ipotesi che non esistano costi di transazione e che al tempo $s_0 = 0$ siano noti i prezzi $v(s_0, s_i)$ degli zcb unitari con scadenza $s_i \in S$, il tasso a breve, r_i, nel modello di BDT è il tasso di uno zcb unitario con scadenza s_{i+1},

$$r_i \equiv r(s_i, s_{i+1}) = \frac{1}{v(s_i, s_{i+1})} - 1 \tag{29}$$

[3]Per la valutazione di opzioni put di tipo europeo sono disponibili soluzioni in forma chiusa in alcuni modelli stocastici della struttura per scadenza dei tassi di interesse. Ben note sono le soluzioni nei modelli gaussiani (Jamshidian [10]), e quella nel modello di Cox, Ingersoll e Ross [4]. Per una trattazione completa si veda Hull [8].

[4]Altri autori hanno proposto modelli in cui l'evoluzione della struttura a termine dipenda anche da un tasso a lungo periodo e dalla correlazione fra queste due quantità (Brennan e Schwartz [3], Nielsen e Ronn [16]). Per una discussione completa si veda Hull e White [9].

[5]Un albero binomiale ricombinante ha la caratteristica che il tasso ottenuto dalla sequenza aleatoria *up-down* è uguale a quello ottenuto dalla sequenza *down-up*, ciò riduce notevolmente la complessità computazionale.

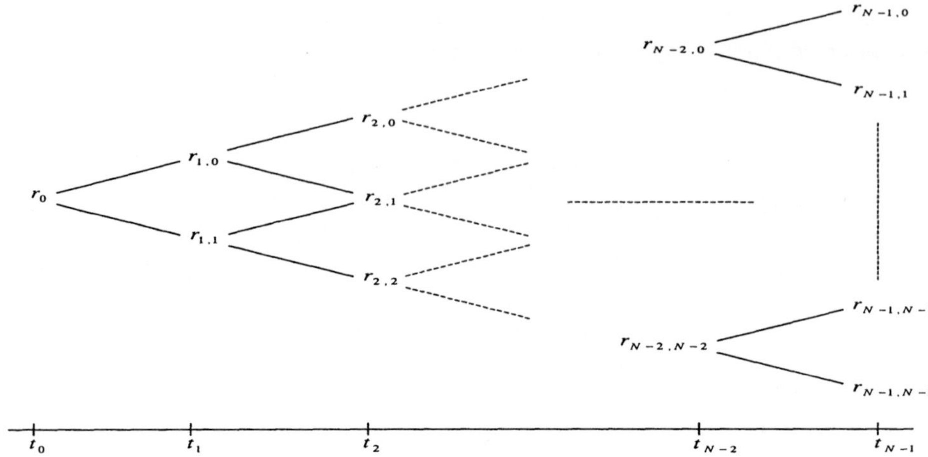

Figure 3: Dinamica binomiale del tasso a breve.

Il processo stocastico del tasso a breve, $\{r_i\}_{i=0}^{N-1}$, schematizzato in Figura 3, deve essere determinato in modo che sia consistente con la struttura a termine dei tassi d'interesse e delle volatilità osservate nel mercato all'istante di valutazione.
Nell'ipotesi che la probabilità di transizione per il primo periodo, $p(s_1)$, sia diversa da zero assegnata la volatilità $\sigma(s_1)$, le seguenti relazioni caratterizzano l'andamento del tasso a breve:

$$\sigma^2(s_1) \equiv V[\log r_1 | r_0] = p(s_1)(1-p(s_1)) \left[\log \frac{r_{1,0}}{r_{1,1}}\right]^2 \tag{30}$$

da cui,

$$r_{1,0} = r_{1,1} \exp\left\{\frac{\sigma(s_1)}{\sqrt{p(s_1)(1-p(s_1))}}\right\} = r_{1,1} g(\sigma(s_1), p(s_1)). \tag{31}$$

Ipotizzando che le probabilità di transizione siano costanti, data la struttura ricombinante del processo binomiale, si dimostra che :

$$r_{i,j} = r_{i,i} \cdots g(\sigma(s_i), p)^{i-j} \quad \forall j = 0, \ldots, i;\ i = 0, \ldots, N-1. \tag{32}$$

Si consideri ora uno zcb unitario con scadenza s_2. Il suo valore attuale dovrà essere uguale al valore atteso di una lira scontata in s_0, quindi:

$$\begin{aligned} v(s_0, s_2) &= \frac{1}{1+r_0}\left[\frac{p}{1+r_{1,0}} + \frac{1-p}{1+r_{1,1}}\right] \\ &= \frac{1}{1+r_0}\left[\frac{p}{1+r_{1,1} g(\sigma(s_1), p)} + \frac{1-p}{1+r_{1,1}}\right]. \end{aligned} \tag{33}$$

Per valori costanti di $p \in (0,1)$, $r_{1,1}$ è l'unica soluzione positiva dell'equazione (33).
Determinati i valori di $r_{1,1}$ ed $r_{1,0}$ è possibile impostare per il periodo successivo un'equazione analoga alla precedente. Ricordando la (32) e che il valore in s_0 di uno

zcb unitario con scadenza s_3 deve essere uguale al valore atteso di una lira scontata in s_0, si ha,

$$v(s_0, s_3) = E_p\left[\prod_{n=0}^{2}\frac{1}{1+r_n}|r_0\right]$$

$$(34)\qquad = \frac{1}{1+r_0}\left[\frac{p}{1+r_{1,0}}\left(\frac{p}{1+r_{2,2}g^2(\sigma(s_2),p)}+\frac{1-p}{1+r_{2,2}g(\sigma(s_2),p)}\right)\right.$$
$$\left.+\frac{1-p}{1+r_{1,1}}\left(\frac{p}{1+r_{2,2}g(\sigma(s_2),p)}+\frac{1-p}{1+r_{2,2}}\right)\right].$$

La soluzione della (34), per l'unica incognita $r_{2,2}$, fornisce il tasso base per movimenti "all-down" del tasso a breve. Noto $r_{2,2}$ ed utilizzando la relazione (32), è possibile determinare, per ogni stato $\{j = 0, 1, 2\}$, il corrispondente valore del tasso a breve.[6]

Iterando il processo per ogni $s_i \in S$ si otterà l'evoluzione del tasso a breve $\{r_i\}_{i=0}^{N-1}$ consistente con le informazioni prevalenti nel mercato.

Nei successivi paragrafi 3.1 e 3.2 considereremo mutui a tasso fisso ed indicizzati con rate semestrali posticipate. La struttura di riferimento per il calcolo della rata è quella illustrata nell'Introduzione. Lo *spread* δ, che tiene conto del rischio di insolvenza da parte del mutuatario, è fissato ad un valore costante.

La dinamica del tasso a breve è stata adattata utilizzando la struttura a termine dei titoli di stato tedeschi al 31-8-1998.[7] L'intervallo, Δ, fra due istanti temporali è stato posto pari ad un mese. La scelta riguardo il parametro Δ dipende da un *trade-off* fra efficienza computazionale e precisione dei prezzi delle opzioni. Per orizzonti temporali lunghi non è opportuno considerare discretizzazioni inferiori ad un mese. Nel caso in esame l'orizzonte temporale copre 30 anni, il numero di nodi è dato da $N(N+1)/2$, dove $N = 360$, per un numero totale di nodi pari a 64980.

La volatilità è stata stimata su base storica utilizzando un database mensile di strutture a termine dal 1990 al 1998. La metodologia utilizzata per le stime è simile a quella proposta da Riskmetrics [18]. In particolare la volatilità è ottenuta pesando tramite un parametro di "smoothing", λ, la volatilità ed il rendimento nel periodo precedente:

$$(35)\qquad \sigma_{t+1} = \sqrt{\lambda\sigma_t^2 + (1-\lambda)\bar{r}_t^2},$$

dove $\bar{r}_t^2$ è la variazione logaritmica del tasso fra $t-1$ e t.

Comunque, la scelta delle volatilità dipende anche dalla sensibilità del decisore. In alcuni casi le aspettative degli operatori sono differenti dalle stime ottenute su dati storici. Per tale motivo il decisore preferirà effettuare un'analisi con volatilità fissata a certi livelli e basare le proprie valutazioni su scenari "worst-case". Le valutazioni numeriche riportate nei paragrafi successivi sono state condotte considerando anche livelli di volatilità costanti al $10\%, 20\%, 30\%$ e 40%.

Il modello di BDT e la valutazione delle opzioni sono stati implementati su un PC con processore Intel Pentium II e 64 Mbyte di RAM. Il codice è stato scritto in C++ e

[6]Diversi autori hanno proposto implementazioni efficienti dal punto di vista computazionale del modello di BDT. Si veda ad esempio, Jamshidian [11], Bjerksund e Stensland [1].

[7]Non è stato possibile reperire dati per il mercato italiano.

compilato con l'opzione -O. Il tempo di calcolo per la determinazione del prezzo di ogni opzione non supera i 3-4 secondi.

3.1. La valutazione dell'OdRA. Come visto nel paragrafo 2.1, l'OdRA è assimilabile ad una opzione americana con prezzo d'esercizio pari al debito residuo D_t. Nella figura 4 è mostrato un campione della griglia binomiale (12 mesi) per un mutuo ventennale. Ogni mese è identificato dal suo numero progressivo riportato in verticale e dall'indice i; gli stati sono rappresentati sull'asse orizzontale e indicati con l'indice j, con $j = 0, 1, \cdots, i$. Ogni nodo sarà pertanto identificato dalla coppia (i, j). In corrispodenza di ogni nodo è trascritto il tasso annuo equivalente,

$$r^*_{i,j} = (1 + r_{i,j})^{12}. \tag{36}$$

nonchè il valore dell'opzione.

Si è ipotizzato che il mutuatario possa esercitare l'opzione un attimo dopo il pagamento della rata. Per tale motivo in corrispondenza del sesto e dodicesimo mese sono riportati le quantità necessarie per la valutazione dell'opzione: (i) il prezzo dell'opzione O_{t_k}, (ii) la differenza fra il valore scontato delle rate future ed il debito residuo $V_{t_k} - D_{t_k}$, (iii) il valore temporale dell'OdRA o_{t_k}, (nella Figura 4, $k = 1, 2$) .

Il mutuatario eserciterà l'opzione al tempo t_k se la variabile $V_{t_k} - D_k$ risulterà maggiore di o_{t_k}. Infatti, nella valutazione, è necessario tenere conto del valore temporale dell'opzione ovvero della possibilità che i tassi scendano ulteriormente nei periodi successivi e l'OdRA sia quindi più vantaggiosa nel futuro. Come si evince dalla Figura 4, nel nodo $(6, 1)$ il valore intrinseco $V_{t_1} - D_1 = 1.442.639, 37$ è maggiore del valore temporale $o_{t_1} 1.096.460, 57$, quindi il valore dell'opzione sarà posto uguale ad $1.442.639, 37$. Il contrario avviene nel nodo $(6, 2)$ dove sarà più conveniente esercitare l'opzione nei periodi successivi in quanto $o_{t_k} > (V_{t_k} - D_k)$.

Il prezzo dell'opzione è chiaramente un costo per l'istituto bancario. Tale costo è da addebitare ai mancati pagamenti futuri nel caso in cui il mutuatario decida di rifinanziarsi ad un tasso più basso, e quindi eserciti l'opzione che ha implicitamente acquistato alla stipula del mutuo.

Nei nodi in cui non sono pagate rate, e quindi non è possibile esercitare l'opzione, sono riportati solamente il tasso a breve $r^*_{i,j}$ ed il valore attuale dell'opzione ottenuto scontando i valori relativi al periodo successivo,

$$O_{i,j} = \frac{1}{2} \frac{O_{i+1,j+1} + O_{i+1,j}}{1 + r^*_{i,j}} \tag{37}$$

Nel nodo $(0, 0)$ si osserverà il prezzo dell'opzione che, in questo caso, è pari a 397.534 lire, circa lo 0.4% del capitale preso in prestito.

Nella Figura 5 sono rappresentati i risultati di alcune simulazioni. In particolare, si può notare la tipica relazione convessa fra il prezzo dell'opzione e la volatilità, dall'altro la dipendenza del prezzo dell'OdRA dalla durata del mutuo. Il grafico è stato ottenuto utilizzando volatilità costanti pari al 10%, 20%, 30%, 40%; l'etichetta "storica" indica il caso in cui la volatilità è stata stimata basandosi sulle serie storiche della struttura a termine.

Nella Tabella 1 sono invece riportati alcuni valori dell'OdRA per diversi livelli dello spread δ.

12	**5.32**	**5.44**	**5.57**	**5.7**	**5.84**	**5.99**	**6.15**	**6.32**	**6.49**	**6.67**	**6.86**	**7.06**	**7.27**
	4245252.77	**3459599.21**	**2655451.35**	**1832528.62**	**990562.18**	**315353.3**	**139219.92**	**63631.62**	**29987.74**	**13884.99**	**6220.79**	**2662.24**	**1080.55**
	4245252.77	**3459599.21**	**2655451.35**	**1832528.62**	**990562.18**	**129296.52**	**-751508.83**	**-1652077.57**	**-2572614.71**	**-3513304.65**	**-4474309.11**	**-5455765**	**-6457782.18**
	3364402.98	**2640952.11**	**1911702.76**	**1223953.78**	**669315.91**	**315353.3**	**139219.92**	**63631.62**	**29987.74**	**13884.99**	**6220.79**	**2662.24**	**1080.55**
11	5.42	5.54	5.67	5.8	5.94	6.09	6.25	6.41	6.58	6.76	6.94	7.14	
	3835105.09	3043470.91	2233438.54	1404752.12	649739.58	226138.78	100900.63	46561.07	21816.79	9996.62	4415.96	1860.33	
10	5.56	5.68	5.8	5.93	6.07	6.21	6.36	6.51	6.67	6.84	7.01		
	3423416.98	2626022.07	1810338.36	1022191.53	435735.76	162678.01	73342.4	34004.48	15818.78	7165.47	3119.92		
9	5.74	5.85	5.96	6.08	6.21	6.34	6.47	6.61	6.75	6.91			
	3010325.35	2207421.61	1409261.35	725286.94	297667.11	117390.34	53385.57	24775.16	11427.8	5113.27			
8	5.93	6.02	6.13	6.23	6.34	6.46	6.57	6.7	6.82				
	2596055.45	1799307.89	1061851.87	508833.58	206437.34	84930.91	38867.41	18001.03	8223.79				
7	6.11	6.19	6.28	6.37	6.46	6.56	6.66	6.76					
	2186556.28	1423236.15	781253.94	355746.68	144903.52	61562.6	28277.31	13038.96					
6	**6.26**	**6.33**	**6.4**	**6.48**	**6.55**	**6.63**	**6.71**						
	2313629.49	**1442639.37**	**565483.17**	**248981.42**	**102672.55**	**44673.23**	**20543.34**						
	2313629.49	**1442639.37**	**551853.67**	**-358951.59**	**-1289981**	**-2241419**	**-3213425**						
	1795529.12	**1096460.57**	**565483.17**	**248981.42**	**102672.55**	**44673.23**	**20543.34**						
5	6.36	6.42	6.48	6.54	6.59	6.65							
	1868227.22	998718.25	405046.06	174874.63	73270.26	32428.46							
4	6.38	6.42	6.47	6.51	6.56								
	1425896.76	698146.37	288406.02	123402.59	52561.97								
3	6.33	6.37	6.41	6.44									
	1056450.8	490673.08	204811.1	87512.3									
2	6.27	6.31	6.34										
	769539.75	345923.83	145393.14										
1	6.26	6.29											
	554839.58	244377.55											
0	6.26												
	397534.33												
	0	**1**	**2**	**3**	**4**	**5**	**6**	**7**	**8**	**9**	**10**	**11**	**12**

Figure 4: Un campione della griglia binomiale. L'opzione può essere esercitata solamente dopo il pagamento della rata. Per il periodo 6 e 12, e per ogni stato, sono riportati (in grassetto) rispettivamente, il prezzo dell'opzione O_{t_k}, la quantità $V_{t_k} - D_k$, ed il valore attuale del prezzo futuro dell'opzione, o_{t_k}.

	Spread δ		
	2%	4%	6%
5 anni	0,110%	0,106%	0,102%
10 anni	0,242%	0,225%	0,209%
15 anni	0,448%	0,400%	0,357%
20 anni	0,650%	0,562%	0,486%
25 anni	0,847%	0,708%	0,592%
30 anni	1,002%	0,815%	0,664%

Table 1: Valore dell'OdRA in funzione dello spread δ e della durata del mutuo. I valori sono espressi in percentuale del capitale preso in prestito $S = 100.000.000$.

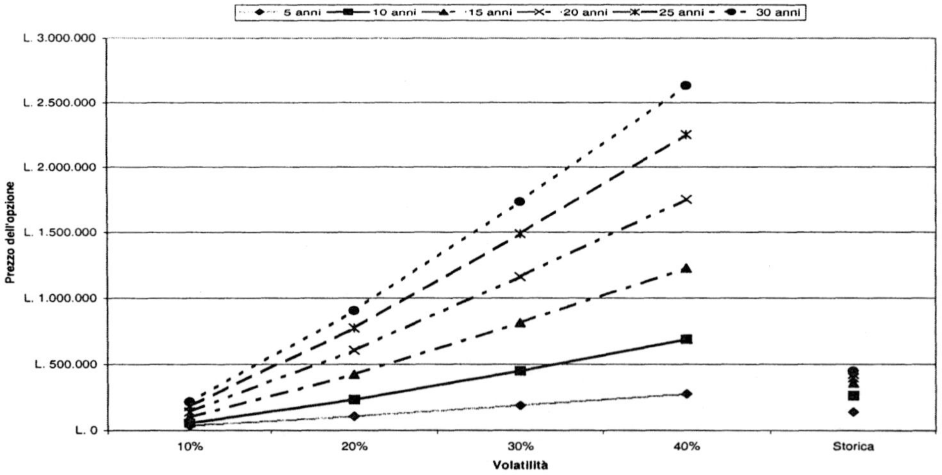

Figure 5: Prezzo dell'OdRA in funzione della volatilità. Il livello dello spread δ è pari al 3% per tutte le scadenze mostrate nel grafico.

Periodo (in mesi)	Opzione Cap
6	0
12	0
18	2,65
24	747,65
30	9262,43
36	24588,35
42	39078,3
48	49481,93
54	53935,34
60	51856,45
66	47087,23
72	41337,92
78	35372,05
84	29627,9
90	24234,22
96	19240,76
102	14572,38
108	10026,46
114	5311,56
Totale	455763,7

Table 2: Il *cap* è equivalente ad un portafoglio di opzioni europee con scadenze in corrispondenza del pagamento delle rate.

3.2. La valutazione del cap. Nel paragrafo 2.2 si è descritto il meccanismo per la determinazione del prezzo *cap* nel caso di mutuo indicizzato. Nelle simulazioni è stato considerato un mutuo con rate semestrali a quote capitale costanti. La quota interessi è calcolata sulla base del tasso osservato nei sei mesi precedenti,

$$r(t_{k-1}, t_k) = \frac{1}{v(t_{k-1}, t_k)} - 1. \tag{38}$$

Tale tasso è ottenuto scontando al tempo t_{k-1} sulla griglia binomiale il prezzo di uno zcb unitario che scade in t_k.
Nella Tabella 3 sono riportati i prezzi delle singole opzioni europee, $P_{0,k}$ per $k = 1, 2, \cdots, m$, nonchè il prezzo del cap,

$$V_{cap} = \sum_{k=1}^{m} P_{0,k}. \tag{39}$$

Il tasso del *cap* è stato posto ad un punto percentuale sopra il tasso prevalente. Come si può notare le prime opzioni hanno un prezzo nullo o quasi, infatti, in tali periodi, la dinamica del tasso a breve si mantiene al di sotto del tasso del *cap* e pertanto le opzioni non producono alcun payoff positivo.
Nella Tabella 3 sono riportati i prezzi del *cap*, in percentuale del capitale preso in prestito, per diversi livelli del *cap rate* e durata del mutuo. I prezzi più alti si ottengono

	Cap Rate				
	6,47%	6,97%	7,47%	8,47%	9,47%
5 anni	0,634%	0,226%	0,065%	0,002%	0%
10 anni	2,213%	1,102%	0,456%	0,044%	0,003%
15 anni	3,774%	2,080%	0,993%	0,166%	0,027%
20 anni	5,329%	3,157%	1,686%	0,415%	0,109%
25 anni	6,738%	4,183%	2,392%	0,717%	0,229%
30 anni	7,688%	4,875%	2,880%	0,941%	0,326%

Table 3: Prezzi del cap in funzione del tasso limite e della durata del mutuo. I prezzi sono in percentuale del capitale mutuato $S = 100.000.000$, mentre il tasso corrente, con spread $\delta = 3\%$, è pari al $6,47\%$.

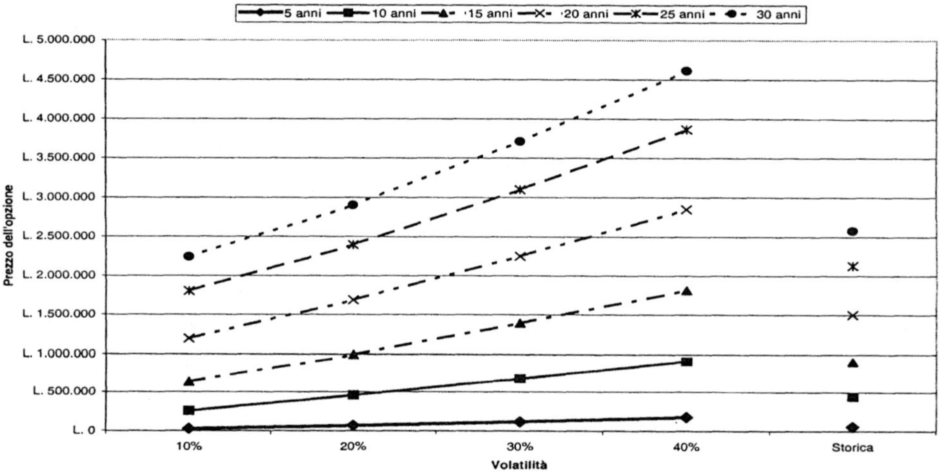

Figure 6: Prezzo del cap in funzione della volatilità. Il livello dello spread δ è pari al 3% per tutte le scadenze mostrate nel grafico.

per *cap rate* molto vicini al valore corrente del tasso annuale, ovvero, per opzioni "at the money" $(6,47\%)$. Nella Figura 6 sono riportati i prezzi del *cap* in funzione della volatilità.

4. Conclusioni. In questo articolo si è proposto uno schema di riferimento generale per la determinazione del prezzo di opzioni implicite nei prestiti a lungo periodo. Si è visto come tali opzioni possano essere facilmente valutate una volta che la dinamica del tasso a breve sia opportunatamente modellizzata.

Per quanto nel lavoro siano state trattate soltanto due tipologie di opzioni, è importante evidenziare che l'analisi può essere estesa ad altri tipi di opzioni senza modificare significativamente la struttura del modello. Tale aspetto è molto importante in quanto fornisce alle istituzioni bancarie un potente strumento per ampliare la propria offerta arricchendo i prodotti con opzioni diverse a secondo delle esigenze dei clienti. Nello stesso tempo permette di determinare il valore delle opzioni analizzate in modo da consentire una impostazione corretta della gestione del rischio di tasso d'interesse implicito nelle operazioni finanziarie discusse nel lavoro.

Appendice A. Nell'ipotesi di esistenza di una misura equivalente di martingala (ciò implica, da un punto di vista finanziario, l'assenza di opportunità di arbitraggio) si può dimostrare la (22) in maniera formale. Infatti, il valore di mercato del debito residuo è dato da,

$$V_{t_j} = \sum_{k=j+1}^{m} E^*_{t_j}\Big[\exp\Big(-\int_{t_j}^{t_k} r(u)du\Big) R_k\Big], \tag{40}$$

dove il valore atteso condizionato è calcolato rispetto alla misura equivalente di martingala e $\{r(t)\}$ rappresenta il processo del tasso *spot*. Sostituendo la (20) nella (40) si ottiene,

$$V_{t_j} = \sum_{k=j+1}^{m} D_{k-1} E^*_{t_j}\Big[\frac{\exp\big(-\int_{t_j}^{t_k} r(u)du\big)}{v(t_{k-1},t_k)}\Big] + \sum_{k=j+1}^{m} (C_k - D_{k-1}) v(t_j,t_k). \tag{41}$$

Dato che

$$E^*_{t_j}\left[\frac{\exp\big(-\int_{t_j}^{t_k} r(u)du\big)}{v(t_{k-1},t_k)}\right] = E^*_{t_j}\Big[\frac{\exp\big(-\int_{t_j}^{t_{k-1}} r(u)du\big)}{v(t_{k-1},t_k)} E^*_{t_{k-1}}\big[\exp\big(-\int_{t_{k-1}}^{t_k} r(u)du\big)\big]\Big], \tag{42}$$

e per definizione

$$E^*_{t_{k-1}}\big[\exp\big(-\int_{t_{k-1}}^{t_k} r(u)du\big)\big] = v(t_{k-1},t_k), \tag{43}$$

la (41) può essere riscritta nel modo seguente,

$$E^*_{t_j}\left[\frac{\exp\big(-\int_{t_j}^{t_k} r(u)du\big)}{v(t_{k-1},t_k)}\right] = v(t_j,t_{k-1}). \tag{44}$$

Sostituendo la (44) nella (41) si ottiene la relazione dell'indebitamento residuo,

$$V_{t_j} = \sum_{k=j+1}^{m} \Big[D_{k-1} v(t_j,t_{k-1}) + (C_k - D_{k-1}) v(t_j,t_k)\Big]. \tag{45}$$

Appendice B. In questa appendice mostreremo come costruire un albero binomiale per il tasso a breve secondo il modello di BDT.

Ipotizzeremo uno scadenzario discreto S con $s_0 = 0$ ed $s_5 = 5$; Δ è stato posto pari ad un anno. Nella Tabella 4 sono riportate la struttura dei tassi e delle volatilità osservate nel mercato per lo scadenzario S.

Data la struttura dei rendimenti $i(s_0, s_k)$, i prezzi in s_0 degli zcb unitari con scadenza $s_k \in S$ sono definiti dalla seguente relazione,

$$v(s_0, s_k) = \frac{1}{(1 + i(s_0, s_k))^{s_k - s_0}} \tag{46}$$

Scadenza	Rendimento	Volatilità
1	10%	20%
2	11%	19%
3	12%	17,2%
4	12,5%	15,3%
5	13%	13,5%

Table 4: Struttura a termine dei rendimenti e delle volatilità.

k	Prezzi in $s_0 = 0$
1	$v(0,1) = \frac{1}{1+i(0,1)} = \frac{1}{1+0,1} = 0,909$
2	$v(0,2) = \frac{1}{(1+i(0,2))^2} = \frac{1}{(1+0,11)^2} = 0,812$
3	$v(0,3) = \frac{1}{(1+i(0,3))^3} = 0,712$
4	$v(0,4) = \frac{1}{(1+i(0,4))^4} = 0,624$
5	$v(0,5) = \frac{1}{(1+i(0,5))^5} = 0,543$

Table 5: Prezzi degli zcb unitari

Nella Tabella 5 sono riportati i prezzi degli zcb unitari per la struttura in Tabella 4. Nel caso di uno zcb con scadenza fra un anno, qualunque sia lo stato, il suo valore a fine anno deve essere pari ad 1. Escludendo la possibilità di arbitraggi non-rischiosi, deve essere,

$$v(0,1) = 0,909 = \frac{\frac{1}{2}1 + \frac{1}{2}1}{1+r} \tag{47}$$

da cui si ricava facilmente il valore del tasso a breve per il primo periodo $r_0 = 0,10$. Seguendo la stessa logica e ricordando l'equazione (33), si può determinare il tasso base $r_{1,1}$. Se si considera infatti uno zcb che scade in $s_2 = 2$, l'evoluzione del prezzo è rappresentato nella Figura 7. In questo caso deve essere

$$v(0,2) = B = \frac{\frac{1}{2}B_u + \frac{1}{2}B_d}{1+r_0} \tag{48}$$

dove il tasso a breve r_0 è stato determinato in precedenza, mentre

$$B_u = \frac{\frac{1}{2}1 + \frac{1}{2}1}{1+r_{1,0}}, \qquad B_d = \frac{\frac{1}{2}1 + \frac{1}{2}1}{1+r_{1,1}}.$$

Sfruttando la relazione (32), si può esprimere il tasso $r_{1,0}$ in funzione di $r_{1,1}$ e della volatilità $\sigma(s_1) = \sigma(1)$, quindi $r_{1,0} = r_{1,1} g(\sigma(1), p)$, dove $g(\sigma(1), p) = exp(2\sigma(1)) = 1,462$. Sostituendo nella (48) le quantità B_u e B_d si ottiene una equazione dove l'unica incognita è $r_{1,1}$,

$$0,812 = \frac{\frac{1}{2}\left[\left(\frac{1}{1+1,462 r_{1,1}}\right) + \left(\frac{1}{1+r_{1,1}}\right)\right]}{1+0,1} \tag{49}$$

da cui si ricava $r_{1,1} = 0,0979$ e $r_{1,0} = 0,1432$. Iterando il procedimento per le successive scadenze è possibile ottenere i valori del tasso uniperiodale che servono per completare l'albero binomiale riportato in Figura 8. Dal punto di vista computazionale, la procedura descritta nell'esempio non è efficiente. Nelle simulazioni numeriche riportate nella sezione 3.1 è stato utilizzato l'algoritmo proposto da Jamshidian [11].

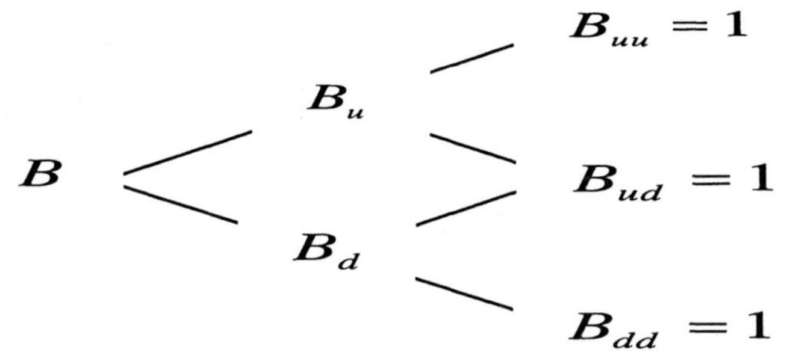

Figure 7: Evoluzione binomiale del prezzo di uno zcb con scadenza in $s_2 = 2$.

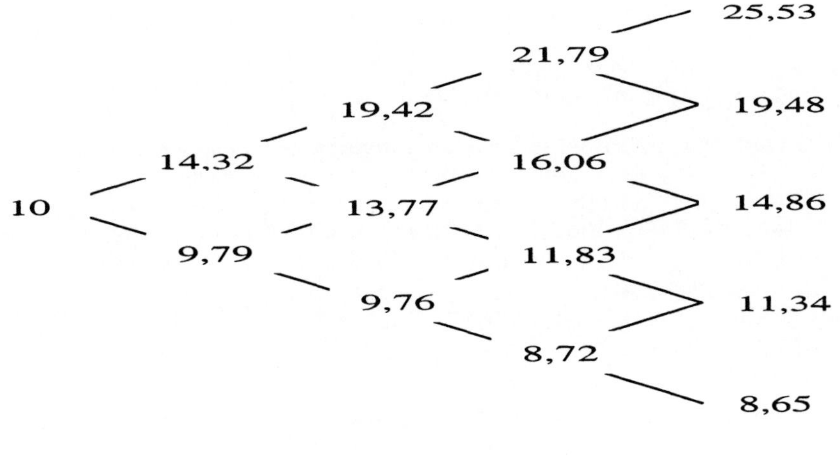

Figure 8: Dinamica del tasso a breve consistente con la struttura a termine dei rendimenti e volatilità.

Riferimenti bibliografici

[1] P. Bjerksund, G. Stensland: *Implementation of the Black-Derman-Toy interest rate model*, J. of Fixed Income (Sept. 1996), 67-75

[2] F. Black, E. Derman, W. Toy: *A one-factor model of interest rates and its application to treasury bond options*, Financial Analysts J. (Jan./Febr. 1990), 33-39

[3] M. J. Brennan, E. S. Schwartz: *An equilibrium model of bond pricing and a test of market efficiency*, J. of Financial and Quantitative Analysis **17** (1982), 301-329

[4] J. Cox, J. Ingersoll, S. Ross: *A theory of the term structure of interest rates*, Econometrica **53** (1985), 385-407

[5] D. Heath, R. Jarrow, A. Morton: *Bond pricing and the term structure of interest rates: a discrete time approximation*, J. of Financial and Quantitative Analysis **25** (1990), 419-440

[6] D. Heath, R. Jarrow, A. Morton: *Bond pricing and the term structure of interest rates: a new methodology for contingent claim valuation*, Econometrica **60** (1992), 77-105

[7] T. Ho, S. Lee: *Term structure movements and pricing interest rate contingent claims*, J. of Finance (Dec. 1986), 1011-1029

[8] J. Hull: *Options, Futures, and other Derivative Securities*, Prentice-Hall, New Jersey 1993

[9] J. Hull, A. White: *New ways with the yield curve*, in: *From Black-Scholes to Black Holes: New Frontiers in Options*, Risk, Finex 1992, p. 107-110

[10] F. Jamshidian: *Bond and option evaluation in the Gaussian interest rate model*, Technical Report, Merrill Lynch 1990

[11] F. Jamshidian: *Forward induction and construction of yield curve diffusion models*, J. of Fixed Income (June 1991), 62-74

[12] R. Jarrow, D. Lando, S. Turnbull: *A Markov model for the term structure of credit risk spreads*, Review of Financial Studies **10, 2** (1997), 481-523

[13] C. Mari: *Price and risk valuation of floating-rate bonds under mean reverting spot rate dynamics*, in: Atti XXI Convegno AMASES, C. Colombo, Roma 1997, p. 435-451

[14] F. Moriconi: *Matematica Finanziaria*, Il Mulino, Bologna 1994

[15] S. S. Nielsen, E. I. Ronn: *A two-factor model for the valuation of the T-Bond futures contract's embedded options*, in: *Advances in Fixed-Income Valuation Modeling and Risk Management* [Ed.: F. J. Fabozzi], New Hope, PA 1997, p. 135-152

[16] S. S. Nielsen, E. I. Ronn: *The valuation of default risk in corporate bonds and interest rate swaps*, in: *Advances in Futures and Options Research*, JAI Press, New York 1997, p. 175-196

[17] L. Peccati: *Matematica per la Finanza Aziendale*, Editori Riuniti, Milano 1994

[18] O. Vasicek: *An equilibrium characterization of the term structure*, J. of Financial Economics **5** (1977), 177-188

Andrea Consiglio, Dipartimento di Organizzazione Aziendale e Amministrazione Pubblica, Università della Calabria, I-87036 Arcavacata di Rende (CS), Italy; a.consiglio@unical.it

Massimo Costabile, Dipartimento di Organizzazione Aziendale e Amministrazione Pubblica, Università della Calabria, I-87036 Arcavacata di Rende (CS), Italy; massimo.costabile@unical.it

Carlo Mari, Dipartimento di Organizzazione Aziendale e Amministrazione Pubblica, Università della Calabria, I-87036 Arcavacata di Rende (CS), Italy; karlmari@tin.it

Ivar Massabò, Dipartimento di Organizzazione Aziendale e Amministrazione Pubblica, Università della Calabria, I-87036 Arcavacata di Rende (CS), Italy; i.massabo@unical.it

Progress in Nonlinear Differential Equations
and Their Applications, Vol. 40

Continuity of Near-Duality Maps and Characterizations of Ideal Spaces of Measurable Functions

GUNTER DIRR, MARTIN VÄTH

Dedicated to Alfonso Vignoli on the occasion of his 60th birthday

Summary: We show that the duality map of any normed linear space is "almost" upper and lower semicontinuous and admits "almost" a continuous selection. The result is applied to characterize ideal spaces of vector-valued functions.

Keywords: Continuous selection of duality maps, semicontinuity of duality maps, spaces of measurable functions, Köthe spaces, ideal spaces, Banach function spaces.

Classification: 46B10, 47H04, 46E30, 46E40, 28B20.

1. Semicontinuity of near-duality maps. The duality map $\mathcal{F}$ of a normed linear space X is a (multivalued) map from X into its dual space X^* which maps $x \in X$ into the set of all functionals $f \in X^*$ which satisfy $f(x) = \|x\|^2$ and $\|f\| = \|x\|$. The duality map $\mathcal{F}$ is well-studied, see e. g. [4]. However, the continuity properties of $\mathcal{F}$ are rather disappointing: In general, $\mathcal{F}$ is not lower semicontinuous and does not admit a continuous selection, even if X has finite dimension, as can be seen for $X = \mathbb{R}^2$ with the max-norm:

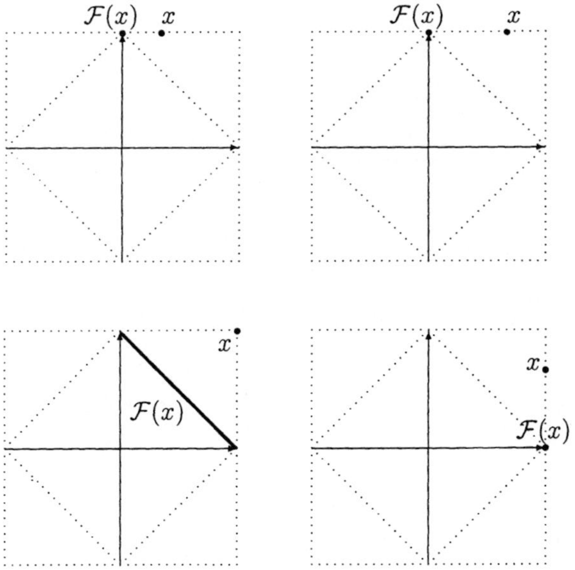

In infinite dimensions it may additionally happen that $\mathcal{F}$ is not upper semicontinuous either, as we shall see later.

For this reason, we allow slight perturbations by an error $\rho \geq 0$, and define the *near-duality map* $\mathcal{F}_\rho$ by letting $\mathcal{F}_\rho(x)$ consist of all linear functionals $f \in X^*$ which satisfy $f(x) = \|x\|^2$ and $\|f\| \leq \|x\|(1+\rho)$. In case $\rho = 0$, we get the above duality map $\mathcal{F}_0 = \mathcal{F}$. For $\rho > 0$, we also define the *open near-duality map* $\mathcal{F}_\rho^o$ at a point $x \neq 0$ as the set of all $f \in X^*$ which satisfy $f(x) = \|x\|^2$ and $\|f\| < \|x\|(1+\rho)$ (and $\mathcal{F}_\rho^o(0) = \{0\}$).

Geometrically, the image $\mathcal{F}_\rho(x)$ is the intersection of the closed ball of the dual space with radius $\|x\|(1+\rho)$ (and center 0) and the hyperplane which passes through $\mathcal{F}(x)$ and which is "orthogonal" to the line through 0 and x. For $\mathcal{F}_\rho^o(x)$ one has to take the corresponding open ball.

One may think of the mentioned hyperplane as a "tangential plane" to the ball of X^* with radius $\|x\|$ (and center 0) which cuts the ray from 0 through x "orthogonally".

The following figures show in $X = \mathbb{R}^2$ with the max-norm and $\rho \approx \frac{1}{3}$ the images $\mathcal{F}_\rho(x)$ and $\mathcal{F}_\rho^o(x)$ for various values of x: In this case the hyperplanes become lines; for $\mathcal{F}_\rho^o(x)$ the endpoints of the line segments are missing:

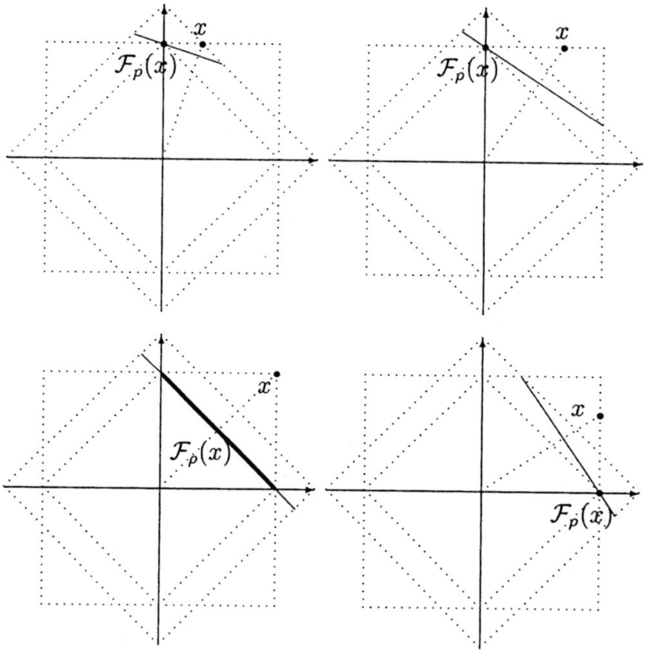

We shall see that $\mathcal{F}_\rho$ and $\mathcal{F}_\rho^o$ are always lower semicontinuous for $\rho > 0$ (even more holds true). However, $\mathcal{F}_\rho$ and $\mathcal{F}_\rho^o$ are not always upper semicontinuous. It turns out that the following notion from [3] is the "right" definition to describe the continuity behaviour of $\mathcal{F}_\rho$ and $\mathcal{F}_\rho^o$:

We call a multivalued map F *ε-δ-upper semicontinuous at x_0* if, given $\varepsilon > 0$, there exists $\delta > 0$ such that $F(x_0 + \delta B) \subseteq F(x_0) + \varepsilon B$, where B denotes the closed unit ball in the corresponding space. Similarly, we call F *ε-δ-lower semicontinuous at x_0* if, given $\varepsilon > 0$, there exists $\delta > 0$ such that $F(x_0) \subseteq F(x) + \varepsilon B$ for each $x \in x_0 + \delta B$. In [1] the same property is called "upper/lower semicontinuous in the ε sense".

Each upper semicontinuous map F is ε-δ-upper semicontinuous, but the converse is not true. We also have the dual fact that each ε-δ lower semicontinuous map F is lower semicontinuous, and the converse is not true. In both cases, the converse holds, if F takes compact values. For the proofs of these statements, see e. g. [2] or [1, 1.1].

The Hahn-Banach extension theorem implies that the duality map $\mathcal{F} = \mathcal{F}_0$ (and consequently also each $\mathcal{F}_\rho$ and each $\mathcal{F}_\rho^\circ$) takes only nonempty values. However, the proof of the Hahn-Banach theorem uses the axiom of choice. Unless explicitly stated, we will not make use of that axiom. Instead, we will assume a weaker form of the axiom of choice throughout, the so-called *axiom of dependent choices* [7] which allows (recursive or nonrecursive) countable choices. Under this weakening of the axiom of choice, the Hahn-Banach extension theorem still can be proved in separable spaces X [6, p. 183]; but for general spaces, no proof can be given.

More precisely, the following facts can not be disproved: It may happen that $X^* = \{0\}$ [8]. Moreover, it may happen for any $\rho \geq 0$ that there exists a space X and some $x_0 \in X$ with $\mathcal{F}_\rho(x_0) = \emptyset$ although $\mathcal{F}_c(x) \neq \emptyset$ for each $x \in X$ and each $c > \rho$:

Example 1.1. Given $\rho \geq 0$, consider the space X of all bounded sequences $x = (\xi_n)_n$, endowed with the norm

$$\|x\| = \sup_n |\xi_n| + \rho \limsup_{n\to\infty} |\xi_n|.$$

We consider X as a space of functions on the measure space $\mathbb{N}$ (with the counting measure). Recall [11] that the associate space X' of X consists of all sequences $y = (\eta_n)_n$ with finite norm

$$\|y\|_{X'} = \sup_{\|x\|\leq 1} \sum_{n=1}^{\infty} |\eta_n||\xi_n|.$$

The space X' coincides with the space l_1 of absolutely summable sequences (with the usual (sum-)norm): One may see this quickly, if one applies [12, Theorem 3.4.5] to find $X' = X''' = (X_L)'$ where X_L denotes the Lorentz space of X which evidently coincides with the space l_∞ of all bounded sequences with the sup-norm. According to the results of [11] it thus can not be disproved without the axiom of choice that X^* may (in the canonical way) be identified with l_1. For $x_0 = (1+\rho)^{-1}(1-\frac{1}{2}, 1-\frac{1}{3}, 1-\frac{1}{4}, \dots) \in X$ the latter implies that all functionals $f \in X^*$ with $f(x_0) = 1$ satisfy $\|f\| > 1 + \rho$, i. e. $\mathcal{F}_\rho(x_0) = \emptyset$.

On the other hand, $\mathcal{F}_c(x) \neq \emptyset$ for each $x \in X$ and each $c > \rho$: Given $x = (\xi_n)_n \in X\setminus\{0\}$ and $c > \rho$, we have $\lambda = (1+\rho)/(1+c) < 1$. Fix some n_0 with $|\xi_{n_0}| \geq \|x\|_\infty \lambda$, and define a functional $f \in X^*$ by $f((\eta_n)_n) = \|x\|^2 \eta_{n_0}/\xi_{n_0}$. Then $f(x) = \|x\|^2$, and $\|f\| = \|x\|^2/|\xi_{n_0}| \leq \|x\|(1+\rho)\|x\|_\infty/\xi_{n_0} \leq \|x\|(1+\rho)/\lambda = \|x\|(1+c)$, i. e. $f \in \mathcal{F}_c(x)$.

We emphasize once more that, in the presence of the axiom of choice, the assumption of the following result is satisfied for $c = 0$:

Theorem 1.1. *Let $c \geq 0$ be such that $\mathcal{F}_c(x) \neq \emptyset$ for each $x \in X$. Then for each $\rho > c$ the maps $\mathcal{F}_\rho$ and $\mathcal{F}_\rho^\circ$ are ε-δ-upper and ε-δ-lower semicontinuous.*

Proof. Concerning the ε-δ-lower semicontinuity at each $x_0 \in X$, we show even more: We prove that for each $\varepsilon > 0$ there is some $\delta = \delta(c, \rho, \varepsilon, \|x_0\|) > 0$ with the following property: For any $f_0 \in \mathcal{F}_\rho(x_0) \supseteq \mathcal{F}_\rho^o(x_0)$ and any $x \in x_0 + \delta B$ there is some $g \in \mathcal{F}_\rho^o(x) \subseteq \mathcal{F}_\rho(x)$ with $\|g - f_0\| < \varepsilon$.

For $x_0 = 0$, we must have $f_0 = 0$, and so even any $g \in \mathcal{F}_\rho^o(x)$ satisfies $\|g - f_0\| \leq \delta(1+\rho) < \varepsilon$ if only $\delta < \varepsilon/(1+\rho)$. Thus, let $x_0 \neq 0$. To simplify our formulas, we introduce a size $\delta_0 = \delta_0(c, \rho, \varepsilon, \|x_0\|) > 0$ which will be determined later; δ will be chosen in dependence of δ_0. Observe that $\|x - x_0\| < \delta$ implies for each $f_0 \in \mathcal{F}_\rho(x_0)$ that $|\,\|x_0\|^2 - f_0(x)| = |f_0(x_0 - x)| \leq \|x_0\|(1+\rho)\delta$ and in particular $|f_0(x)| \geq \|x_0\|^2 - \|x_0\|(1+\rho)\delta$. For sufficiently small $\delta > 0$ the last expression is positive, and moreover, we have for any $x \in x_0 + \delta B$ and any $f_0 \in \mathcal{F}_\rho(x_0)$ the estimate

$$\begin{aligned} \left|\frac{\|x\|^2}{f_0(x)} - 1\right| &= \frac{|(\|x\| + \|x_0\|)(\|x\| - \|x_0\|) + (\|x_0\|^2 - f_0(x))|}{|f_0(x)|} \\ &\leq \frac{(2\|x_0\| + \delta)\delta + \|x_0\|(1+\rho)\delta}{\|x_0\|^2 - \|x_0\|(1+\rho)\delta} < \delta_0. \end{aligned} \tag{1}$$

Without loss of generality, assume that $\delta < \|x_0\|/2$ and $\delta < \|x_0\|\delta_0/2$. Given $x \in x_0 + \delta B$, fix some $f \in \mathcal{F}_c(x)$, and put

$$\lambda = \frac{(1+\delta_0)^2(1+\rho) - (1+\rho)}{(1+\delta_0)^2(1+\rho) - (1+c)}$$

and

$$g = (1-\lambda)\frac{\|x\|^2}{f_0(x)} f_0 + \lambda f.$$

Then we have $g(x) = \|x\|^2$. Moreover, observing that $\|x\| \geq \|x_0\| - \delta > \|x_0\|/2 > 0$ and so

$$\|x_0\| \leq \|x\| + \delta < \|x\| + \delta_0\|x_0\|/2 < \|x\|(1+\delta_0),$$

we find in view of (1) that

$$\frac{\|x\|\|x_0\|}{f_0(x)} \leq (1+\delta_0)\frac{\|x\|^2}{f_0(x)} \leq (1+\delta_0)^2.$$

Hence, we have the estimate

$$\begin{aligned} \|g\| &\leq (1-\lambda)\frac{\|x\|^2}{f_0(x)}\|x_0\|(1+\rho) + \lambda\|x\|(1+c) \\ &< \|x\|\left((1-\lambda)(1+\delta_0)^2(1+\rho) + \lambda(1+c)\right) = \|x\|(1+\rho). \end{aligned}$$

Thus, $g \in \mathcal{F}_\rho^o(x) \subseteq \mathcal{F}_\rho(x)$. Moreover, by (1),

$$\begin{aligned} \|g - f_0\| &\leq |(1-\lambda)\frac{\|x\|^2}{f_0(x)} - 1|\,\|f_0\| + \lambda\|x\|(1+c) \\ &\leq [\delta_0 + \lambda(1+\delta_0)]\,\|x_0\|(1+\rho) + \lambda(\|x_0\| + \delta)(1+c) =: C. \end{aligned}$$

Observing that $\delta < \|x_0\|/2$ and

$$\lambda \leq \frac{(1+\delta_0)^2(1+\rho) - (1+\rho)}{(1+\rho) - (1+c)} = \frac{1+\rho}{\rho - c}(2\delta_0 + \delta_0^2),$$

we may now choose $\delta_0 > 0$ such that $C < \varepsilon$. Then $\|g - f_0\| < \varepsilon$ for δ satisfying the earlier mentioned estimates.

To prove the ε-δ-upper semicontinuity of $\mathcal{F}_\rho$ and $\mathcal{F}_\rho^o$ at a point $x \neq 0$, we apply the above statement: Observes that, given $\varepsilon > 0$, one may choose the corresponding value $\delta = \delta(c, \rho, \varepsilon, a)$ in this statement such that δ depends continuously on $a \in (0, \infty)$: For this choice, the minimum δ_1 of $\delta(c, \rho, \varepsilon, a)$ for $a \in [\|x\|/2, 3\|x\|/2]$ is positive. With $\delta_2 = \min\{\delta_1, \|x\|/2\}$, we then have $\mathcal{F}_\rho(x + \delta_2 B) \subseteq \mathcal{F}_\rho^o(x) + \varepsilon B$. Indeed, given $x_0 \in x + \delta_2 B$ and $f \in \mathcal{F}_\rho(x_0)$, we have $x \in x_0 + \delta(c, \rho, \varepsilon, \|x_0\|)B$, and so there is some $g \in \mathcal{F}_\rho^o(x)$ with $\|f_0 - g\| < \varepsilon$. □

There is some relation between $\mathcal{F}_\rho$ and $\mathcal{F}_\rho^o$:

Proposition 1.1. *The sets $\mathcal{F}_\rho(x)$ and $\mathcal{F}_\rho^o(x)$ are always convex. Moreover, $\mathcal{F}_\rho(x)$ is closed, and if $\mathcal{F}_\rho^o(x) \neq \emptyset$, then $\mathcal{F}_\rho(x)$ is the closure of $\mathcal{F}_\rho^o(x)$.*

Proof. The only nontrivial statement is the inclusion $\mathcal{F}_\rho(x) \subseteq \overline{\mathcal{F}_\rho^o(x)}$. Thus, let $f \in \mathcal{F}_\rho(x)$ be given. Fix some $g \in \mathcal{F}_\rho^o(x)$. The convex combinations $f_n = (1 - n^{-1})f + n^{-1}g$ belong to $\mathcal{F}_\rho^o(x)$ and converge to f. Hence, $f \in \overline{\mathcal{F}_\rho^o(x)}$. □

Without the axiom of choice, it does not follow from $\mathcal{F}_\rho(x) \neq \emptyset$ that $\mathcal{F}_\rho^o(x) \neq \emptyset$ (formally, for $\rho = 0$, this conclusion fails even under the axiom of choice, if one chooses the natural definition $\mathcal{F}_0^o(x) := \emptyset$ for $x \neq 0$):

Example 1.2. Consider in the space X of Example 1.1 the point $x = (1+\rho)^{-1}(1, 1, \ldots) \in X$. Recall that it can not be disproved that X^* is equal to l_1. But the latter implies that each $f \in X^*$ with $f(x) = 1$ satisfies $\|f\| \geq 1 + \rho$, i. e. $\mathcal{F}_\rho^o(x) = \emptyset$. On the other hand, the functional $f((\xi_n)_n) = (1 + \rho)\xi_1$ belongs to $\mathcal{F}_\rho(x)$.

Now we are in a position to prove our main result:

Theorem 1.2. *Let $c \geq 0$ be such that $\mathcal{F}_c(x) \neq \emptyset$ for each $x \in X$. Then for each $\rho > c$ the maps $\mathcal{F}_\rho$ and $\mathcal{F}_\rho^o$ are lower semicontinuous; moreover, $\mathcal{F}_\rho$ is upper semicontinuous if and only if X has finite dimension. More precisely, if X does not have finite dimension, then $\mathcal{F}_\rho$ $(\rho > c)$ is upper semicontinuous at $x_0 \in X$ only if $x_0 = 0$. The same holds for $\mathcal{F}_\rho^o$ $(\rho > c)$ in all spaces X which are not one-dimensional.*

Proof. The first statements are immediate consequences of Theorem 1.1 and the remarks before. The fact that $\mathcal{F}_\rho^o$ $(\rho > c)$ is not upper semicontinuous at $x_0 \neq 0$ is almost trivial, since $\mathcal{F}_\rho^o$ is homogeneous and since $\mathcal{F}_\rho^o(x_0)$ is contained in a smallest open ball U around 0: Each neighborhood of x_0 contains a point of the form $x = \lambda x_0$ with $\lambda > 1$, and the set $\mathcal{F}_\rho^o(x) = \lambda \mathcal{F}_\rho^o(x_0)$ is not contained in U.

Now, assume that X does not have finite dimension. We prove that $\mathcal{F}_\rho$ is not upper semicontinuous (and simultaneously show the same for $\mathcal{F}_\rho^o$ again) at each $x_0 \in X \setminus \{0\}$. Since $\mathcal{F}_c$ does not take empty values, we may identify X in the canonical way with a subspace of X^{**}, after passing to an equivalent norm. In particular, X^{**} can not have finite dimension, and so X^* can not have finite dimension either. Fix some $f_0 \in \mathcal{F}_c(x_0)$. Then X^* is the direct sum of span$\{f_0\}$ and $Y = \{g \in X^* : g(x_0) = 0\}$: Indeed, span$\{f_0\} \cap Y = \{0\}$, and each $f \in X^*$ may be written as $f = g + \lambda f_0$ where $\lambda = f(x_0)/f_0(x_0)$ and $g = f - \lambda f_0 \in Y$. In particular, Y does not have finite dimension. A recursive application of Riesz' lemma shows that there exists a sequence $g_n \in Y$ with $\|g_n\| = 1$ such that $\|g_n - g_k\| \geq \frac{1}{2}$ $(k < n)$. Given $\rho > c$, fix a sequence $\lambda_n \in (\frac{1}{2}, 1)$ with $\lambda_n \to 1$, and put

$$f_n = \lambda_n f_0 + \frac{\rho - c}{2}\|x_0\| g_n$$

and $x_n = \lambda_n x_0$. Then

$$f_n(x_n) = \lambda_n^2 f_0(x_0) = \lambda_n^2 \|x_0\|^2 = \|x_n\|^2$$

and

$$\|f_n\| \leq \lambda_n \|f_0\| + \frac{\rho - c}{2}\|x_0\| < (1 + c)\|x_n\| + (\rho - c)\|x_n\|,$$

and so $f_n \in \mathcal{F}_\rho^o(x_n) \subseteq \mathcal{F}_\rho(x_n)$. Observe that $f_n(x_0) = \lambda_n f_0(x_0) \neq \|x_0\|^2$, and so $f_n \notin \mathcal{F}_\rho(x_0)$. In particular, Proposition 1.1 implies that $\delta_n = \operatorname{dist}(f_n, \mathcal{F}_\rho(x_0)) = \operatorname{dist}(f_n, \mathcal{F}_\rho^o(x_0)) > 0$. Choose a null sequence $r_n \in (0, \delta_n)$ with $r_n < \|x_0\|(\rho - c)/15$, and let $B_n \subseteq X^*$ denote the closed ball with center f_n and radius r_n. Since

$$\begin{aligned}\|f_n - f_k\| &= \|(\lambda_n - \lambda_k) f_0 + \frac{\rho - c}{2}\|x_0\|(g_n - g_k)\| \\ &\geq \frac{\rho - c}{4}\|x_0\| - |\lambda_n - \lambda_k|\,\|f_0\|\end{aligned}$$

for $n \neq k$, and since $|\lambda_n - \lambda_k| \to 0$ as $n, k \to \infty$, we may assume without loss of generality that $\|f_n - f_k\| \geq \|x_0\|(\rho - c)/5$ $(n \neq k)$. In particular, all balls B_n have a pairwise distance of at least $\|x_0\|(\rho - c)/15$, and so their union is closed. By construction, $V = X^* \setminus \bigcup B_n$ is a neighborhood of $\mathcal{F}_\rho(x_0) = \overline{\mathcal{F}_\rho^o(x_0)}$. However, since $x_n \to x_0$, we find in each neighborhood of x_0 some point x_n such that $f_n \in \mathcal{F}_\rho^o(x_n) \subseteq \mathcal{F}_\rho(x_n)$ is not contained in V. Hence, $\mathcal{F}_\rho$ and $\mathcal{F}_\rho^o$ are not upper semicontinuous at x_0. □

If we assume the axiom of choice, the only question concerning the semicontinuity of $\mathcal{F}_\rho$ and $\mathcal{F}_\rho^o$ which has not been answered yet is what happens in case $\rho = 0$ for the duality map $\mathcal{F} = \mathcal{F}_0$. If X has finite dimension, we already saw that $\mathcal{F}$ need not be lower semicontinuous, but the following result implies that $\mathcal{F}$ must always be upper semicontinuous in this case:

Theorem 1.3. *Assume either the axiom of choice or that X is separable. Then all of the maps $\mathcal{F}_\rho$ $(\rho \geq 0)$ are upper semicontinuous with respect to the weak* topology.*

Proof. If $\mathcal{F}_\rho$ is not upper semicontinuous at some $x_0 \neq 0$, there is a weak*-open set $U \supseteq \mathcal{F}_\rho(x_0)$ and a sequence $x_n \to x_0$ such that each $\mathcal{F}_\rho(x_n) \setminus U$ contains an element f_n. Observe that $r = \limsup \|f_n\| \leq \|x_0\|(1 + \rho)$. In particular, $(f_n)_n$ is bounded in X^*. If X is separable, the sequence f_n thus contains a weakly* convergent subsequence. If we assume the axiom of choice, the ball $2rB \subseteq X^*$ is weakly* compact. Hence, in both cases the following is true (in the second case apply e. g. [13, Corollary 8.3]): There is some $f \in X^*$ such that each weak*-neighborhood of f contains f_n for infinitely many n. In particular, for any $\varepsilon > 0$ and any $x \in X$ there are infinitely many n with $|(f - f_n)(x)| < \varepsilon$. For infinitely many n we have additionally $\|f_n\| \leq r + \varepsilon$, thus $|f(x)| = |(f - f_n)(x)| + |f_n(x)| \leq \varepsilon + (r + \varepsilon)\|x\|$. Since x and $\varepsilon > 0$ have been arbitrary, we may conclude that $\|f\| \le r \le \|x_0\|(1+\rho)$. An analogous argument with $x = x_0$ shows that for any $\varepsilon > 0$ we have

$$|f(x_0) - f_n(x_n)| = |(f - f_n)(x_0) + f_n(x_0 - x_n)| \le \varepsilon + (r+\varepsilon)\|x_0 - x_n\|$$

for infinitely many n. Since $\|x_0 - x_n\| \to 0$ and $f_n(x_n) = \|x_n\|^2 \to \|x_0\|^2$, we may conclude that $f(x_0) = \|x_0\|^2$. Thus, we have proved $f \in \mathcal{F}_\rho(x_0) \subseteq U$, and so U is a weak*-open neighborhood of f. By the choice of f, this implies $f_n \in U$ for infinitely many n, contradicting the definition of f_n. □

Corollary 1.1. *If X has finite dimension, then the duality map $\mathcal{F}_0$ is upper semicontinuous.*

In infinite dimensions, Theorem 1.3 is the best we can say in general: There exists a separable Banach space X such that $\mathcal{F}_0$ is not upper semicontinuous, not even ε-δ-upper semicontinuous. Moreover, if we assume the axiom of choice, this even holds with respect to the weak topology on X^*:

Example 1.3. Put $X = L_1([0,1])$, and for $0 \le t \le 1$, define $x(t) \in X$ by

$$x(t)(s) = \begin{cases} -1 & \text{if } s \le t, \\ 1 & \text{if } s > t. \end{cases} \tag{2}$$

With the canonical identification $X^* = L_\infty([0,1])$ we have $\mathcal{F}(x(t)) = \{x(t)\}$ (for details see e. g. [12, Example 2.1.2]). But although $x : [0,1] \to L_1 = X$ is continuous, the map $x : [0,1] \to L_\infty$ (which corresponds to $\mathcal{F} \circ x$) is discontinuous at each point. Since $\mathcal{F}(x(t))$ is compact, we may conclude that $\mathcal{F}$ is not even ε-δ-upper semicontinuous at each $x(t)$. Moreover, if we assume the axiom of choice, the map $x : [0,1] \to L_\infty$ is discontinuous, if we equip $L_\infty = X^*$ with the weak topology: In fact, given $t_0 \in (0,1)$, it is easily verified that

$$f(y) = \lim_{s \to t_0^+} y(s) + \lim_{s \to t_0^-} y(s)$$

defines a bounded linear functional on the subspace of all $y \in L_\infty$ for which the limits exist (after modifying y on an appropriate null set). By Hahn-Banach's theorem, we may extend f to an element of X^*. But since $|f(x(t)) - f(x(t_0))| = 2$ for all $t \ne t_0$, the function x is not weakly continuous at t_0, as claimed.

Corollary 1.2. *Assume the axiom of choice. Then for each $\rho > 0$ the maps $\mathcal{F}_\rho$ and $\mathcal{F}_\rho^o$ admit a continuous selection.*

Proof. In view of Theorem 1.2 and Proposition 1.1, the statement for $\mathcal{F}_\rho$ is just an application of Michael's selection theorem [10]. Since each continuous selection of $\mathcal{F}_{\rho/2}$ is a continuous selection of $\mathcal{F}_\rho^o$, the result follows. □

Observe that the proof of Michael's selection theorem is based on Stone's result that metric spaces are paracompact which in turn makes essential use of the axiom of choice (actually, a well-order on X is required).

2. Characterization of ideal spaces. We give now an application of Corollary 1.2. Actually, this application was the main motivation for the above results.

Let $(Y, |\cdot|)$ be a normed linear space, and S be a (σ-additive positive) measure space. We call a function $x : S \to Y$ (totally or strongly Bochner) measurable, if for each subset $E \subseteq S$ of finite measure there exists a sequence of simple functions which converges to x on E a. e. (in the sense of the Lebesgue extension of S). If Y is a Banach space, this is the classical definition used in e. g. [5]. Consider a normed linear space $(X, \|\cdot\|)$ of (classes of) measurable functions $x : S \to Y$. Then X is called a *preideal space*, if $x \in X$ and $|z(s)| \leq |x(s)|$ for a measurable function $z : S \to Y$ imply $z \in X$ and $\|z\| \leq \|x\|$. If X is complete, it is called an *ideal space*. Although this definition is rather general, most results which are known for Lebesgue-Bochner spaces $L_p(S, Y)$ carry over to preideal spaces, see e. g. [12,14-16].

On the one hand, the name "ideal space" is explained by the fact that in the case $Y = \mathbb{R}$ the preideal spaces are precisely the order ideals (see e. g. [9,15]) in the space of measurable functions. On the other hand, one may also explain the name by the fact that the unit ball of X is invariant under (pointwise) multiplication with functions from the unit ball of L_∞. This second explanation also makes sense in case $Y \neq \mathbb{R}$, if one replaces L_∞ by $L_\infty(S, \mathcal{L}(Y))$ where $\mathcal{L}(Y)$ denotes the space of bounded linear operators in Y with the operator norm: It is trivial that the unit ball of any preideal space X is invariant under multiplication with functions from the unit ball of $L_\infty(S, \mathcal{L}(Y))$. But it is not clear whether also the converse holds: Does this property imply that X is a preideal space?

If the answer to this question is positive, we say that Y has the *characterization property*. It has been proved constructively (i. e. without the axiom of choice) in [12, Theorem 2.1.3] that Y has the characterization property if S is σ-finite and if $\mathcal{F}_\rho(x) \neq \emptyset$ for each $\rho > 0$. Of course, if we assume the axiom of choice, the latter is automatically satisfied. We shall prove now by completely different methods that the axiom of choice implies that it is not necessary that S be σ-finite:

Theorem 2.1. *Assume the axiom of choice. Then each normed linear space Y has the characterization property.*

It follows from [12, Theorem 2.1.2] that the statement of Theorem 2.1 may equivalently be reformulated as:

Proposition 2.1. *Assume the axiom of choice. Let $\mathcal{F}_\rho$ denote the near-duality map of a normed linear space Y, and let S be a measure space. If $x : S \to Y$ is measurable, then $\mathcal{F}_\rho \circ x$ admits a measurable selection for each $\rho > 0$.*

Proposition 2.1 is of course an immediate consequence of Corollary 1.2. For the reader's convenience, let us briefly recall how Proposition 2.1 implies Theorem 2.1:

Proof of Theorem 2.1. Let X be a normed linear space of measurable functions $x : S \to Y$ such that the unit ball is invariant under (pointwise) multiplication of functions from the unit ball of $L_\infty(S, \mathcal{L}(Y))$. Let $x \in X$ and a measurable function $z : S \to Y$ with $|z(s)| \leq |x(s)|$ be given. Given $\varepsilon > 0$, let f be a measurable selection of $\mathcal{F}_\varepsilon \circ x$, and put

$$y(s) = \begin{cases} z(s)|x(s)|^{-2}f(s) & \text{if } x(s) \neq 0, \\ 0 & \text{if } x(s) = 0. \end{cases}$$

Then $y/(1+\varepsilon)$ belongs to the unit ball of $L_\infty(S, \mathcal{L}(Y))$, and so $w(s) = \frac{y(s)}{1+\varepsilon}\frac{x(s)}{\|x\|}$ belongs to the unit ball of X. Hence, $z(s) = w(s)(1+\varepsilon)\|x\|$ belongs to X, and $\|z\| \le (1+\varepsilon)\|x\|$. Since $\varepsilon > 0$ was arbitrary, we even have $\|z\| \le \|x\|$. Thus, X is a preideal space. □

The reader should be warned that the statement of Proposition 2.1 may fail for $\rho = 0$, so that it is indeed necessary to use near-duality maps instead of duality maps for the proof of Theorem 2.1:

Example 2.1. Let $S = [0,1]$, $Y = L_1(S)$, and $x : S \to Y$ be given by (2). With the canonical identification $Y^* = L_\infty(S)$, Example 1.3 implies that the only possible selection of $\mathcal{F}_0 \circ x$ is x, considered as a mapping from S into $Y^* = L_\infty$. But although $x : S \to L_1$ is continuous and thus measurable, the mapping $x : S \to L_\infty$ is not measurable: Otherwise, almost all values $x(t)$ would belong to a separable subset of L_∞ [5, III.6., Theorem 10]. But this is not true: Since $|x(t) - x(\tau)|_{L_\infty} = 2$ for $t \ne \tau$, the image $x(T)$ is not separable for uncountable $T \subseteq S$ (one does not need the uncountable axiom of choice for this conclusion, see e. g. [12, Lemma A.1.1]).

References

[1] J.-P. AUBIN, A. CELLINA: *Differential Inclusions*, Springer, Berlin 1984

[2] YU. G. BORISOVICH, B. D. GEL'MAN, A. D. MYSHKIS, V. V. OBUKHOVSKIĬ: *Multivalued maps* [in Russian], Itogi Nauki Tekhniki **19** (1982), 127–230; Engl. transl.: J. Soviet Math. **24** (1984), 719–791

[3] YU. G. BORISOVICH, B. D. GEL'MAN, A. D. MYSHKIS, V. V. OBUKHOVSKIĬ: *Introduction to the Theory of Multivalued Maps* [in Russian], Izd. Voronezh. Gos. Univ., Voronezh 1986

[4] K. DEIMLING: *Nonlinear Functional Analysis*, Springer, Berlin 1985

[5] N. DUNFORD, J. T. SCHWARTZ: *Linear operators I*, Int. Publ., New York, 1966

[6] H. G. GARNIR, M. DE WILDE, J. SCHMETS: *Analyse fonctionnelle. Théorie constructive des espaces linéaires à semi-normes, Tome I: Théorie générale*, Birkhäuser, Basel 1968

[7] T. J. JECH: *The Axiom of Choice*, North-Holland Publ. Company, Amsterdam, 1973

[8] W. A. J. LUXEMBURG, M. VÄTH: *The existence of nontrivial bounded functionals implies the Hahn-Banach extension theorem*, submitted

[9] W. A. J. LUXEMBURG, A. C. ZAANEN: *Riesz Spaces I*, North-Holland Publ. Company, Amsterdam 1971

[10] E. MICHAEL: *Continuous selections I*, Ann. Math. **63** (1956), 361–381

[11] M. VÄTH: *The dual space of L_∞ is L_1*, Indag. Math., **9** (1998), no. 4, 619–625

[12] M. VÄTH: *Ideal Spaces*, Lect. Notes Math. **1664**, Springer, Berlin 1997

[13] B. VON QUERENBURG: *Mengentheoretische Topologie*, Springer, Berlin 1979

[14] A. C. ZAANEN: *Integration*, North-Holland Publ. Company, Amsterdam 1967

[15] A. C. ZAANEN: *Introduction to Operator Theory in Riesz spaces*, Springer, Berlin 1997

[16] P. P. ZABREĬKO: *Ideal spaces of functions I* [in Russian], Vestnik Jaroslav. Univ. **8** (1974), 12–52

GUNTHER DIRR, Universität Würzburg, Mathematisches Institut, Am Hubland, D-97074 Würzburg, Germany; `dirr@mathematik.uni-wuerzburg.de`

MARTIN VÄTH, Universität Würzburg, Mathematisches Institut, Am Hubland, D-97074 Würzburg, Germany; `vaeth@mathematik.uni-wuerzburg.de`

Progress in Nonlinear Differential Equations
and Their Applications, Vol. 40

A Spectral Theory for Semilinear Operators and its Applications

WENYING FENG, JEFF WEBB

Dedicated to Alfonso Vignoli on the occasion of his 60th birthday

Summary: By generalizing the concept of $(0, k)$-epi mappings to that of $(0, L, k)$-epi mappings, we introduce the definition of spectrum for a semilinear operator pair (L, N), where L is a Fredholm operator of index zero, N is a nonlinear operator. We prove that the new spectrum has similar properties with that of nonlinear operators previously studied in [8]. Applications of the new theory are made to some semilinear operator equations.

Keywords: Fredholm operator, semilinear operator, spectrum, L-compact, L-k-set contraction.

Classification: 47H12, 47H09.

1. Introduction. The notion of spectrum is particularly important for linear operators but spectral ideas have also been used in the study of nonlinear operator equations, for example, [4,5,9]. This has led to the development of various theories of spectrum for nonlinear operators which attempt to preserve the useful properties of the linear case, see for example [1,15].

An important contribution was made by Furi, Martelli and Vignoli, [6]. Their theory did preserve many useful properties, it coincided with the usual concept in the linear case, the spectrum depended on the asymptotic behaviour of the functions and had some stability properties but, in general, did not contain eigenvalues. Based on this study, some new ideas were used by Feng [8] to define a spectrum for nonlinear operators which shared some of the important and useful properties of the FMV-spectrum but also contains eigenvalues.

The aim of this paper is to extend the theory of [8] to a semilinear operator pair (L, N), where L is a linear operator which is Fredholm of index zero, and N is nonlinear, which situation arises frequently in applications to differential equations.

The spectrum of the semilinear operator pair (L, N), $\sigma(L, N)$, will be defined by generalizing the concept of $(0, k)$-epi mappings [16] to that of $(0, L, k)$-epi mappings. Some properties of $(0, L, k)$-epi mappings, such as existence results, normalization property, localization property, homotopy property, will be proved. These results generalize the results of [7], [12] and [16]. We prove that $\sigma(L, N)$ has similar properties to that of the spectrum of nonlinear operators of [8], for example, it contains eigenvalues and is closed. When L is the identity map, this spectrum reduces to the spectrum defined in [8]. Following a decomposition of the spectrum given by Furi, Martelli and Vignoli [6], we also decompose $\sigma(L,N)$ into various parts $\sigma_\delta(L,N), \sigma_m(L,N), \sigma_\omega(L,N)$ and $\sigma_\pi(L,N)$ and obtain several results concerning these parts. In our final section some applications of this theory are given which extend some existence results for semilinear operator equations.

2. Preliminaries. We first give some definitions and fix notation that will be used.

Let E and F be two Banach spaces and let $\Omega \subset E$ be an open bounded subset. The notion of p-epi mapping was introduced by Furi, Martelli and Vignoli [7]. Later, this concept was generalized to (p,k)-epi mapping in [16], which utilises the notion of k-set contractive maps. For a bounded set A, $\alpha(A)$ denotes the Kuratowski (set) measure of non-compactness of A. A continuous map $f : D(f) \subset E \to F$ is said to be a k-set contraction if $\alpha(f(A)) \leq k\alpha(A)$ for every bounded $A \subset D(f)$. For further details see [2,6] and references therein.

Definition 2.1. A continuous mapping $f : \overline{\Omega} \to F$ is said to be *p-admissible* $(p \in F)$ if $f(x) \neq p$ for $x \in \partial\Omega$.

Definition 2.2. A 0-admissible mapping $f : \overline{\Omega} \to F$ is said to be $(0,k)$-epi if for each k-set contraction $h : \overline{\Omega} \to F$ with $h(x) \equiv 0$ on $\partial\Omega$ the equation $f(x) = h(x)$ has a solution in Ω. Similarly, a p-admissible mapping $f : \overline{\Omega} \to F$ is said to be (p,k)-epi if the map f_p defined by $f_p(x) = f(x) - p$, $x \in \overline{\Omega}$, is $(0,k)$-epi.

It was proved in [16] that the (p,k)-epi mappings have 'existence', 'boundary dependence', 'normalization', 'localization' and 'homotopy' properties similar to those of topological degree theory, as was known for p-epi mappings [6].

Analogous to the concept of k-set contraction [2], the concept of L-k-set contraction for semilinear operators was defined in [10]. Suppose that $L : \mathrm{dom}(L) \subset E \to F$ is a closed, linear Fredholm operator of index zero, assume that $\ker(L) \neq \{0\}$ and $\mathrm{dom}(L)$ is dense in E. Write $E = \ker(L) \oplus E_1$, and $F = F_0 \oplus \mathrm{im}(L)$ and let $P : E \to \ker(L)$, $Q : F \to F_0$ be projections. Also let L_P denote the invertible operator L restricted to $\mathrm{dom}(L) \cap E_1$ into $\mathrm{im}(L)$. Write $K_P = {L_P}^{-1}$, $K_{PQ} = K_P(I - Q)$, let $\Pi : F \to F/\mathrm{im}(L)$ be the quotient map, and let $\Lambda : F/\mathrm{im}(L) \to \ker(L)$ be the linear isomorphism (see [10]).

Definition 2.3. Suppose that $\mathrm{dom}(L) \cap \Omega \neq \emptyset$ and $N : \overline{\Omega} \to F$ is a nonlinear mapping. N is said to be *L-compact* if

1. $\Pi N : \overline{\Omega} \to \mathrm{coker}(L)$ is continuous and $\Pi N(\overline{\Omega})$ bounded.
2. $K_{PQ}N : \overline{\Omega} \to E$ is compact.

N is said to be a *L-k-set contraction* if

1. $\Pi N : \overline{\Omega} \to \mathrm{coker}(L)$ is continuous and $\Pi N(\overline{\Omega})$ bounded.
2. $K_{PQ}N : \overline{\Omega} \to E$ is a k-set contraction.

Given a continuous map f from a subset $D(f)$ of E to F, the following three numbers $\alpha(f)$, $\omega(f)$ and $d(f)$ were studied in [6].

$$\alpha(f) = \inf\{k \geq 0 : \alpha(f(A)) \leq k\alpha(A) \text{ for every bounded } A \subset D(f)\},$$

$$\omega(f) = \sup\{k \geq 0 : \alpha(f(A)) \geq k\alpha(A) \text{ for every bounded } A \subset D(f)\},$$

$$d(f) = \liminf_{\|x\|\to\infty,\, x\in D(f)} \frac{\|f(x)\|}{\|x\|}.$$

We shall also use the following number

$$m(f) = \sup\{k \geq 0 : \|f(x)\| \geq k\|x\| \text{ for all } x \in E\}.$$

If $f(x) \neq 0$, for every $x \neq 0$, we let

$$\begin{aligned}\nu_r(f,0) = \inf\{k \geq 0, &\text{ there exists a } k\text{-set contraction } g : B_r \to E \\ &\text{with } g \equiv 0 \text{ on } \partial B_r \text{ such that } f(x) = g(x) \text{ has no solution in } B_r\},\end{aligned}$$

where $B_r \subset E$ is the closed ball with radius r centred at 0. The *measure of solvability* [8] of f at 0 is then $\nu(f) = \inf\{\nu_r(f,0), r > 0\}$.

3. $(0, L, k)$**-epi mappings.** The following concept is a generalization of the $(0, k)$-epi mappings of [16].

Definition 3.1. Let $L : \operatorname{dom}(L) \subset E \to F$ be a closed Fredholm operator of index zero. Let $\Omega \subset E$ be a bounded open set and $N : \overline{\Omega} \to F$ be a continuous map. Suppose $0 \notin (L-N)(\operatorname{dom}(L) \cap \partial\Omega)$. We say that $L - N$ is a $(0, L, k)$-*epi mapping* in $\operatorname{dom}(L) \cap \overline{\Omega}$ if for every continuous bounded map $h : \overline{\Omega} \to F$ such that

1. h is a L-k-set contraction;

2. $h(x) \equiv 0$ for $x \in \partial\Omega$,

the equation $Lx - Nx = h(x)$ has at least one solution in $\operatorname{dom}(L) \cap \Omega$.

The following property is clear.

Property 3.2 (Existence result). *Suppose $L - N$ is $(0, L, k)$-epi for some $k \geq 0$. Then the equation $Lx - Nx = 0$ has a solution in* $\operatorname{dom}(L) \cap \Omega$.

We prove some further properties.

Property 3.3 (Normalization property). *Suppose that L is invertible and N is a continuous L-k-set contraction with $k < 1$, $(L - N)(\operatorname{dom}(L) \cap \partial\Omega) \neq 0$. If $0 \in \Omega$, then $L - N$ is $(0, L, k_1)$-epi for every k_1 with $k + k_1 < 1$.*

Proof. Let $h : \overline{\Omega} \to F$ be a L-k_1-set contraction with $k_1 + k < 1$. Suppose that h is continuous and bounded with $h(x) \equiv 0$ on $\partial\Omega$. The equality $\ker(L) = \{0\}$ implies that

$\Lambda\Pi = 0$, $Q = 0$ and $P = 0$. Hence $L^{-1} : F \to \mathrm{dom}(L)$ is a bounded linear operator and $L^{-1}(N+h) : \overline{\Omega} \to E$ is a $(k_1 + k)$-set contraction. Define $h_1 : E \to E$ by

$$h_1(x) = \begin{cases} L^{-1}(Nx + h(x)), & \text{if } x \in \mathrm{dom}(L) \cap \overline{\Omega}, \\ 0, & \text{if } x \notin \mathrm{dom}(L) \cap \overline{\Omega}. \end{cases}$$

Let

$$M = \sup\{\|h_1(x)\| : x \in \mathrm{dom}(L) \cap \overline{\Omega}\} \text{ and } B = \{x \in E : \|x\| \leq M\}.$$

Then $h_1 : B \to B$ is a $(k_1 + k)$-set contraction and so has a fixed point $x_0 \in B$. If $x_0 \notin \mathrm{dom}(L) \cap \overline{\Omega}$, $h_1(x_0) = 0$ implies $x_0 = 0$. This contradicts $0 \in \Omega$. So $x_0 \in \mathrm{dom}(L) \cap \overline{\Omega}$ and $Lx_0 - Nx_0 = h(x_0)$. If $x_0 \in \mathrm{dom}(L) \cap \partial\Omega$, then $h(x_0) = 0$ and $Lx_0 - Nx_0 \neq 0$, a contradiction. Hence $x_0 \in \mathrm{dom}(L) \cap \Omega$ and $L - N$ is $(0, L, k_1)$-epi on $\mathrm{dom}(L) \cap \overline{\Omega}$. □

Property 3.4 (Localization property). *If $L - N : \mathrm{dom}(L) \cap \overline{\Omega} \to F$ is $(0, L, k)$-epi and $(L-N)^{-1}(0) \subset \mathrm{dom}(L) \cap \Omega_1$, where $\Omega_1 \subset \Omega$ is an open set. Then $L - N$ restricted to $\mathrm{dom}(L) \cap \overline{\Omega_1}$ is a $(0, L, k)$-epi map.*

Property 3.5 (Homotopy property). *Let $L - N : \mathrm{dom}(L) \cap \overline{\Omega} \to F$ be $(0, L, k)$-epi and $h : [0,1] \times \overline{\Omega} \to F$ be a L-k_1-set contraction such that $0 \leq k_1 \leq k < 1$ and $h(0, x) = 0$ for all $x \in \overline{\Omega}$. Furthermore, suppose that $Lx - Nx + h(t, x) \neq 0$ for all $x \in \mathrm{dom}(L) \cap \partial\Omega$ and all $t \in [0,1]$. Then $Lx - Nx + h(1, x) : \mathrm{dom}(L) \cap \overline{\Omega} \to F$ is $(0, L, k - k_1)$-epi.*

Proof. Let $g : \overline{\Omega} \to F$ be a continuous L-$(k - k_1)$-set contraction with $g(x) \equiv 0$ on $\partial\Omega$. Let

$$S = \{x \in \mathrm{dom}(L) \cap \overline{\Omega}, Lx - Nx = g(x) - h(t, x), \text{ for some } t \in [0,1]\}.$$

Then S is bounded. We will prove that S is closed. Suppose that $x_n \in S$, $x_n \to x_0$ ($n \to \infty$). Let $t_n \in [0,1]$ satisfy $Lx_n - Nx_n = g(x_n) - h(t_n, x_n)$. Then $\{t_n\}$ has a convergent subsequence $t_{n_k} \to t_0$. Since N, g, h are continuous operators, we have , (as $n_k \to \infty$),

$$L(x_{n_k}) = N(x_{n_k}) + g(x_{n_k}) - h(t_{n_k}, x_{n_k}) \to N(x_0) + g(x_0) - h(t_0, x_0).$$

L closed implies that $x_0 \in \mathrm{dom}(L)$ and $N(x_0) + g(x_0) - h(t_0, x_0) = L(x_0)$. Hence $x_0 \in S$ and S is closed. Since g is a L-k-set contraction, there exists $x' \in \mathrm{dom}(L) \cap \Omega$ such that $Lx' - Nx' = g(x')$, so $x' \in S$ and S is not empty. Moreover, $S \cap \partial\Omega = \emptyset$. By Urysohn's Lemma, there exists a continuous function $0 \leq \phi \leq 1$, such that

$$\phi(x) = \begin{cases} 1, & x \in S, \\ 0, & x \in \partial\Omega. \end{cases}$$

Let

$$\overline{h}(x) = g(x) - h(\phi(x), x), \ x \in \overline{\Omega}.$$

Then $\overline{h}$ is a L-k-set contraction with $\overline{h}(x) \equiv 0$ on $\partial\Omega$. Thus the equation $Lx - Nx = \overline{h}(x)$ has a solution $x'' \in \mathrm{dom}(L) \cap \Omega$. Thus $x'' \in S$, so that $\phi(x'') = 1$, and

$$L(x'') - N(x'') + h(1, x'') = g(x'').$$

Hence $Lx - Nx + h(1, x)$ is $(0, L, k - k_1)$-epi. □

4. Regular maps and the spectrum of semilinear operators. In the following, subspaces E_1, F_0 and maps $P, Q, L_P, K_P, K_{PQ}, \Pi, \Lambda$ will be as defined in section 2. We also suppose that $N : E \to F$. To introduce the semilinear spectrum, we first make the following definition.

Definition 4.1. For $\lambda \in \mathbb{C}$, let $f_\lambda(L, N) : E \to E$ be defined by

$$f_\lambda(L, N)(x) = \lambda(I - P)x - (\Lambda\Pi + K_{PQ})Nx.$$

We will use f_λ for $f_\lambda(L, N)$ when no confusion is likely.

Definition 4.2. For $\lambda \in \mathbb{C}$, the operator $\lambda L - N$ is said to be *regular* if $\omega(f_\lambda) > 0$, $m(f_\lambda) > 0$ and $\nu(f_\lambda) > 0$. The resolvent set is defined by

$$\rho(L, N) = \{\lambda \in \mathbb{C} : \lambda L - N \text{ is regular}\}.$$

The spectrum is the set $\sigma(L, N) = \mathbb{C} \setminus \rho(L, N)$.

To prove properties of the spectrum, we need the following two lemmas.

Lemma 4.3. *Let $h : F/\mathrm{im}(L) \to F_0$ be the natural linear isomorphism and let $J = h\Lambda^{-1}$. Then $\Lambda\Pi + K_{PQ} : F \to \mathrm{dom}(L)$ is a linear isomorphism and $L + JP : \mathrm{dom}(L) \to F$ is invertible with $(L + JP)^{-1} = \Lambda\Pi + K_{PQ}$.*

Proof. Obviously, $J : \ker(L) \to F_0$ is an isomorphism. For $x \in \mathrm{dom}(L)$, suppose that $(L + JP)x = 0$. Then $Lx = -JPx \in F_0$, so $JPx = Lx = 0$, $x \in \ker(L)$ and then $JPx = Jx = 0$. This implies that $x = 0$. Hence $L + JP$ is invertible on $\mathrm{dom}(L)$.

Now, let $y \in F$ and suppose $(\Lambda\Pi + K_{PQ})y = 0$. Then

$$\Lambda\Pi y = -K_P(I - Q)y \in \mathrm{dom}(L) \cap E_1.$$

This implies that $\Lambda\Pi y = 0$ and $(I - Q)y = 0$. Thus $y \in \mathrm{im}(L) \cap F_0$, so that $y = 0$. Therefore, $\Lambda\Pi + K_{PQ}$ is one-to-one. For every $y \in F$, we have

$$(L + JP)(\Lambda\Pi + K_{PQ})y = JP\Lambda\Pi y + (I - Q)y = h\Pi y - Qy + y = y.$$

Hence $L + JP$ is onto. Also, for every $x \in \mathrm{dom}(L)$, we have

$$(\Lambda\Pi + K_{PQ})(L + JP)x = (I - P)x + J^{-1}h\Pi JPx = x.$$

Hence $\Lambda\Pi + K_{PQ}$ is the (bounded) inverse of $L + JP$. □

Lemma 4.4. *Let $y \in F$ and $\lambda \in \mathbb{C}$. Then*

1. $\lambda Lx - Nx = y$ *if and only if* $f_\lambda(x) = (\Lambda\Pi + K_{PQ})y$.

2. $\lambda L - N : \mathrm{dom}(L) \to F$ *is onto if and only if* $f_\lambda : E \to \mathrm{dom}(L)$ *is onto.*

Proof. Let $y \in F$, then

$$\begin{aligned}\lambda Lx - Nx = y &\iff \lambda Lx = Q(Nx+y) + (I-Q)(Nx+y)\\ &\iff \lambda Lx = (I-Q)(Nx+y),\ Q(Nx+y) = 0\\ &\iff \lambda(I-P)x = K_{PQ}(Nx+y),\ \Lambda\Pi(Nx+y) = 0\\ &\iff \lambda(I-P)x = (\Lambda\Pi + K_{PQ})(Nx+y)\\ &\iff f_\lambda(x) = (\Lambda\Pi + K_{PQ})y.\end{aligned}$$

Suppose $\lambda L - N$ is onto. For $x \in \text{dom}(L)$, by Lemma 4.3, $x = (\Lambda\Pi + K_{PQ})y$ for some $y \in F$. Let $x_0 \in \text{dom}(L)$ be such that $\lambda Lx_0 - Nx_0 = y$. Then by 1, $f_\lambda(x_0) = (\Lambda\Pi + K_{PQ})y = x$. Hence, $f_\lambda : E \to \text{dom}(L)$ is onto.

On the other hand, suppose $f_\lambda : E \to \text{dom}(L)$ is onto. Then for $y \in F$, $(\Lambda\Pi + K_{PQ})y \in \text{dom}(L)$. There exists $x_0 \in E$, $f_\lambda(x_0) = (\Lambda\Pi + K_{PQ})y$. Thus $x_0 \in \text{dom}(L)$ and $\lambda Lx_0 - Nx_0 = y$. Hence $\lambda L - N$ is onto. □

Relative to the measure of solvability of f at 0, $\nu(f)$, we have the following definition.

Definition 4.5. Let $r > 0$ and define

$$\begin{aligned}\nu_{Lr}(\lambda L - N, 0) = \inf\{k \geq 0 :\ &\text{there is a } L\text{-}k\text{-set contraction } g : B_r \to F,\\ &g \equiv 0 \text{ on } \partial B_r, \text{ and } \lambda Lx - Nx = g(x) \text{ has no solution in } B_r \cap \text{dom}(L)\}.\end{aligned}$$

Let $\nu_L(\lambda L - N) = \inf\{\nu_{Lr}(\lambda L - N, 0), r > 0\}$. The number $\nu_L(\lambda L - N)$ will be called the *measure of solvability* of $\lambda L - N$ at 0.

It is clear that $\nu_L(\lambda L - N) > 0$ if and only if there exists $\varepsilon > 0$ such that $\lambda L - N$ is $(0, L, \varepsilon)$-epi on $\text{dom}(L) \cap B_r$ for every $r > 0$.

We now establish some properties of regular mappings.

Proposition 4.6. *If $\lambda L - N$ is regular, then $\omega(\lambda L - N) > 0$, $m(\lambda L - N) > 0$ and $\nu_L(\lambda L - N) > 0$. If L is bounded, then $\lambda L - N$ is regular if and only if the above conditions are satisfied.*

Proof. Firstly, we have $\omega(f_\lambda) = \omega(K_P(\lambda L) - K_P(I-Q)N)$ and so

$$\omega(K_P)\omega(\lambda L - N) \leq \omega(f_\lambda) \leq \alpha(K_P)\omega(\lambda L - N).$$

Since $\alpha(K_P) \leq \|K_P\| < +\infty$, if $\omega(f_\lambda) > 0$ then $\omega(\lambda L - N) > 0$. If L is bounded, then $\text{dom}(L) = E$, $\omega(K_P) > 0$. Hence $\omega(\lambda L - N) > 0$ implies $\omega(f_\lambda) > 0$. Next, $f_\lambda = H^{-1}(\lambda H(I-P) - N)$ and $H(I-P) = L$, (recall $H = L + JP$). So $\|H^{-1}(\lambda L - N)x\| \geq m(f_\lambda)\|x\|$. Since H^{-1} is bounded,

$$\|(\lambda L - N)x\| \geq \frac{m(f_\lambda)}{\|H^{-1}\|}\|x\|.$$

Thus $m(f_\lambda) > 0$ implies $m(\lambda L - N) > 0$. When L is bounded, we have

$$\|H\|\,\|f_\lambda(x)\| \geq \|(\lambda L - N)x\| \geq m(\lambda L - N)\|x\|.$$

Thus $m(\lambda L - N) > 0$ implies that $m(f_\lambda) > 0$.
Now suppose $\nu(f_\lambda) > 0$. Thus there exists $\varepsilon > 0$ such that f_λ is $(0,\varepsilon)$-epi on every B_r with $r > 0$. Let $h : B_r \to F$ be a bounded continuous L-ε-set contraction with $h(x) \equiv 0$ on ∂B_r. Then $(\Lambda\Pi + K_{PQ})h : B_r \to E$ is continuous,

$$\alpha((\Lambda\Pi + K_{PQ})h) = \alpha(K_{PQ}h) \leq \varepsilon,$$

and $(\Lambda\Pi + K_{PQ})h(x) \equiv 0$ on ∂B_r. By the definition of a ε-epi map, the equation

$$f_\lambda(x) = (\Lambda\Pi + K_{PQ})h(x)$$

has a solution $x_0 \in B_r$, $\|x_0\| < r$. Therefore

$$\lambda(I-P)x_0 = (\Lambda\Pi + K_{PQ})(h(x_0) + N(x_0)) = H^{-1}(h(x_0) + N(x_0)).$$

Thus $\lambda L x_0 - N x_0 = h(x_0)$. Since $H^{-1} : F \to \mathrm{dom}(L)$, we have

$$\lambda(I-P)x_0 \in H^{-1}F \subset \mathrm{dom}(L),$$

so that $x_0 \in \mathrm{dom}(L) \cap \{x : \|x\| < r\}$. Therefore $\lambda L - N$ is $(0, L, \varepsilon)$-epi on $\mathrm{dom}(L) \cap B_r$, thus $\nu_L(\lambda L - N) > 0$. Suppose that L is bounded and there exists $\varepsilon > 0$ such that $\lambda L - N$ is $(0, L, \varepsilon)$-epi on $\mathrm{dom}(L) \cap B_r$. Let $h : B_r \to F$ be a continuous map with $\alpha(h) \leq \varepsilon$ and such that $h(x) \equiv 0$ on ∂B_r. Then $\alpha(H^{-1}Hh) \leq \varepsilon$, so Hh is a L-ε-set contraction. By definition, $\lambda Lx - Nx = Hh(x)$ has a solution $x_0 \in \mathrm{dom}(L) \cap O_r = O_r$. By Lemma 4.4, $f_\lambda(x_0) = h(x_0)$. Hence f_λ is $(0,\varepsilon)$-epi on B_r. □

Notation. Let $\Sigma(L, N)$ denote the set

$$\left\{\lambda \in \mathbb{C} : \liminf_{\|x\| \to \infty} \frac{\|f_\lambda(L,N)(x)\|}{\|x\|} = 0\right\}.$$

Proposition 4.7. *Let $\lambda \in \mathbb{C}$, $\lambda \notin \Sigma(L, N)$ and $\nu_L(\lambda L - N) > 0$. Then $\lambda L - N$ is onto.*

Proof. Let $y \in F$ and define

$$S = \{x \in \mathrm{dom}(L) \cap E, \lambda Lx - Nx = ty, \text{ for some } t \in [0,1]\}.$$

We assert that S is bounded. In fact, if there exist $x_n \in S$, $\|x_n\| \to \infty$ as $n \to \infty$, let $t_n \in [0,1]$ be such that $\lambda L x_n - N x_n = t_n y$. By Lemma 4.4, we have

$$\lambda(I-P)x_n - (\Lambda\Pi + K_{PQ})N(x_n) = t_n(\Lambda\Pi + K_{PQ})y.$$

Then

$$\frac{\|\lambda(I-P)x_n - (\Lambda\Pi + K_{PQ})N(x_n)\|}{\|x_n\|} = \frac{\|t_n(\Lambda\Pi + K_{PQ})y\|}{\|x_n\|} \to 0.$$

This contradicts $\lambda \notin \Sigma(L, N)$.
Now let $R > 0$ be such that $S \subset B_R$. Then

$$\lambda Lx - Nx - ty \neq 0 \text{ for } x \in \mathrm{dom}(L) \cap \partial B_R.$$

By Property 3.5, $\lambda L - N - y$ is $(0, L, \varepsilon)$-epi for some $\varepsilon > 0$ on $\mathrm{dom}(L) \cap B_R$. By the existence result, there exists $x_0 \in \mathrm{dom}(L) \cap B_R$ such that $\lambda L x_0 - N x_0 = y$. Thus $\lambda L - N$ is onto. □

Propositions 4.7 and 4.6 yield the following result.

Theorem 4.8. *Assume that $\lambda \in \rho(L, N)$, then $\lambda L - N$ is onto.*

We also have:

Theorem 4.9. *All eigenvalues of (L, N) are in the spectrum of (L, N).*
Proof. Suppose $0 \neq x_0 \in \mathrm{dom}(L)$ and $\lambda L x_0 - N x_0 = 0$. Then by Lemma 4.4, $f_\lambda(x_0) = 0$. Hence $m(f_\lambda) = 0$ and $\lambda \in \sigma(L, N)$. □

We now prove that the spectrum of semilinear operators is closed.

Theorem 4.10. *$\rho(L, N)$ is an open set and $\sigma(L, N)$ is closed.*
Proof. Suppose $\lambda_1 \in \rho(L, N)$. Let λ_2 be such that

$$|\lambda_2 - \lambda_1| < \min\{\omega(f_{\lambda_1}), \nu(f_{\lambda_1}), m(f_{\lambda_1})/\|I - P\|\}.$$

Then $f_{\lambda_2} = f_{\lambda_1} + (\lambda_2 - \lambda_1)(I - P)$. So,

$$\omega(f_{\lambda_2}) \geq \omega(f_{\lambda_1}) - \alpha((\lambda_2 - \lambda_1)(I - P)) = \omega(f_{\lambda_1}) - |\lambda_2 - \lambda_1| > 0.$$

Also, we have

$$\|f_{\lambda_2}(x)\| \geq \|f_{\lambda_1}(x)\| - |\lambda_2 - \lambda_1| \|(I - P)x\| \geq (m(f_{\lambda_1}) - |\lambda_2 - \lambda_1| \|(I - P)\|)\|x\|.$$

Thus $m(f_{\lambda_2}) > 0$. Let $h(t, x) = t(\lambda_2 - \lambda_1)(I - P)x$, $t \in [0, 1]$. Then $\alpha(h) \leq |\lambda_2 - \lambda_1| < \nu(f_{\lambda_1})$ and $h(0, x) = 0$ for $x \in E$. If $f_{\lambda_1}(x) + t(\lambda_2 - \lambda_1)(I - P)x = 0$, then

$$m(f_{\lambda_1})\|x\| \leq \|f_{\lambda_1}(x)\| \leq |\lambda_2 - \lambda_1| \|I - P\| \|x\|.$$

Hence $x = 0$. Since f_{λ_1} is $(0, \varepsilon)$-epi on every ball for some $\varepsilon > 0$, by the homotopy property of (p, k)-epi maps, f_{λ_2} is $(0, \varepsilon - \alpha(h))$-epi. Thus $\nu(f_{\lambda_2}) > 0$, $\lambda_2 L - N$ is regular and $\lambda_2 \in \rho(L, N)$. □

In [10], a concept of regularity was given for a map $L - N$ when N is also linear. The following theorem shows that regular maps according to Definition 4.2 when N is linear coincides with that given in [10].

Theorem 4.11. *Suppose N is a linear operator. Then $\lambda L - N$ is regular if and only if $\lambda L - N$ has a continuous inverse.*

Proof. If $\lambda L - N$ is regular, it is one-to-one, onto and there exists $m > 0$ such that $\|(\lambda L - N)(x)\| \geq m\|x\|$. So $(\lambda L - N)^{-1}$ is continuous. Conversely, suppose that $\lambda L - N$ has a continuous inverse. There exists $x_0 \in E$ such that $f_\lambda(x_0) = 0$, and then by Lemma 4.4, $\lambda L x_0 - N x_0 = 0$. Thus $x_0 = 0$ so that f_λ is one-to-one.

Let $x' \in E$ and $y_0 = Hx'$. Since $\lambda L - N$ is onto, there exists $x_0 \in \mathrm{dom}(L)$ such that $\lambda L x_0 - N x_0 = y_0$. This implies that $f_\lambda(x_0) = H^{-1} y_0 = x'$. Hence, $f_\lambda : E \to E$ is onto.

Since f_λ is continuous, f_λ^{-1} is also continuous. By Proposition 3.2.1 of [6], $\omega(f_\lambda) > 0$. Let $m' > 0$ be such that $\|f_\lambda^{-1}(x)\| \leq m'\|x\|$ for every $x \in E$. Then $\|f_\lambda(x)\| \geq (1/m')\|x\|$. Thus $m(f_\lambda) > 0$.

Furthermore, for an arbitrary bounded subset S of F, let $A = f_\lambda^{-1}(S)$. Then A is a bounded subset of E and

$$\omega(f_\lambda)\alpha(A) \leq \alpha(f_\lambda(A)) \implies \alpha(f_\lambda^{-1}(S)) \leq \frac{\alpha(S)}{\omega(f_\lambda))}.$$

Hence $f_\lambda : O_r \to F$ is continuous, one-to-one and $1/\omega(f_\lambda)$-proper (where $O_r = \{x \in E : \|x\| < r\}$). Since f_λ^{-1} is continuous, $f_\lambda(O_r)$ is an open set. By Theorem 2.3 of [16], f_λ is $(0,k)$-epi for $0 < k < \omega(f)$. Hence $\nu(f_\lambda) > 0$. By the above arguments, $\lambda L - N$ is regular. □

Corollary 4.12. *If N is a linear operator and $L+N$ is regular, then $L+N$ is $(0,L,\varepsilon)$-epi for all $\varepsilon < \omega(f)$.*

5. Decomposition of the spectrum. In this section, we shall discuss various parts of the spectrum. As in [6], we shall use the following symbols:

$$\sigma_\delta(L,N) = \{\lambda : \nu(f_\lambda) = 0\}, \tag{5.1}$$

$$\sigma_m(L,N) = \{\lambda : m(f_\lambda) = 0\}, \tag{5.2}$$

$$\sigma_\omega(L,N) = \{\lambda : \omega(f_\lambda) = 0\}, \tag{5.3}$$

$$\sigma_\pi(L,N) = \sigma_m(L,N) \cup \sigma_\omega(L,N). \tag{5.4}$$

Theorem 5.1. *Suppose that N is a L-k-set contraction, and N is an odd mapping. If $\lambda \in \sigma(L,N)$ and $|\lambda| > k$ then $\lambda \in \sigma_m(L,N)$.*

Proof. Firstly we have $\omega(f_\lambda) \geq |\lambda| - \alpha(H^{-1}N) = |\lambda| - k > 0$. Assume that $m(f_\lambda) > 0$. Then for every $x \in E$, $\|f_\lambda(x)\| \geq m(f_\lambda)\|x\|$. Let $r > 0$ and $B_r = \{x : \|x\| \leq r\}$. Suppose $g : B_r \to E$ is an ε-set contraction with $\varepsilon < |\lambda| - k$ and $g(x) \equiv 0$ on ∂B_r. We shall prove that $f_\lambda(x) = g(x)$ has a solution $x_0 \in B_r \setminus \partial B_r$. Let

$$h(x) = Px + \frac{1}{\lambda}(\Lambda\Pi + K_{PQ})Nx + \frac{g(x)}{\lambda}.$$

Then

$$\alpha(h) \leq \frac{k + \alpha(g)}{|\lambda|} \leq \frac{k+\varepsilon}{|\lambda|} < 1.$$

Hence h is a k_1-set contraction with $k_1 < 1$. For $x \in \partial B_r$, $(I-h)(x) = (1/\lambda)f_\lambda(x) \neq 0$. Also $h|_{\partial B_r}$ is odd. So, the degree, $d(I-h, O_r, 0) \neq 0$. Therefore there exists $x_0 \in B_r \setminus \partial B_r$ such that

$$x_0 = Px_0 + \frac{1}{\lambda}(\Lambda\Pi + K_{PQ})Nx_0 + \frac{g(x_0)}{\lambda}.$$

Hence f_λ is $(0,\varepsilon)$-epi on B_r. By the definition, $\lambda \in \rho(L,N)$. This contradicts $\lambda \in \sigma(L,N)$. So, we must have $m(f_\lambda)=0$. □

In [10], it was proved that if N is linear and L-compact, $\mu \in \mathbb{C}$ is not a regular value of (L,N), then μ is an eigenvalue of (L,N). The following result generalizes this.

Theorem 5.2. *Assume that N is a L-k-set contraction, N is odd and positively homogeneous. If $\lambda \in \sigma(L,N)$ and $|\lambda|>k$, then λ is an eigenvalue of (L,N).*

Proof. By Theorem 5.1, we know that $\lambda \in \sigma_m(L,N)$. So, for $n \in \mathbb{N}$, there exists $x_n \in E$ such that

$$\|\lambda(I-P)x_n - H^{-1}N(x_n)\| < \frac{\|x_n\|}{n}.$$

Since N is positively homogeneous, we have

$$\|\lambda(I-P)\frac{x_n}{\|x_n\|} - H^{-1}N\Big(\frac{x_n}{\|x_n\|}\Big)\| \to 0,\ (n\to\infty).$$

Let

$$A=\bigcup_{n=1}^{\infty}\left\{\frac{x_n}{\|x_n\|}\right\}.$$

Then A is bounded and $\omega(f_\lambda)\alpha(A)\le \alpha(f_\lambda(A))=0$. Since N is a L-k-set contraction and $|\lambda|>k$, $\omega(f_\lambda)>0$. This implies that $\alpha(A)=0$, so we may assume that $x_n/\|x_n\|\to x_0$ (for a subsequence). Then $\|x_0\|=1$ and $\lambda(I-P)x_0 - H^{-1}N(x_0)=0$. By Lemma 4.4, we obtain $\lambda Lx_0 - Nx_0 = 0$. Thus λ is an eigenvalue of (L,N). □

It is known that the boundary of the spectrum of a linear operator is contained in its approximate point spectrum. In the nonlinear case, the boundary of $\sigma_{fmv}(f)$ is contained in $\sigma_\pi(f)$ [6]. Thus if $\lambda \in \partial\sigma_{fmv}(f)$, then $d(\lambda - f)=0$ or $\omega(\lambda - f)=0$. To prove a similar result in the semilinear case for the boundary of the spectrum, we first prove the following result.

Proposition 5.3. *$\sigma_m(L,N)$ and $\sigma_\omega(L,N)$ are closed sets. Let*

$$U=\{\lambda : \lambda\in\sigma_\delta(L,N)\setminus\sigma_\pi(L,N) \ \ \textit{and}\ \ \lambda L - N \ \ \textit{is not onto}\}.$$

Then U is an open set.

Proof. (1) Suppose that $\lambda_n \in \sigma_m(L,N)$ and $\lambda_n\to\lambda$. We shall show that $\lambda\in\sigma_m(L,N)$. Otherwise, there exists $m>0$ such that $\|f_\lambda(x)\|\ge m\|x\|$ for every $x\in E$. Since $f_{\lambda_n}=f_\lambda+(\lambda_n-\lambda)(I-P)$, we have

$$\|f_{\lambda_n}(x)\| \ge (m-|\lambda-\lambda_n|\|I-P\|)\|x\|.$$

For $|\lambda_n-\lambda|<m/\|I-P\|$, we obtain $\lambda_n\notin\sigma_m(L,N)$, a contradiction.

(2) Suppose $\lambda_n\in\sigma_\omega(L,N)$ and $\lambda_n\to\lambda$. If $\omega(f_\lambda)>0$, then from

$$\omega(f_{\lambda_n})=\omega(f_\lambda+(\lambda_n-\lambda)(I-P)x)\ge\omega(f_\lambda)-|\lambda-\lambda_n|,$$

we obtain $\lambda_n\notin\sigma_\omega(L,N)$ when $|\lambda-\lambda_n|<\omega(f_\lambda)$. Hence $\omega(f_\lambda)=0$ and $\lambda\in\sigma_\omega(L,N)$.

(3) Suppose that $\lambda \in U$, then $\lambda \in (\sigma_\pi(L,N))^c$. By (1) and (2), $\sigma_\pi(L,N)$ is closed. So there exists $\delta > 0$ such that for $\mu \in \mathbb{C}$ with $|\mu - \lambda| < \delta$, $\mu \notin \sigma_\pi(L,N)$. Now assume that there exist $\mu_n \notin U$, $\mu_n \to \lambda$. Suppose that $|\mu_n - \lambda| < \delta$, then $\mu_n \notin \sigma_\pi(L,N)$. So $\mu_n \notin \sigma_\delta(L,N) \setminus \sigma_\pi(L,N)$ implies that $\mu_n \notin \sigma_\delta(L,N)$. Then $\mu_n L - N$ is regular, hence is onto. By Lemma 4.4, f_{μ_n} is onto. If $\mu_n \in \sigma_\delta(L,N) \setminus \sigma_\pi(L,N)$, then f_{μ_n} is also onto because $\mu_n \notin U$.
Therefore, for $y \in E$, there exist $x_n \in E$ such that $f_{\mu_n}(x_n) = y$. Then

$$\begin{aligned}\|y\| &= \|\mu_n(I-P)x_n - (\Lambda\Pi + K_{PQ})Nx_n\| \\ &\geq \|f_\lambda(x_n)\| - |\mu_n - \lambda|\|I-P\|\|x_n\| \\ &\geq m\|x_n\| - |\mu_n - \lambda|\|I-P\|\|x_n\|.\end{aligned}$$

Suppose that $|\mu_n - \lambda| < m/2(\|I-P\|)$, then $\|x_n\| \leq 2\|y\|/m$. Thus $\{x_n\}_{n=1}^\infty$ is bounded. Since $\|(I-P)x_n\| \leq \|I-P\|\|x_n\|$, we have

$$f_\lambda(x_n) = y + (\lambda - \mu_n)(I-P)(x_n) \to y \ (n \to \infty).$$

Since

$$\omega(f_\lambda)\alpha(\{x_1, x_2, \ldots\}) \leq \alpha(f_\lambda(\{x_1, x_2, \ldots\})) \leq \alpha(\{f_\lambda(x_1), f_\lambda(x_2), \ldots\}) = 0,$$

and $\omega(f_\lambda) > 0$, we obtain $\alpha(\{x_1, x_2, \ldots\}) = 0$. Thus $\{x_n\}_{n=1}^\infty$ has a convergent subsequence. Let $x_{n_k} \to x_0$, then $f_\lambda(x_{n_k}) \to f_\lambda(x_0) = y$. We have shown that f_λ is onto, so $\lambda L - N$ is onto. This contradicts $\lambda \in U$. By the above argument, there exists $\delta_1 > 0$ such that $|\mu - \lambda| < \delta_1$ implies $\mu \in U$, thus U is open. □

The result for the boundary of the spectrum in the semilinear case can now be given.

Theorem 5.4. *The inclusion*

$$\partial\sigma(L,N) \subseteq \{\sigma_\pi(L,N)\} \cup \{\lambda : \lambda L - N \textit{ is onto}\}$$

holds.

Proof. Suppose that $\lambda \in \partial\sigma(L,N)$. Since $\sigma(L,N)$ is closed, we have $\lambda \in \sigma(L,N)$. Assume that $\lambda \notin \sigma_\pi(L,N)$ and $\lambda L - N$ is not onto. Then $\lambda \in U$. By Proposition 5.3, there exists $\delta > 0$ such that for every $\mu \in \mathbb{C}$ with $|\mu - \lambda| < \delta$, $\mu \in U$. On the other hand, there exist $\mu_n \notin \sigma(L,N)$ with $\mu_n \to \lambda$. When $|\mu_n - \lambda| < \delta$, we have $\mu_n \in U$. This contradiction proves that $\lambda \in \sigma_\pi(L,N)$ or $\lambda L - N$ is onto. □

When N is positively homogeneous operator, we have the following result.

Proposition 5.5. *Suppose N is positively homogeneous. Then*

$$\sigma_m(L,N) \setminus \sigma_\omega(L,N) \subseteq \{\lambda : \lambda \text{ is an eigenvalue of } (L,N)\} \subseteq \sigma_m(L,N).$$

Proof. Let $\lambda \in \sigma_m(L,N) \setminus \sigma_\omega(L,N)$. Then there exist $x_n \in E$, $\|x_n\| = 1$ such that

$$\|\lambda(I-P)x_n - (\Lambda\Pi + K_{PQ})N(x_n)\| < \frac{1}{n} \to 0.$$

So, $\omega(f_\lambda)\alpha(\{x_1,x_2,\ldots\}) \le \alpha(\{f_\lambda(x_1), f_\lambda(x_2),\ldots\}) = 0$. Now $\lambda \notin \sigma_\omega(L,N)$ implies that $\omega(f_\lambda) > 0$. Hence $\alpha(\{x_1,x_2,\ldots\}) = 0$ and $\{x_n\}_{n=1}^\infty$ has a convergent subsequence. Suppose that $x_{n_k} \to x_0$, then $\|x_0\| = 1$ and $f_\lambda(x_0) = 0$. By Lemma 4.4, $\lambda L(x_0) - N(x_0) = 0$. Hence λ is an eigenvalue of (L,N). The other part of the proposition is clear. □

Proposition 5.6. *The inclusion*

$$\mathbb{C} \setminus (\sigma_\pi(L,N) \cup \{\lambda : \lambda L - N \textit{ is onto}\}) \subseteq \sigma_\delta(L,N) \setminus \partial\sigma(L,N)$$

holds. Therefore, $\sigma(L,N) \setminus \partial\sigma(L,N) \subseteq \sigma_\delta(L,N)$.

Proof. Let $\lambda \in \mathbb{C} \setminus (\sigma_\pi(L,N) \cup \{\lambda : \lambda L - N \text{ is onto}\})$, then $\lambda \notin \sigma_\pi(L,N)$ and $\lambda L - N$ is not onto. So $\lambda \in \sigma_\delta(L,N)$. By Theorem 5.4, $\lambda \notin \partial\sigma(L,N)$. Thus $\lambda \in \sigma_\delta(L,N) \setminus \partial\sigma(L,N)$. □

6. Applications of the spectral theory. We apply the above spectral theory to extend some existence results for semilinear operator equations. The first theorem in this section generalizes Corollary 1 in [13], which has been widely applied in the study of differential equations. In [13], the mapping A was assumed to be linear and L-compact, and N was L-compact.

Theorem 6.1. *Let $A : E \to F$ be an odd, positively homogeneous operator, which is a L-k_1-set contraction on the closed unit ball of E and such that* $\ker(L - A) = \{0\}$. *If $0 \in \Omega$ and $N : \Omega \to F$ is a L-k_2-set contraction with*

$$Lx \neq (1-t)Ax + tNx \textit{ for all } t \in [0,1],\ x \in \mathrm{dom}(L) \cap \partial\Omega,$$

then the equation

$$Lx = Nx$$

has at least one solution in $\mathrm{dom}(L) \cap \overline{\Omega}$ *provided that* $2k_1 + k_2 < 1$.

Proof. Since $\ker(L - A) = \{0\}$, 1 is not an eigenvalue of (L,A). By Theorem 5.2, $1 \in \rho(L,A)$. Theorem 5.1 implies that $f_1 = (I - P) - (\Lambda\Pi + K_{PQ})A$ is $(0,\varepsilon)$-epi on $\overline{\Omega}$ for every $0 \le \varepsilon < 1 - k_1$. By Proposition 4.6, $L - A$ is $(0, L, \varepsilon)$-epi on $\overline{\Omega}$ for every $0 \le \varepsilon < 1 - k_1$. Let $h : [0,1] \times \overline{\Omega} \to F$ be defined by $h(t,x) = tAx - tNx$, $t \in [0,1]$. Then $h(0,x) = 0$, h is a L-$(k_1 + k_2)$-set contraction and,

$$Lx - Ax + h(t,x) \neq 0 \text{ for all } x \in \mathrm{dom}(L) \cap \partial\Omega,\ t \in [0,1].$$

By Property 3.5, $L - N$ is $(0, L, \varepsilon)$-epi for $0 < \varepsilon < 1 - 2k_1 - k_2$. Thus there is $x_0 \in \Omega$ such that $Lx_0 = Nx_0$. □

Suppose that $T : E \to F$ is a k-set contraction and is asymptotically linear with asymptotic derivative B. It was proved in Lemma XI.3 of [10] that if $0 \le k \le \omega(L)$, and $\mathrm{im}(T - B) \subset \mathrm{im}(L - B)$, then $Lx - Tx = 0$ has a solution. In the following, by using a different method, we obtain a different condition for the existence of a solution of the equation $\lambda Lx - Tx = 0$.

Theorem 6.2. *Suppose that T is a k-set contraction and is asymptotically linear with asymptotic derivative B. Let $\lambda \in \mathbb{C}$ with $|\lambda| > 3k/\omega(L)$. Then $\lambda Lx - Tx = 0$ has a solution provided λ is not an eigenvalue of (L, B).*

Proof. Suppose that λ is not an eigenvalue of (L, B). Since T is a k-set contraction, B is also a k-set contraction with $\alpha(B) = k < |\lambda|\omega(L)$, [17]. So $\lambda L - B$ is a Fredholm operator of index zero [10]. Let $P : E \to \ker(\lambda L - B)$ be the projection, $J : F_1 \to \ker(\lambda L - B)$ be the linear isomorphism, where F_1 is a subspace of F with $F_1 \oplus \mathrm{im}(\lambda L - B) = F$. Suppose that $T = B + R$ and let

$$S = \{x : \lambda Lx - Bx - J^{-1}Px + t[J^{-1}Px - R(x)] = 0, t \in [0,1]\}.$$

Case (1). Assume that there exist $x_n \in S$ with $\|x_n\| \to \infty$. Let $y_n = x_n/\|x_n\|$ and $t_n \in [0,1]$ be such that

$$\lambda Ly_n - By_n - J^{-1}Py_n + t_nJ^{-1}Py_n \to 0, \ (n \to \infty).$$

As $J^{-1}P$ is compact, we can assume that $J^{-1}Py_n - t_nJ^{-1}Py_n \to y_0 \in F_1$. Then $\lambda Ly_n - By_n \to y_0$. Since $|\lambda| > \alpha(B)/\omega(L)$, we obtain $\omega(\lambda L - B) > 0$ and $\alpha(\{y_1, y_2, \ldots\}) = 0$. Suppose that $y_{n_k} \to x_0$. Then $\|x_0\| = 1$ and $\lambda Ly_{n_k} \to y_0 + B(x_0)$. Since L is closed, $\lambda Lx_0 - Bx_0 = y_0 \in F_1$. This ensures that $\lambda Lx_0 - Bx_0 = 0$. Hence λ is an eigenvalue of (L, B). This contradicts our assumption.

Case (2). Suppose that there exists $r > 0$ such that for all $x \in \partial B_r$,

$$\lambda Lx - Bx - J^{-1}Px + t[J^{-1}Px - Rx] \neq 0, \ t \in [0,1].$$

Since $\lambda L - B - J^{-1}P$ is one-to-one and $J^{-1}P$ is a linear compact operator, by Theorem 5.2, $1 \in \rho(\lambda L - B, J^{-1}P)$. Thus $\lambda L - B - J^{-1}P$ is regular. Let $L_1 = \lambda L - B$ and $f_1(x) = (I - P)x - (\Lambda\Pi + K_{PQ})J^{-1}Px$. By Theorem 4.11, f_1 is $(0, \varepsilon)$-epi for any $\varepsilon < \omega(f_1) = 1$. Also by Proposition 4.6, $L_1 - J^{-1}P$ is $(0, L_1, \varepsilon)$-epi for every $\varepsilon < 1$. Let $H(x,t) = t[J^{-1}Px - R(x)]$, and note that $\alpha(H) \leq \alpha(R)$. Since $|\lambda| > 3k/\omega(L)$, we have

$$\alpha(R) = \alpha(T - B) \leq 2k < |\lambda|\omega(L) - k \leq |\lambda|\omega(L) - \alpha(B) \leq \omega(\lambda L - B) = \omega(L_1).$$

Hence H is a L_1-k-set contraction with $k = \alpha(R)/\omega(L_1) < 1$ [10]. Now suppose that $\alpha(R)/\omega(L_1) < \varepsilon_1 < 1$, then $L_1 - J^{-1}P$ is $(0, L_1, \varepsilon_1)$-epi. Again by Property 3.5, we obtain that

$$L_1x - J^{-1}Px + H(x,1) = \lambda L - T$$

is $(0, L_1, \varepsilon)$-epi for some $\varepsilon > 0$ on B_r. Hence there exists $x_0 \in E$ such that $\lambda Lx_0 - Tx_0 = 0$. We are done. □

In [14], Mawhin gave some existence theorems of Leray-Schauder type. The following theorem generalizes Theorem 2.2 of that paper.

Theorem 6.3. *Let Ω be an open bounded connected and convex subset of E with $0 \in \Omega$. Let $N : \overline{\Omega} \to F$ be a L-k_1-set contraction. Let $A : E \to F$ be a linear L-k_2-set contraction and $h : \overline{\Omega} \to F$ be a L-compact map such that*

(1) $h(\partial\Omega) \subset (L + A)(\mathrm{dom}(L) \cap \Omega)$.

(2) $\ker(L + A) = \{0\}$.

(3) $Lx + (1-t)(Ax - hx) + tNx \neq 0$ *for* $x \in \partial\Omega \cap \mathrm{dom}(L)$ *and* $t \in (0,1)$.

(4) $k_1 + 2k_2 < 1$.

Then the equation $Lx + Nx = 0$ *has at least one solution in* $\mathrm{dom}(L) \cap \overline{\Omega}$.

Proof. The condition $\ker(L+A) = \{0\}$ implies that $1 \in \rho(L,A)$ since A is a linear L-k_2-set contraction with $k_2 < 1$. So by Corollary 4.12, $L + A$ is $(0, L, k)$-epi for every $k < \omega(f)$, where $f = (I - P) - (\Lambda\Pi + K_{PQ})A$ and $\omega(f) > 1 - k_2$. Let $h_1(t,x) = th(x)$. Then $h_1(t,x) : [0,1] \times \overline{\Omega} \to F$ is a L-compact map. Assume that there exists $x_0 \in \partial\Omega \cap \mathrm{dom}(L)$ such that

$$Lx_0 + Ax_0 - t_0 h(x_0) = 0.$$

Then

$$Lx_0 + Ax_0 = t_0 h(x_0) = t_0(L+A)x_1,$$

where $x_1 \in \mathrm{dom}(L) \cap \Omega$. Thus $x_0 = t_0 x_1$, $t_0 \in [0,1]$. This contradicts the connectedness and convexity of Ω. Hence for $x \in \partial\Omega \cap \mathrm{dom}(L)$, we have $Lx + Ax \neq th(x)$. By Property 3.5, we obtain that $L + A - h$ is $(0, L, k)$-epi for every $k < \omega(f)$ on $\mathrm{dom}(L) \cap \overline{\Omega}$. Let $h_2 : [0,1] \times \overline{\Omega} \to F$ be defined by

$$h_2(t,x) = t(Ax - h(x)) + tNx.$$

Then h_2 is a L-$(k_2 + k_1)$-set contraction and $k_2 + k_1 < 1 - k_2 < \omega(f)$. The assumption 3 and Property 3.5 imply that $L + N : \mathrm{dom}(L) \cap \overline{\Omega} \to F$ is $(0, L, \varepsilon)$-epi for some $\varepsilon > 0$. Hence there exists $x_0 \in \mathrm{dom}(L) \cap \Omega$ which is a solution of the equation $Lx + Nx = 0$. □

Remark 6.4. In Theorem 6.3, let $k_1 = k_2 = 0$, $h : \overline{\Omega} \to F$ be defined by $h(x) = z$ with $z \in (L+A)(\mathrm{dom}(L) \cap \Omega)$. Then this Theorem reduces to Theorem 2.2 of [14].

References

[1] J. Appell, M. Dörfner: *Some spectral theory for nonlinear operators*, Nonlin. Anal. TMA **28** (1997), 1955-1976

[2] K. Deimling: *Nonlinear Functional Analysis*, Springer, Berlin 1985

[3] D. E. Edmunds, J. R. L. Webb: *Remarks on nonlinear spectral theory*, Boll. Unione Mat. Ital. **B-2** (1983), 377-390

[4] S. Fučik, J. Nečas, J. Souček, V. Souček: *Spectral Analysis of Nonlinear Operators*, Lect. Notes Math. **346**, Springer, Berlin 1973

[5] M. Furi, A. Vignoli: *A nonlinear spectral approach to surjectivity in Banach spaces*, J. Funct. Anal. **20** (1975), 304-318

[6] M. Furi, M. Martelli, A. Vignoli: *Contributions to the spectral theory for nonlinear operators in Banach spaces*, Annali Mat. Pura Appl. **118** (1978), 229-294

[7] M. Furi, M. Martelli, A. Vignoli: *On the solvability of nonlinear operator equations in normed spaces*, Annali Mat. Pura Appl. **124** (1980), 321-343

[8] W. FENG: *A new spectral theory for nonlinear operators and its applications*, Abstr. Appl. Anal. **2** (1997), 163-183

[9] W. FENG, J. R. L. WEBB: *Surjectivity results for nonlinear mappings without oddness conditions*, Comment. Math. Univ. Carolinae **38** (1997), 15-28

[10] R. E. GAINES, J. MAWHIN: *Coincidence Degree and Nonlinear Differential Equations*, Lect. Notes Math. **568**, Springer, Berlin 1977

[11] J. IZE, I. MASSABÒ, J. PEJSACHOWICZ, A. VIGNOLI: *Nonlinear multiparametric equations: structure and topological dimension of global branches of solutions*, in: Proc. Sympos. Pure. Math. **45**, Amer. Math. Soc., Providence RI 1986, p. 529-540

[12] M. MARTELLI: *Positive eigenvectors of wedge maps*, Annali Mat. Pura Appl. **145** (1986), 1-32

[13] J. MAWHIN: *Topological degree methods in nonlinear boundary value problems*, in: NSF-CBMS Regional Conference Series in Math. **40**, Amer. Math. Soc., Providence RI 1979, p. 1-122

[14] J. MAWHIN: *Topological degree and boundary value problems for nonlinear differential equations*, in: Topological Methods for Ordinary Differential Equations [Ed.: P. M. FITZPATRICK, M. MARTELLI, J. MAWHIN, R. NUSSBAUM], Lect. Notes Math. **1537**, Springer, Berlin 1991, p. 74-142

[15] R. H. MARTIN: *Nonlinear Operators and Differential Equations in Banach Spaces*, John Wiley & Sons, New York 1976

[16] E. U. TARAFDAR, H. B. THOMPSON: *On the solvability of nonlinear noncompact operator equations*, J. Austral. Math. Soc. **43** (1987), 103-126

[17] J. F. TOLAND: *Topological methods for nonlinear eigenvalue problems*, Battelle, Advanced Studies Center, Mathematics Report **77**, Geneva 1973

WENYING FENG, Trent University, AMINSS Program, Peterborough, K9J 7B8, Canada; `wfeng@trentu.ca`

JEFF WEBB, University of Glasgow, Department of Mathematics, Glasgow, G12 8QW, Scotland UK; `jrlw@maths.gla.ac.uk`

Progress in Nonlinear Differential Equations
and Their Applications, Vol. 40

Feedback Stability of Closed Sets for Nonlinear Control Systems

LECH GÓRNIEWICZ, PAOLO NISTRI

Dedicated to Alfonso Vignoli on the occasion of his 60th birthday

Summary: Using methods from the multivalued analysis we show the existence of feedback controls which "stabilizes" a given closed set $K \subseteq \mathbb{R}^n$ satisfying a suitable regularity property with respect to the dynamics of a nonlinear control system. For this, we study the structure and the properties of the external contingent Bouligand cone to K and we use a suitable selection theorem of the regulation map associated to the system.

Keywords: Proximate retract, φ-convex set, sleek set, external contingent Bouligand cone, regulation map, feedback control.

Classification: 93C15, 93B52, 32D45.

Acknowledgement: This work is supported in part by GNAFA-CNR (Italy) and by Polish KBN grant.

0. Introduction. In this paper we consider an autonomous nonlinear control system described by the differential equations

$$\dot{x} = f(x, u), \tag{1}$$

where f is a continuous function. The state variable x belongs to $\mathbb{R}^n$ and the control variable u belongs to a set $U(x) \subset \mathbb{R}^m$ depending on x. The assumptions on the multivalued map $x \mapsto U(x)$ will be specified later.

The control problem for system (1) that we consider in this paper is in the class of control problems involving sliding manifolds. Specifically, the problem that we consider here can be formulated as follows.

- Given a "sufficiently" smooth nonempty closed set $K \subset \mathbb{R}^n$, a suitable neighborhood I of K, and an initial state $x_0 \in I$, we want, by means of a control u which takes values in $U(x)$, $x \in I$, to steer in finite time and then hold the state x of system (1) in a prescribed ε-neighborhood of K.

The set K represents the required behaviour of the controlled dynamics. In the case when K can be defined as set of zeros of a continuously differentiable map $s : \mathbb{R}^n \to \mathbb{R}^m$, $m \leq n$, $s = (s_j)_{j=1}^m$ and one uses feedback control laws $u = u(x)$ which are discontinuous along the surfaces $s_j(x) = 0$, usually $u_j(x) = -\operatorname{sgn} s_j(x)$, $j = 1, 2, \ldots, m$, then this is the classical nonlinear variable structure control problem for which there is a very broad literature, see e.g. the monographs [16] and [17].

Another approach for solving the proposed problem is based on the theory of singularly perturbed ordinary differential equations and was proposed in [7]. It consists in defining a dynamical feedback controller as the solution of a differential equation containing a small parameter $\varepsilon > 0$. This equation is directly derived from the dynamics f and the function s. In fact, the states corresponding to such controls enter any prescribed neighborhood of K and remain there for all the future times. One of the advantages of this approach is that these controls are absolutely continuous and so the corresponding dynamics (1) does not require any regularization, for instance in the sense of Filippov or Krasowski, as in the case of discontinuous feedback controls. Furthermore, the so-called chattering phenomenon, which is one of the main drawbacks of discontinuous feedback controls, is eliminated.

In this paper we aim at solving the above problem in a different way, that is we will use tools and methods of multivalued analysis. In particular, we will consider the external contingent Bouligand cone $\widetilde{T}_K(x)$ to the set K, for $x \in I$, and the selection theory for multivalued maps. Indeed, multivalued analysis is of great relevance for the study of control problems: viability, invariance, and stability problems for system (1), among many others, are widely investigated by methods of multivalued analysis, see the monographs [1-4,14] and the paper [12]. One of the main difficulties here is to establish the lower semicontinuity of the external contingent Bouligand cone $\widetilde{T}_K(x)$, at any $x \in I$, in order to use a selection result from [5] for a suitably defined regulation map associated to (1). This permits us to derive the existence of a trajectory of (1) starting from x_0 and having the required property. We also would like to point out that we do not require any particular conditions on (1), like f to be affine in u, in order to ensure the lower semicontinuity of the considered regulation map. In fact, the employed selection result does not require such a condition. Therefore, the advantages of our approach are related to the generality of the control system (1) and of the control set $U(x)$.

The paper is organized as follows. In Section 1, for a large class of sets K, Proposition 1 provides the lower semicontinuity of the cone-valued map $x \mapsto \widetilde{T}_K(x)$ on I. For this, we prove a suitable representation for the cone $\widetilde{T}_K(x)$ which is known in the case when K is a convex set. In Propositions 2 and 3 under different assumptions we show the lower semicontinuity of suitable cone-valued selections $\widetilde{T}_{K,\gamma}(x)$ of $\widetilde{T}_K(x)$ which allow us to define the suitable regulation map for (1) as the intersection of $\widetilde{T}_{K,\gamma}(x)$ with the set of velocities $F(x) := f(x, U(x))$. Finally, in Section 2 we show how to solve the proposed problem. Indeed, by using both a selection result of [5] for the regulation map and a result for the approximation in graph of multivalued maps of [11], we prove in Theorem 2 the existence of a trajectory $x = x(t)$, $t \geq 0$, $x(0) = x_0 \in I$, of (1) corresponding to a control $u(t) \in U(x(t))$ which reaches in finite time a prescribed neighborhood of K and remains there for all future times. Moreover, we outline other possible controllability results which can be obtained by means of the proposed approach.

1. External contingent Bouligand cone. In this section we establish some relevant properties of the external contingent Bouligand cone for a class of nonempty, closed sets of $\mathbb{R}^n$. These properties will be employed in the next section for solving the control problem for (1) stated above.

We start with some definitions.

Definition 1. Let $K \subset \mathbb{R}^n$ be nonempty. Let $x \in \overline{K}$, the closure of K. The *contingent Bouligand cone* $T_K(x)$ is defined by

$$T_K(x) := \{v \in \mathbb{R}^n : \liminf_{\tau \to 0+} \frac{d_K(x+\tau v)}{\tau} = 0\}.$$

Here $d_K(\cdot)$ is the distance function defined as $d_K(x) := \inf_{y \in K} d(x,y)$ where $d(x,y) = |x-y|$ and $|\cdot|$ denotes the Euclidean norm in $\mathbb{R}^n$. It turns out that $T_K(x)$ is a closed cone. If the cone-valued map $x \mapsto T_K(x)$ is lower semicontinuous at any $x \in K$, the set K is said to be *sleek*. Finally, the *polar cone* of $T_K(x)$ is the normal cone to K, denoted by $N_K(x)$.

Definition 2. Let $K \subset \mathbb{R}^n$ be a nonempty closed set. Denote by $\pi_K(\cdot)$ the *metric projection* on K defined by

$$\pi_K(x) := \{y \in K : |x-y| = d_K(x)\}.$$

We assume the following condition on K.

(C) There exists an open neighborhood I of K in $\mathbb{R}^n$ such that, for any $x \in I \setminus K$, the metric projection $\pi_K(x)$ is singlevalued.

Remark 1. In [10, Theorem 4.8] for a nonempty closed set $K \subset \mathbb{R}^n$ satisfying (C) it is shown that $x \mapsto \pi_K(x)$ is continuous in I and locally Lipschitz in $I \setminus K$ together with $\operatorname{grad} d_K(x)$. Furthermore, in [10, Theorem 4.18] it is proved, in an implicit way, that condition (C) is equivalent to the following property:

(Γ) There exists a continuous function $\varphi : K \times K \to \mathbb{R}_+$ such that for all $x, y \in K$ and $v \in N_K(x)$

$$\langle v, y-x \rangle \le \varphi(x,y)|v|\,|y-x|^2.$$

Here $\langle \cdot,\cdot \rangle$ denotes the inner product in $\mathbb{R}^n$. A nonempty closed set satisfying this property is called *φ-convex set* (see [6]). In [8] it is shown that a φ-convex set of an infinite dimensional Hilbert space satisfies (C) with $x \mapsto \pi_K(x)$ continuous in I. Furthermore, a nonempty closed set K of a Banach space which satisfies (C) with $\pi_K : I \to K$ continuous (metric retraction) is called *proximate retract* (see e.g. [5,13,15]). Therefore, in a finite dimensional space, the class of φ-convex sets coincides with that of proximate retracts. It is still an open question if a proximate retract in an infinite dimensional Hilbert space is φ-convex.

Remark 2. Note that any φ-convex set K is sleek (see [9]). Moreover, convex sets as well as sets with $C^{1,1}$-boundary are φ-convex.

For any $x \in I$, consider the set $K(x)$ defined by

$$K(x) := K + d_K(x)B_1,$$

where $B_1 \subset \mathbb{R}^n$ denotes the closed unit ball in $\mathbb{R}^n$. We pose the following problem.

(P) Show that the cone-valued map defined in I by

$$x \mapsto T_{K(x)}(x)$$

is lower semicontinuous at any $x \in I$, i.e. the set $K(x)$ is sleek at any $x \in I$.

To solve problem (P) is one of the aims of this section.

Remark 3. Note that, by [3, Theorem 4.1], we have that if $x \mapsto T_{K(x)}(x)$ is lower semicontinuous at x then the cone $T_{K(x)}(x)$ is convex.

The first step to solve (P) is to prove the following result which is known in the case when K is convex [3, p. 141].

Lemma 1. *Assume that K is a nonempty closed set satisfying condition* (C). *Then*

$$T_{K(x)}(x) = T_K(\pi_K(x)) + T_{d_K(x)B_1}(x - \pi_K(x)), \quad x \in I.$$

Proof. Obviously if $x \in K$ there is nothing to prove. Therefore let $x \in I \setminus K$ and assume that $v \in T_{K(x)}(x)$, this means that there exist sequences $\tau_n \to 0+$, $v_n \to v$ such that $x_n := x + \tau_n v_n \in K(x)$, i.e. $d_{K(x)}(x_n) = 0$. By property (C) and Remark 1, there exists $l = l(x) > 0$ such that $|y - y_n| \leq l|x_n - x|$ for $n \in \mathbb{N}$ sufficiently large with $y = \pi_K(x)$ and $y_n = \pi_K(x_n)$. On the other hand for $n \in \mathbb{N}$ there exists $b_n \in B_1$ such that $x_n = y_n + |x - y|b_n$, since $|x_n - y_n| = d_K(x_n) \leq d_K(x)$. In fact, from $x_n \in K(x)$ it follows that there exist $\widehat{y}_n \in K$ and $\widehat{b}_n \in B_1$ such that $x_n = \widehat{y}_n + d_K(x)\widehat{b}_n$, thus $d_K(x_n) = d_K(\widehat{y}_n + d_K(x)\widehat{b}_n) \leq d_K(\widehat{y}_n) + d_K(x)|\widehat{b}_n| \leq d_K(x)$. Furthermore, by passing to a subsequence if necessary, we have that $b_n \to b$, where $b = \dfrac{(x - y)}{|x - y|}$. Rewrite $x_n = x + \tau_n v_n$ as follows

$$x + \tau_n v_n = y + |y_n - y|\frac{y_n - y}{|y_n - y|} + (x - y) + |x - y|\,|b_n - b|\frac{b_n - b}{|b_n - b|}, \tag{2}$$

where $b_n := \dfrac{x_n - y_n}{|x - y|}$. Observe that

$$y_n = y + |y_n - y|\frac{y_n - y}{|y_n - y|} \in K$$

with $|y_n - y| \to 0+$ and

$$|x - y|b_n = (x - y) + |x - y|\,|b_n - b|\frac{b_n - b}{|b_n - b|} \in d_K(x)B_1$$

with $|x - y|\,|b_n - b| \to 0+$. Then

$$w_0 := \lim_{n\to\infty} \frac{y_n - y}{|y_n - y|} \in T_K(y), \quad z_0 := \lim_{n\to\infty} \frac{b_n - b}{|b_n - b|} \in T_{d_K(x)B_1}(x - y).$$

On the other hand from (2) we have

$$v_n = \frac{|y_n - y|}{\tau_n}\frac{y_n - y}{|y_n - y|} + \frac{|x - y|\,|b_n - b|}{\tau_n}\frac{b_n - b}{|b_n - b|}, \tag{3}$$

with $\frac{|y_n - y|}{\tau_n} \leq l|v_n| \leq M$ for n sufficiently large. Thus, by passing to a subsequence if necessary, we have

$$\frac{|y_n - y|}{\tau_n} \to \alpha_0 \geq 0 \quad \text{and} \quad \frac{|x-y|\,|b_n - b|}{\tau_n} \to \beta_0 \geq 0$$

obtaining from (3)

$$v = \alpha_0 w_0 + \beta_0 z_0 \in T_K(y) + T_{d_K(x)B_1}(x-y).$$

Viceversa, let us prove now that from $w \in T_K(y)$ and $z \in T_{d_K(x)B_1}(x-y)$, where $y = \pi_K(x)$ and $x \in I \setminus K$, it follows that $w + z \in T_{K(x)}(x)$. Observe that, without loss of generality, we can assume that $|z| = |x-y|$. Assume $w \in T_K(y)$ then there exist sequences $\hat{\tau}_n \to 0+$ and $w_n \to w$ such that $y + \hat{\tau}_n w_n \in K$. We show now the existence of a sequence $z_n \to z$, such that $|z_n| = |x-y|$ and

$$(x-y) + \hat{\tau}_n z_n \in d_K(x)B_1.$$

In fact, for any $z \in T_{d_K(x)B_1}(x-y)$, with $|z| = |x-y|$, there exists a sequence $z_n \to z$ such that $|z_n| = |x-y|$ for any $n \in \mathbb{N}$. Therefore, for any sequence $\tau_n \to 0+$, we have that

$$(x-y) + \tau_n z_n \in d_K(x)B_1,$$

for n sufficiently large. In particular,

$$(x-y) + \hat{\tau}_n z_n \in d_K(x)B_1.$$

In conclusion,

$$y + \hat{\tau}_n w_n + (x-y) + \hat{\tau}_n z_n = x + \hat{\tau}_n(w_n + z_n) \in K(x)$$

with $\hat{\tau}_n \to 0+$, thus $w + z \in T_{K(x)}(x)$. This completes the proof. □

Remark 4. As a consequence of Lemma 1 and the closedness of $T_{K(x)}(x)$ we have that, in our case, the set

$$T_K(\pi_K(x)) + T_{d_K(x)B_1}(x - \pi_K(x)), \quad x \in I,$$

is closed.

Definition 3. Let $\gamma > 0$ and K as in Lemma 1. For any $x \in I \setminus K$, we define the closed cone $\tilde{T}_{K,\gamma}(x)$ as follows

$$\tilde{T}_{K,\gamma}(x) := \{v \in \mathbb{R}^n : \liminf_{\tau \to 0+} \frac{d_K(x+\tau v) - d_K(x)}{\tau} \leq -\gamma\}.$$

We assume that for γ sufficiently small $\tilde{T}_{K,\gamma}(x)$ is nonempty. For simplicity, we denote by $\tilde{T}_K(x)$ the external contingent Bouligand cone $\tilde{T}_{K,0}(x)$. Observe that if $x \in K$ then $\tilde{T}_{K,\gamma}(x)$ reduces to $T_K(x)$.

Lemma 2. *The equality $\tilde{T}_K(x) = T_{K(x)}(x)$ holds for any $x \in I$.*

Proof. Let $v \in T_{K(x)}(x)$, i.e. $\liminf_{\tau \to 0+} \frac{d_{K(x)}(x+\tau v)}{\tau} = 0$. On the other hand,

$$d_{K(x)}(x+\tau v) = \inf_{z \in K(x)} d(x+\tau v, z),$$

where $z = \zeta + d_K(x)b$, $\zeta \in K$, $b \in B_1$. Thus

$$d_{K(x)}(x+\tau v) \geq d_K(x+\tau v) - d_K(x),$$

since $d(x+\tau v, \zeta + d_K(x)b) \geq d(x+\tau v, \zeta) - d_K(x)$, for any $\zeta \in K$ and any $b \in B_1$. In conclusion

$$0 \geq \liminf_{\tau \to 0+} \frac{d_K(x+\tau v) - d_K(x)}{\tau},$$

which implies $v \in \widetilde{T}_K(x)$.

Viceversa, assume that $v \in \widetilde{T}_K(x)$ hence there exist sequences $\tau_n \to 0+$, $v_n \to v$ such that

$$d_K(x+\tau_n v_n) \leq d_K(x). \tag{4}$$

Let $y_n = \pi_K(x+\tau_n v_n)$, thus $|x - y_n + \tau_n v_n| = d_K(x+\tau_n v_n)$ and by (4) we get $|x - y_n + \tau_n v_n| \leq |x-y|$ for any $n \in \mathbb{N}$. This implies that $x + \tau_n v_n \in K(x)$, since $x + \tau_n v_n \in y_n + d_K(x)B_1$. Consequently, we obtain $d_{K(x)}(x+\tau_n v_n) = 0$ for any $n \in \mathbb{N}$, hence $v \in T_{K(x)}(x)$. This completes the proof. □

Proposition 1. *Assume that $K \subset \mathbb{R}^n$ is a nonempty closed set which satisfies condition* (C). *Then the cone-valued map $x \mapsto \widetilde{T}_K(x)$ is lower semicontinuous at any $x \in I$ with nonempty, closed, convex values.*

Proof. The proof easily follows from Lemmas 1 and 2, taking into account Remarks 1 and 3. □

Definition 4. Let $\varphi : \mathrm{Dom}(\varphi) \to \mathbb{R}$ be a function and $x \in \mathrm{Dom}(\varphi)$. We define the *contingent epiderivative* $D_\uparrow\varphi(x)(v)$ of φ at x in the direction v as follows

$$D_\uparrow\varphi(x)(v) := \liminf_{\substack{\tau \to 0+ \\ v' \to v}} \frac{\varphi(x+\tau v') - \varphi(x)}{\tau}.$$

As pointed out in Remark 1, if K is a nonempty closed set satisfying condition (C), then $x \mapsto d_K(x)$ is continuously differentiable in $I \setminus K$. Therefore, in this case we have that $D_\uparrow d_K(x)(v) = \langle \mathrm{grad}\, d_K(x), v\rangle$ for any $x \in I \setminus K$ and $v \in \mathbb{R}^n$. Furthermore, if K is any nonempty closed subset of $\mathbb{R}^n$ and $x_0 \in \mathbb{R}^n \setminus K$ is such that $x \mapsto d_K(x)$ is differentiable at x_0, then $\pi_K(x)$ is a singleton (see [10]). Finally, since the distance function $x \mapsto d_K(x)$, $x \in \mathbb{R}^n$, is Lipschitz of constant 1, from [3, Proposition 6.1.7] it follows that for any $x \in \mathbb{R}^n$

(i) $D_\uparrow d_K(x)(v) = \liminf_{\tau \to 0+} \frac{d_K(x+\tau v) - d_K(x)}{\tau}$, for any $v \in \mathbb{R}^n$.

(ii) $|D_\uparrow d_K(x)(v)| \leq |v|$, for any $v \in \mathbb{R}^n$.

In particular from (ii) it follows that d_K is contingently epidifferentiable, i.e. $D_\uparrow d_K(x)(0) = 0$ at any $x \in \mathbb{R}^n$, or equivalently $D_\uparrow d_K(x)(v) > -\infty$ at any $x \in \mathbb{R}^n$ whenever $v \in \mathbb{R}^n$. Finally, observe that if the epigraph of d_K : $E_p(d_K)$ is sleek with $\text{Dom}(d_K) = D$, i.e. $x \mapsto T_{E_p(d_K)}(x, d_K(x))$ is lower semicontinuous at any $x \in D$, then the cone-valued map $x \mapsto E_p(D_\uparrow d_K(x))$ is also lower semicontinuous with closed, convex values at any $x \in D$; in fact $T_{E_p(d_K)}(x, d_K(x)) = E_p(D_\uparrow d_K(x))$, see [3, Proposition 6.1.4].

The following two results are in the spirit of those of [1, Section 9.4.4].

Proposition 2. *Assume that $K \subset \mathbb{R}^n$ is a nonempty closed set which satisfies condition* (C). *Moreover, assume that for γ sufficiently small and any $x \in I\setminus K$ there exists $\overline{v} \in \mathbb{R}^n$ such that $D_\uparrow d_K(x)(\overline{v}) < -\gamma$. Then the cone-valued map $x \mapsto \widetilde{T}_{K,\gamma}(x)$, $x \in I \setminus K$, is lower semicontinuous with nonempty, closed, convex values.*

Proof. For $x \in I \setminus K$, let us define the map

$$\widehat{T}_{K,\gamma}(x) := \{v \in \mathbb{R}^n : D_\uparrow d_K(x)(v) < -\gamma\}.$$

By assumption this set is nonempty. We shall show that the graph of $\widehat{T}_{K,\gamma}$ is open, where $\text{Dom}\,\widehat{T}_{K,\gamma} = I \setminus K$. For this, let $(x, v) \in \text{Graph}\,\widehat{T}_{K,\gamma}$, for any sequence $(x_n, v_n) \to (x, v)$ we have that

$$\lim_{n\to\infty} D_\uparrow d_K(x_n)(v_n) = D_\uparrow d_K(x)(v) < -\gamma.$$

Furthermore, the set $\widehat{T}_{K,\gamma}(x)$ is convex, since $D_\uparrow d_K(x)(v) = \langle \text{grad}\, d_K(x), v\rangle$. Finally, consider the map

$$x \mapsto \widetilde{T}_K(x) \cap \widehat{T}_{K,\gamma}(x) = \widehat{T}_{K,\gamma}(x), \quad x \in I \setminus K.$$

This map is lower semicontinuous, since it is the intersection of the lower semicontinuous map $x \mapsto \widetilde{T}_K(x)$ (Proposition 1) with a map having open graph. Furthermore, it is convex valued and so $x \mapsto \widetilde{T}_{K,\gamma}(x) = \overline{\widehat{T}_{K,\gamma}(x)}$ is also lower semicontinuous. As a consequence $x \mapsto \widetilde{T}_{K,\gamma}(x)$ is lower semicontinuous at any $x \in I$. □

Finally, we can also prove the following result which provides the same conclusion of Proposition 2 under different assumptions.

Proposition 3. *Let $K \subset \mathbb{R}^n$ be a nonempty closed set. Let I be a neighborhood of K. Assume that $E_p(d_K)$ is sleek with* $\text{Dom}\, d_K = I$, *and assume that for $\gamma > 0$ sufficiently small and any $x \in I \setminus K$ there exists $\overline{v} \in \mathbb{R}^n$ such that $D_\uparrow d_K(x)(\overline{v}) < -\gamma$. Then the cone-valued map $x \mapsto \widetilde{T}_{K,\gamma}(x)$, $x \in I \setminus K$, is lower semicontinuous with nonempty, closed, convex values.*

Proof. Let $x_n \to x$ and $v \in \widetilde{T}_{K,\gamma}(x)$, hence $(v, -\gamma) \in E_p(D_\uparrow d_K(x))$. By the lower semicontinuity of the map $x \mapsto E_p(D_\uparrow d_K(x))$ there exist sequences $v_n \to v$ and $\varepsilon_n \geq 0$, $\varepsilon_n \to 0$ such that

$$(v_n, -\gamma - \varepsilon_n) \in E_p(D_\uparrow d_K(x_n)).$$

Let $\overline{v} \in \widetilde{T}_{K,\gamma}(x)$, hence $\overline{u} = -\gamma - D_\uparrow d_K(x)(\overline{v}) > 0$ and so $(\overline{v}, -\gamma - \overline{u}) \in E_p(D_\uparrow d_K(x))$. Therefore there exist sequences $\overline{v}_n \to \overline{v}$ and $\varepsilon'_n > 0$, $\varepsilon'_n \to 0$ such that

$$(\overline{v}_n, -\gamma - \overline{u} + \varepsilon'_n) \in E_p(D_\uparrow d_K(x_n)).$$

Define now $w_n = (1-\theta_n)v_n + \theta_n \overline{v}_n$, where $\theta_n = \dfrac{\varepsilon_n}{2(\varepsilon_n + \varepsilon_n')} \in [0,1]$. We have that

$$(w_n, -\gamma - \varepsilon_n/2) = (1-\theta_n)(v_n, -\gamma - \varepsilon_n) + \theta_n(\overline{v}_n, -\gamma - \overline{u} + \varepsilon_n).$$

Since $E_p(D_\uparrow d_K(x_n))$ is convex it follows that $(w_n, -\gamma - \varepsilon_n/2) \in E_p(D_\uparrow d_K(x_n))$, that is $D_\uparrow d_K(x_n)(v_n) \leq -\gamma - \varepsilon_n/2$ and so $w_n \in \widetilde{T}_{K,\gamma}(x_n)$ with $w_n \to v$. Moreover, $\widetilde{T}_{K,\gamma}(x)$ is convex. Indeed, the fact that $E_p(d_K)$ is sleek implies the upper semicontinuity of the map $(x,v) \mapsto D_\uparrow d_K(x)(v)$ for $x \in I \setminus K$ and $v \in \mathbb{R}^n$. Now, for $v_1, v_2 \in \widehat{T}_{K,\gamma}(x)$, define $E = \{\theta \in [0,1] : D_\uparrow d_K(x)(\theta v_1 + (1-\theta)v_2) < -\gamma\}$. It is nonempty by assumption, moreover it is easy to verify that E is both closed and open relatively to $[0,1]$ and so $E = [0,1]$. □

Remark 5. To the best knowledge of the authors the characterization of the nonempty, closed sets K for which $E_p(d_K)$ is sleek is an open question.

2. Control problems.

In this section we consider the autonomous nonlinear control system (1). We assume the following conditions:

(A_0) $f : \overline{I} \times \mathbb{R}^m \to \mathbb{R}^n$ is a continuous map and I is an open set of $\mathbb{R}^n$. The control parameter u belongs to the set $U(x)$ depending on $x \in I$. The set-valued map $x \mapsto U(x) \subset \mathbb{R}^m$ is upper semicontinuous with nonempty, compact, convex values.

We aim at solving the following control problem for system (1).

(CP) Given a nonempty closed set $K \subset \mathbb{R}^n$ satisfying condition (C). Given $x_0 \in I$ with I being a neighborhood of K assigned by condition (C). We want to show the existence of a trajectory $x = x(t)$, $t \geq 0$, $x(0) = x_0 \in I \setminus K$, of the control system (1) corresponding to a control $u(t) \in U(x(t))$, $t \geq 0$, which reaches a prescribed neighborhood of K in finite time and remains in it for all future times.

Observe that here problem (CP) is formulated having in mind Propositions 1 and 2 as tools to solve it. Clearly, if one wants to use Proposition 3 then K and I are the sets satisfying the conditions of Proposition 3; moreover, in this case we assume that K is sleek.

Consider the set of velocities $F(x) := f(x, U(x))$ for $x \in I$. It follows that the map $x \mapsto F(x)$, $x \in I$, is upper semicontinuous with nonempty, compact values. Furthermore we assume the following condition.

(A_1) $F(x)$ is a convex set for any $x \in I$.

Definition 5. For $x \in I$, we define the *regulation map* associated to (1) as follows

$$R_{K,\gamma}(x) := F(x) \cap \widetilde{T}_{K,\gamma}(x).$$

We assume that

(A_2) $R_{K,\gamma}(x)$ is nonempty for any $x \in I$ and for $\gamma > 0$ sufficiently small.

Proposition 4. *Under assumptions (A_0)-(A_2) and the assumptions of Proposition 2, for any $\delta > 0$ there exists a continuous map $g : I \to \mathbb{R}^n$ such that*

(i) $g(x) \in \widetilde{T}_{K,\gamma}(x)$ *for any* $x \in I$;

(ii) g *is a* δ*-approximation in graph of the map* F. *That is* Graph $(g + \delta B_1) \subseteq$ Graph $(F) + \delta B_1$.

Proof. Under our assumption the regulation map $x \mapsto R_{K,\gamma}(x)$, $x \in I$, has nonempty, compact, convex values. Thus the proof is a direct consequence of Lemma 5.1 and Remark 5.2 of [5]. □

The following result provides the relevant behaviour of the trajectories of the dynamical system $\dot{x} = g(x)$.

Theorem 1. *Under the assumptions of Proposition 4, any solution of the initial value problem*

$$\text{(5)} \qquad \begin{cases} \dot{x} = g(x) \\ x(0) = x_0, \quad x_0 \in I \setminus K, \end{cases}$$

reaches the set K *in finite time* $t_0 > 0$ *and remains in it for all* $t \geq t_0$. *Moreover* $t_0 \leq d_K(x_0)/\gamma$.

Proof. Let $x = x(t)$ be any solution of the initial value problem (5). Let $t \geq 0$ and $d(t) := d_K(x(t))$. Observe that $t \mapsto d_K(x(t))$ is an absolutely continuous function. For $\tau > 0$ we can write $x(t+\tau) = x(t) + \tau\dot{x}(t) + \tau\alpha(\tau)$, where $\alpha(\tau) \to 0$ as $\tau \to 0$. We have that

$$\dot{d}(t) = \lim_{\tau \to 0+} \frac{d_K(x(t+\tau)) - d_K(x(t))}{\tau}$$
$$= \lim_{\tau \to 0+} \frac{d_K(x(t) + \tau\dot{x}(t) + \tau\alpha(\tau)) - d_K(x(t))}{\tau} \leq -\gamma.$$

Since $\dot{x}(t) = g(x(t)) \in \widetilde{T}_{K,\gamma}(x(t))$. Thus for $t \geq 0$ we have proved that $\dot{d}(t) \leq -\gamma$, namely

$$d(t) \leq d(0) - \gamma t = d_K(x_0) - \gamma t.$$

Let $t_0 > 0$ such that $d(t_0) = 0$ and $d(t) > 0$ for all $0 \leq t < t_0$, then $t_0 \leq \dfrac{d_K(x_0)}{\gamma}$. □

As a straightforward consequence of [11, Theorem 1, p. 87] and [3, Theorem 8.2.10] we can derive the following.

Theorem 2. *For any* $\varepsilon > 0$ *there exists* $\delta > 0$ *such that if* $g : I \to \mathbb{R}^n$ *is a continuous* δ*-approximation in graph of the multivalued map* F *then, for any solution* $y = y(t)$, $t \geq 0$, *of the initial-value problem* (5), *there exists a solution* $x = x(t)$, $t \geq 0$, *of the initial-value problem*

$$\begin{cases} \dot{x} \in F(x) \\ x(0) = x_0 \end{cases}$$

such that for any time interval $[0, a]$ *we have*

$$\max_{t \in [0,a]} |x(t) - y(t)| \leq \varepsilon.$$

Moreover, there exists a measurable control $u(t) \in U(x(t))$, *such that* $\dot{x}(t) = f(x(t), u(t))$ *for a.a.* $t \geq 0$.

It is now evident that, as consequence of Theorems 1 and 2, the trajectory $x = x(t)$, $t \geq 0$, reaches in finite time the ε-neighborhood of K and remains in it for all the future times. Hence, problem (CP) is solved.

Remark 6. If $x \mapsto F(x)$ is lower semicontinuous, then the celebrated Michael selection theorem ensures the existence of a continuous selection $g(x)$ of the regulation map $x \mapsto R_{K,\gamma}(x)$, $x \in I$, and so we can take $x(t) = y(t)$, $t \geq 0$. Note that, under our assumptions, $x \mapsto F(x)$ is lower semicontinuous if $x \mapsto U(x)$ is lower semicontinuous.

Remark 7. Observe that, if we assume in Theorem 1 the conditions of Proposition 1, then by means of the above arguments we can show the existence of a trajectory $x = x(t)$, $t \geq 0$, of the control system which does not leave a prescribed ε-neighborhood of I. Finally, observe that the condition $F(x) \subseteq \tilde{T}_{K,\gamma}(x)$, or $F(x) \subseteq \tilde{T}_K(x)$, allows us to prove the same behaviour for all the trajectories of the control system.

References

[1] J. P. Aubin: *Viability Theory*, Birkhäuser, Basel 1991

[2] J. P. Aubin, A. Cellina: *Differential Inclusions*, Springer, Berlin 1984

[3] J. P. Aubin, H. Frankowska: *Set-Valued Analysis*, Birkhäuser, Basel 1990

[4] J. P. Aubin, J. Ekeland: *Applied Nonlinear Analysis*, John Wiley & Sons, New York 1981

[5] H. Ben-El-Mechaiekh, W. Kryszewski: *Equilibria of set-valued maps of nonconvex domains*, Trans. Amer. Math. Soc. **349** (1997), 4159-4179

[6] A. Canino: *On φ-convex sets and geodesics*, J. Diff. Eqns. **75** (1988), 118-157

[7] A. Cavallo, G. De Maria, P. Nistri:, *Some control problems solved via a sliding manifold approach*, Diff. Eqns. Dynam. Systems **1** (1993), 295–310

[8] G. Colombo, V. V. Goncharov: *A class of non-convex sets with locally well-posed projection*, Report SISSA Trieste No. 78/99/M, 1-6

[9] G. Colombo, V. V. Goncharov: *Variational properties of closed sets in Hilbert spaces*, Report SISSA Trieste No. 57/99/M, 1-16

[10] H. Federer: *Curvature measures*, Trans. Amer. Math. Soc. **93** (1959), 418-491

[11] A. F. Filippov: *Differential Equations with Discontinuous Righthand Sides*, Kluwer Academic Publishers, New York 1988

[12] L. Górniewicz, P. Nistri: *An invariance problem for control systems with deterministic uncertainty*, in: Topology in Nonlinear Analysis **35** [Ed.: K. Gęba, L. Górniewicz], Banach Center Publications (1996), 193-205

[13] L. Górniewicz, P. Nistri, V. Obukhovskii: *Differential inclusions on proximate retracts of Hilbert spaces*, Int. J. Nonlin. Diff. Eqns. TMA **3** (1997), 13-26

[14] H. KISIELEWICZ: *Differential Inclusions and Optimal Control*, PNN-Polish Scientific Publishers, Warszawa; Kluwer Academic Publishers, New York 1991

[15] S. PLASKACZ: *Periodic solutions of differential inclusions on compact subsets of* $\mathbb{R}^n$, J. Math. Anal. Appl. **148** (1990), 202-212

[16] V. I. UTKIN: *Sliding Modes and Their Applications in Variable Structure Systems* [in Russian], MIR, Moscow 1978

[17] V. I. UTKIN: *Sliding Modes in Control Optimization*, Springer, Berlin 1997

LECH GÓRNIEWICZ, Faculty of Mathematics and Informatics, Nicholas Copernicus University, Chopina 12/18, PL-87-100 Toruń, Poland; gorn@mat.uni.torun.pl

PAOLO NISTRI, Dipartimento di Ingegneria dell'Informazione, Facoltà di Ingegneria, Università di Siena, Via Roma 56, I-53100 Siena, Italy; pnistri@dii.unisi.it

Progress in Nonlinear Differential Equations
and Their Applications, Vol. 40

Two Mechanical Systems and Equivariant Degree

JORGE IZE

A Alfonso por sus 60 años y por los 26 de amistad

Summary: In this paper, the equivariant degree for orthogonal maps is applied to two spring-pendulum systems. For large amplitudes of a vertical oscillation of the spring, one obtains bifurcation of oscillations where the pendulum has a movement in space with a periodicity combining space and time.

Keywords: Equivariant degree, orthogonal maps, spring-pendulum.

Classification: 58B05, 34C25, 47H15, 54F45, 55Q91, 58E09.

Acknowledgement: This paper was partially written while the author was visiting the University of Rome, Tor Vergata. Support from the CNR and Conacyt G 25427E is acknowledged. The author wishes to thank his colleagues C. Garza for pointing out to him some of the references, and A. Olvera for the numerical work supporting the results.

In this paper, I shall consider two spring-pendulum systems and show how the equivariant degree for orthogonal maps, defined in [8], may be used to show bifurcation from a $\mathbb{S}^1$-orbit to a T^2-orbit.

The first system consists of a spring, moving vertically, with a rigid pendulum suspended at the end. If one pulls the pendulum downwards slightly, one obtains a stable harmonic oscillation. For a stronger pull, this oscillation looses its stability and one has an oscillation of the pendulum in a plane. For a still stronger pull, one gets an oscillation with a triangular pattern in space. Stronger pulls seem to lead to more complicated patterns.

The second apparatus is a pendulum with an elastic shaft. The same succession of patterns is observed and follow the behavior predicted by the application of the equivariant degree.

Most of the papers on these systems concentrate on the bifurcation from the steady position, near a resonance point, and not from a large amplitude oscillation of the spring. Furthermore, all treat the planar movement only.

In Section 1, the equations will be derived. Section 2 gives the study of the linearization of the problem near a vertical oscillation, giving rise to a Mathieu equation in the first system and a singular Hill equation in the second.

In Section 3, I shall recall the relevant results of the equivariant orthogonal degree which will be applied, in Section 4, to the two systems.

1. The two systems. For the first system, the spring has length l_0 at rest, a constant K, is suspended at the origin, with a mass M at the end, i.e. at the point $(0,0,l)$, orienting the $z-$axis downwards. From this mass, one attachs a rigid pendulum, of length r_0, with a mass m at its end, of coordinates (x,y,z). The kinetic and potential energies are

$$\begin{aligned} T &= \frac{1}{2}M\dot{l}^2 + \frac{1}{2}m(\dot{x}^2+\dot{y}^2+\dot{z}^2), \\ K &= \frac{1}{2}K(l-l_0)^2 - Mgl - mgz \end{aligned}$$

with the relation

$$r_0^2 = x^2+y^2+(z-l)^2.$$

Instead of using a Lagrange multiplier for this holonomic relation, we shall write

$$l = z - r = z - (r_0^2 - x^2 - y^2)^{1/2},$$

assuming thus that $0 \le l \le z$, i.e. that the pendulum does not reach the horizontal position. The Euler equations, $d((T-K)_{\dot{x}_j})/dt = (T-K)_{x_j}$, will give

$$\mathcal{M}\begin{pmatrix} \ddot{x} \\ \ddot{y} \\ \ddot{Z} \end{pmatrix} + A\begin{pmatrix} x/r \\ y/r \\ 0 \end{pmatrix} + (B+C)\begin{pmatrix} x/r \\ y/r \\ 1 \end{pmatrix} = 0$$

where $Z = z - r_0 - l_0 - (m+M)g/K$, so that $Z=0$ corresponds to the static solution,

$$\mathcal{M} = \begin{pmatrix} m + Mx^2/r^2 & Mxy/r^2 & Mx/r \\ Mxy/r^2 & m+My^2/r^2 & My/r \\ Mx/r & My/r & m+M \end{pmatrix}$$

$$A = mg, \qquad B = K(Z + r_0 - r),$$

and

$$C = M[(\dot{x}^2+\dot{y}^2)/r + (x\dot{x}+y\dot{y})^2/r^3] = M[(r_0-r)^{\cdot\cdot} - (x\ddot{x}+y\ddot{y})/r].$$

Note that $\det\mathcal{M} = m^2(m+M+M(x^2+y^2)/r^2)$ and that $x=y=0, Z=a\cos(\nu_0 t+\varphi)$ is solution, with $\nu_0 = (K/(m+M))^{1/2}$ is the frequency of the oscillation of the spring with the mass $m+M$. Inverting $\mathcal{M}$ one obtains a similar equation but with A replaced by $(m+M)m/\det\mathcal{M}, B+C$ replaced by $m^2(B+C)/\det\mathcal{M}$.

We shall be looking for periodic solutions of the system, of unknown frequency ν, close to ν_0. Thus, we shall make the change of time scale $\tau = \nu t$, with $x' = dx/d\tau$ and look for $2\pi-$periodic solutions of

$$(I) \qquad f(X,\nu) \equiv \nu^2\mathcal{M}\begin{pmatrix} x'' \\ y'' \\ Z'' \end{pmatrix} + A\begin{pmatrix} x/r \\ y/r \\ 0 \end{pmatrix} + (B+\nu^2 C)\begin{pmatrix} x/r \\ y/r \\ 1 \end{pmatrix} = 0$$

where in C one replaces $\dot{x}$ by x'.

Note again, that $x=y=0, Z(\tau) = a\cos(\nu\tau/\nu_0+\varphi)$ is solution, hence $\nu = \nu_0/n, x = y = 0, Z_n(\tau) = a\cos(n\tau+\varphi)$ gives a solution in the space of 2π-periodic functions.

The mapping $f(X,\nu)$ is continuous from $C^2_{2\pi}$ into $C^0_{2\pi}$, spaces of C^2, respectively C^0, 2π-periodic functions.

Lemma 1.1. (a) *The mapping $f(X,\nu)$ is $\mathbb{S}^1\times\mathbb{S}^1$-orthogonal, that is*

1. $f(T_{\varphi,\psi}X,\nu)=T_{\varphi,\psi}f(X,\nu)$,
2. $\int_0^{2\pi} f(X,\nu)\cdot X'dt=0, \quad \int_0^{2\pi} f(X,\nu)\cdot AXdt=0,$

where $T_{\varphi,\psi}X(\tau)=\mathcal{R}_\psi X(\tau+\varphi), \mathcal{R}_\psi=\begin{pmatrix}\cos\psi & -\sin\psi & 0\\ \sin\psi & \cos\psi & 0\\ 0 & 0 & 1\end{pmatrix}$

and $AX=A(x,y,z)^T=(-y,x,0)^T$.

(b) $f(X,\nu)$ *is also reversible, in the sense that* $f(R_\varepsilon X,\nu)=R_\varepsilon f(X,\nu)$, *where*

$$R_\varepsilon(x(\tau),y(\tau),Z(\tau))=(\varepsilon x(-\tau),\varepsilon y(-\tau),Z(-\tau)), \qquad \varepsilon=\pm1.$$

Proof. The equivariance with respect to the time shift, $X(\tau+\varphi)$, follows from the fact that the system is autonomous. Furthermore, since, in C, one has $X'\cdot X'$ and $X\cdot X'$, this term is invariant under the rotation $\mathcal{R}_\psi$. Also, since $\mathcal{M}=mI+MD$, it is easy to check that $D(\mathcal{R}_\psi X)=\mathcal{R}_\psi D\mathcal{R}_\psi^{-1}X$, hence $D(\mathcal{R}_\psi X)X''=\mathcal{R}_\psi DX''$.

For the orthogonality, one has that $f(X,\nu)\cdot X'=\frac{d}{d\tau}(T+K)=0$, by conservation of energy. One may also verify this orthogonality directly, noticing the form of C and that $(xx'+yy')/r=(r_0-r)'$.

On the other hand, $f(X,\nu)\cdot AX=m(xy''-x''y)$ which integrates to 0. The reversibility is clear. □

Remark 1.1. If (x,y,Z) is solution, this also the case of $(x,-y,Z),(-x,y,Z)$ and (y,x,Z) besides $\mathcal{R}_\psi X$.

The second system consists of an elastic spring, of length at rest r_0 and suspended at the origin (the z-axis is again oriented downwards), with a mass m at its end. One has

$$\begin{aligned} T &= \frac{1}{2}m(\dot{x}^2+\dot{y}^2+\dot{z}^2)\\ K &= \frac{1}{2}K(r-r_0)^2-mgz, \end{aligned}$$

where $r=(x^2+y^2+z^2)^{1/2}$. The Euler equations are then

$$m\ddot{X}+K(r-r_0)X/r-mg(0,0,1)^T=0,$$

with the constant solution $x=y=0, z=r_0+mg/K$.

Let $Z=z-(r_0+mg/K)$, then

$$m\begin{pmatrix}x\\ y\\ Z\end{pmatrix}^{\cdot\cdot}+K\begin{pmatrix}x\\ y\\ z\end{pmatrix}+K(r_0/r)\begin{pmatrix}-x\\ -y\\ (x^2+y^2)/(Z+r+r_0+mg/K)\end{pmatrix}=0.$$

Again $x = y = 0, Z = a\cos(\nu_0 t + \varphi)$, with $\nu_0 = (K/m)^{1/2}$, is solution.
Let us then make the change of variable $\tau = \nu t$, with $x' = dx/d\tau$, and we shall look at 2π-periodic solutions of

$$g(X,\nu) = m^2\nu^2 \begin{pmatrix} x \\ y \\ Z \end{pmatrix}'' + K \begin{pmatrix} x \\ y \\ Z \end{pmatrix} + K(r_0/r) \begin{pmatrix} -x \\ -y \\ (x^2+y^2)/(Z+r+r_0+mg/K) \end{pmatrix} = 0$$

where $r = (x^2 + y^2 + (Z + r_0 + mg/K)^2)^{1/2}$, which has the solution $x = y = 0, \nu = \nu_0/n, Z_n = a\cos(n\tau + \varphi)$.

Lemma 1.2. *The mapping $g(X,\nu)$ is $\mathbb{S}^1 \times \mathbb{S}^1$-orthogonal and reversible.*

Proof. The equivariance with respect to the time shift and the rotation around the z-axis are easy to prove. The orthogonality to X' follows from the conservation of energy and that to AX is immediate. The reversibility follows as in the previous system. □

Remark 1.2. One has also the same relations as in Remark 1.1.

2. Linearization. Let us linearize $f(X,\nu)$ near $x = y = 0, \nu = \nu_0/n, Z_n = a\cos(n\tau + \varphi)$, with $\nu = \nu_0/n + \mu, Z = Z_n + z$:

$$\begin{aligned} (\nu_0/n)^2 x'' &+ (g + KZ_n/(m+M))x/r_0 = 0, \\ (\nu_0/n)^2 y'' &+ (g + KZ_n/(m+M))y/r_0 = 0, \\ z'' + n^2 z &- 2\mu\nu_0^{-1} n^3 Z_n = 0, \end{aligned}$$

or else $L_n X = 0$, where

$$L_n X = X'' + \begin{pmatrix} n^2(\alpha + 2\beta\cos(n\tau+\varphi))x \\ n^2(\alpha + 2\beta\cos(n\tau+\varphi))y \\ n^2 z + \mu\gamma\cos(n\tau+\varphi) \end{pmatrix}$$

with $\alpha = \nu_0^{-2} g/r_0 = g(m+M)/r_0K, \beta = \nu_0^{-2} Ka/2r_0(m+M) = a/2r_0, \gamma = -2\nu_0^{-1} n^3 a$, that is, the two first equations are Mathieu's equation and the third is a resonant harmonic oscillation. The amplitude a of the vertical oscillation plays the role of an extra parameter.
Since L_n is conjugate to the operator with $\varphi = 0$ (due to the equivariance: [6]), it is enough to study the latter case.

Lemma 2.1. *One has $\dim\ker L_n = 2,4,6$, with eigenvectors $\mu = 0, z = \cos n\tau$ or $\sin n\tau$, x and y are Mathieu functions corresponding to analytic curves $\alpha_{k/n}(\beta), \tilde{\alpha}_{k/n}(\beta)$ passing through $\alpha_{k/n}(0) = \tilde{\alpha}_{k/n}(0) = (k/n)^2$. Solutions on $\alpha_{k/n}(\beta)$ are even in τ and those on $\tilde{\alpha}_{k/n}(\beta)$ are odd. Furthermore, $\alpha_{k/n}(-\beta) = \tilde{\alpha}_{k/n}(\beta), \alpha_{k/n}(\beta)$ coincides with $\tilde{\alpha}_{k/n}(\beta)$ if $k/n \neq k_1/2, k_1$ an integer, while $\alpha_{k_1/2}(\beta)$ and $\tilde{\alpha}_{k_1/2}(\beta)$ intersect only at $\beta = 0$.*

Also, $\alpha_{k/n}(\beta)$ and $\tilde{\alpha}_{k/n}(\beta)$ tend to $-\infty$ when $|\beta|$ goes to ∞; $\alpha_0(\beta) = \alpha_0(-\beta) < 0$ and $\tilde{\alpha}_0(\beta)$ does not exist. Moreover, $\alpha_{k/n}(\beta)$ foliates the region between the curves bifurcating from two consecutive half-integers (i.e. these curves do not intersect and are dense). In that region any solution of $L_2X = 0$ (not necessarily periodic) is bounded, while in the complementing region (the Arnold's tongues), the solutions are unbounded, as well as the other solution on the transition curves $\alpha_{k_1/2}(\beta)$ and $\tilde{\alpha}_{k_1/2}(\beta)$.

If $x_n(\tau)$ is a 2π-periodic solution for $\alpha_{k/n}(\beta)$ and $k/n = k_1/n_1$, with k_1 and n_1 relatively prime, then $x_n(\tau) = x_{n_1}(n\tau/n_1)$, in particular, $x_n(\tau)$ is $(2\pi n_1/n)$-periodic. The solutions $x_n(\tau)$ on $\alpha_{k/n}(\beta)$ have $2k$ simple internal zeros and on $\tilde{\alpha}_{k/n}(\beta)$ one has $2k-1$ internal zeros.

Proof. On the space of 2π-periodic functions one needs that the last equation has bounded solutions, hence it cannot be resonant, thus $\mu = 0$ and z is a combination of $\cos n\tau$ and $\sin n\tau$.

For the Mathieu equation, if $x(\tau)$ is a 2π-periodic solution, then so are $x(-\tau), x(\tau) + x(-\tau)$ and $x(\tau) - x(-\tau)$. Hence one may assume that $x(\tau)$ has a definite parity. This can be proved also from the reversibility. Furthermore, for a given (α, β), one has at most one even and one odd solution, from the uniqueness of the initial value problem.

Similarly, $x(\tau + \pi/n)$ is a solution for $(\alpha, -\beta)$ and $x(\tau + 2\pi/n)$ is a solution for (α, β). In particular, $x(\tau + \pi/n) \pm x(-\tau + \pi/n)$ are two linearly independent solutions for $(\alpha, -\beta)$, since of different parity, unless one of them is 0. If, for example, $x(\tau)$ is even and $x(\tau+\pi/n) = x(-\tau+\pi/n)$, then $x(\tau)$ is $2\pi/n$-antiperiodic and hence $4\pi/n$-periodic (this implies that n is even). If the even part of $x(\tau + \pi/n)$ is 0, then the results are interchanged. Thus, if $x(\tau)$ is not $4\pi/n$-periodic, then there are two periodic solutions for $(\alpha, -\beta)$ and also for (α, β). While, if $x(\tau)$ is $4\pi/n$-periodic, then the change of variable $y(\tau) = x(2\tau/n)$ will reduce L_n to the classical Mathieu equation, i.e. with $n = 2$. A complete study of that case, via continuous fractions, Fourier series and numerical results, can be found in the thesis [5], proving the lemma.

Here, we shall recall the proof of the analyticity of $\alpha_{k/n}(\beta)$: Assume that for (α_0, β_0) one has a solution $x_0(\tau)$ of a definite parity. Consider then L_n in the space of periodic functions with that parity.

Take $x = ax_0 + x_1$, with x_1 being L^2-orthogonal to x_0. Then, the Liapunov-Schmidt reduction implies that $L_nx = 0$ is equivalent to a unique analytic solution $x_1(a, \alpha, \beta) = ax_1(1, \alpha, \beta)$, with $x_1(1, \alpha_0, \beta_0) = 0$ and a solution to the bifurcation equation

$$\alpha - \alpha_0 + 2(\beta - \beta_0) \int_0^{2\pi} \cos n\tau (x_0^2 + x_0x_1) d\tau = 0,$$

after normalizing x_0 to have norm 1 in L^2. The implicit function theorem implies that this equation has a unique analytic local solution $\alpha(\beta)$, with

$$\alpha'(\beta_0) = -2 \int_0^{2\pi} \cos\, n\tau x_0^2 d\tau = \int_0^{2\pi} (\alpha_0 x_0^2 - x_0'^2/n^2)/\beta_0 d\tau.$$

This implies the monotonicity of $\alpha(\beta)$ for $\alpha_0 < 0$. The conservation of the number of zeros on $\alpha_{k/n}(\beta)$ is classical and is that of $\cos k\tau$ or $\sin k\tau$, solutions for $\alpha_{k/n}(0)$. It is then easy to see that $\alpha'_{k/n}(0) = 0$, except if $k/n = 1/2$, with $\alpha'_{1/2}(0) = -1 = -\tilde{\alpha}'_{1/2}(0)$.
□

Remark 2.1. One may prove that, for $k/n \neq 1/2$, one has

$$\alpha''_{k/n}(0) = \tilde{\alpha}''_{k/n}(0) = \frac{4}{4(k/n)^2 - 1}.$$

In fact, since x_0 is analytic in β, one has that

$$\alpha''(\beta) = -4 \int_0^{2\pi} \cos n\tau x_0 x_{0\beta} d\tau$$

and, by differentiating the Mathieu equation, that at $\beta = 0$, one has that $x_{0\beta}$ is a solution (of the same parity of x_0) of

$$y'' + k^2 y = -2n^2 \cos n\tau x_0,$$

where one has used that $\alpha(0) = (k/n)^2$ and $\alpha'(0) = 0$. Then, for $x_0 = \cos k\tau/\sqrt{\pi}$, one has that $x_{0\beta}$ is orthogonal to x_0, and

$$x_{0\beta} = \frac{n}{\sqrt{\pi}} \left[\frac{\cos(n+k)\tau}{n+2k} + \frac{\cos(n-k)\tau}{n-2k} \right]$$

and the result follows.

The linearization of $g(X, \nu)$ near $x = y = 0, \nu = \nu_0/n, Z_n = a\cos(n\tau + \varphi)$, with $\nu = \nu_0/n + \mu, Z = Z_n + z$, leads to

$$\begin{aligned}
x'' &+ n^2 x(\alpha + \beta\cos(n\tau + \varphi))/(1 + \beta\cos(n\tau + \varphi)) = 0, \\
y'' &+ n^2 y(\alpha + \beta\cos(n\tau + \varphi))/(1 + \beta\cos(n\tau + \varphi)) = 0, \\
z'' &+ n^2 z - 2nm\nu_0 Z_n \mu = 0,
\end{aligned}$$

where $\alpha = (mg/K)/(r_0 + mg/K)$ and $\beta = a/(r_0 + mg/K)$.
In this linearization, we have taken $r = Z_n + r_0 + mg/K$, i.e. that $|\beta| \leq 1$. Note that $|\beta| = 1$, corresponds to $a = r_0 + mg/K$, i.e. to $z_n = 0$, for $n\tau + \varphi = \pi$, i.e. to a spring totally collapsed.
Note also that $0 < \alpha < 1$. Thus, we shall work in the rectangle $0 \leq \alpha \leq 1, |\beta| \leq 1$. The first two equations are singular Hill's equations, while the third will have a non-resonant solution only for $\mu = 0$ and $z = \varepsilon\cos(n\tau + \psi)$. Note that, due to the equivariance, one has only to study the linearization, M_n, for $\varphi = 0$.

Lemma 2.2. *One has dim ker $M_n = 2, 4$ or 6, on analytic curves $\alpha_{k/n}(\beta), \tilde{\alpha}_{k/n}(\beta)$ passing through $\alpha_{k/n}(0) = \tilde{\alpha}_{k/n}(0) = (k/n)^2$. Solutions on $\alpha_{k/n}(\beta)$ are even in τ and those on $\tilde{\alpha}_{k/n}(\beta)$ are odd. Furthermore, $\alpha_{k/n}(-\beta) = \tilde{\alpha}_{k/n}(\beta), \alpha_{k/n}(\beta)$ coincides with $\tilde{\alpha}_{k/n}(\beta)$ if $k/n \neq 1/2$, while $\alpha_{1/2}(\beta)$ and $\tilde{\alpha}_{1/2}(\beta)$ intersect only at $\beta = 0, \alpha = 1/4$. The region between them is a region of instability, while the region between $\alpha_{1/2}(\beta)$ and $\tilde{\alpha}_{1/2}(-\beta)$ is foliated by the curves $\alpha_{k/n}(\beta)$.*
Numerically $\alpha_{k/n}(\beta)$ goes to 0 when $|\beta|$ goes to 1, if $k/n < 1/2$ and to 1, if $k/n > 1/2$. $\alpha_0(\beta) \equiv 0$, with unique solution $1 + \beta\cos n\tau$, while $\alpha_1(\beta) = 1 = \tilde{\alpha}_1(\beta)$ with solutions $\cos n\tau$ and sin *$n\tau$. The solutions on $\alpha_{1/2}(\beta)$ and $\tilde{\alpha}_{1/2}(\beta)$ are $(2\pi/n)$-antiperiodic and*

if $k/n = k_1/n_1$, with k_1 and n_1 relatively prime, then $x_n(\tau) = x_{n_1}(n\tau/n_1)$, with $x_n(\tau)$ with period $(2\pi n_1/n)$. Finally, $x_n(\tau)$ on $\alpha_{k/n}(\beta)$ has $2k$ simple internal zeros and on $\tilde{\alpha}_{k/n}(\tau)$ one has $2k-1$ internal zeros.

Proof. As in Lemma 2.1, if $x_n(\tau)$ is not $(4\pi/n)$-periodic then M_n has a double kernel at (α,β) and if $x_n(\tau)$ has this period, then one may reduce the study to the case $n = 2$.

The Liapunov-Schmidt argument yields the bifurcation equation

$$h(\alpha,\beta) = \int_0^{2\pi} \frac{(\alpha-\alpha_0+(\beta-\beta_0+\alpha\beta_0-\alpha_0\beta)\cos n\tau)(x_0^2+x_0x_1)}{(1+\beta\cos n\tau)(1+\beta_0\cos n\tau)}d\tau = 0,$$

where x_0 and x_1 have the same meaning as before. Then,

$$h_\alpha(\alpha_0,\beta_0) = \int_0^{2\pi} x_0^2(1+\beta_0\cos n\tau)^{-1}d\tau > 0$$

and

$$h_\beta(\alpha_0,\beta_0) = (1-\alpha_0)\int_0^{2\pi} \cos n\tau x_0^2(1+\beta_0\cos n\tau)^{-2}d\tau.$$

Thus, one has analytic curves $\alpha(\beta)$, for $|\beta|<1$, which must cross the α-axis at some $\alpha(0) = (k/n)^2$, for $0 \le k \le n$.

In particular for $n = 1$, there are only two curves, $\alpha_0(\beta) = 0$ with only even solutions and $\alpha_1(\beta) = 1$ with both parities, with explicit solutions given in the lemma. For $n = 2$, one has an additional curve, for each parity, going through $1/4$ for $\beta = 0$.

This implies that $\alpha_{k/n}(\beta) = \tilde{\alpha}_{k/n}(\beta)$, for $k/n \neq 1/2$, and are symmetric with respect to the α-axis. The fact that $\alpha_{1/2}(\beta)$ and $\tilde{\alpha}_{1/2}(\beta)$ don't cross each other, except at $\beta = 0$, requires a lengthy argument using Fourier series.

Since, for $n = 2$ and $\beta = 0$, one has the solutions $\cos\tau$ and $\sin\tau$, then $\alpha'_{1/2}(0) = -3/8$ and $\tilde{\alpha}'_{1/2}(0) = 3/8$.

The rest of the lemma uses Floquet theory and scaling. □

Remark 2.2. By normalizing x_0 in such a way that $h_\alpha(\alpha_0,\beta_0) = 1$, then $\alpha'(\beta) = -h_\beta$. Hence, for $k/n \neq 1/2$ for which $\alpha'(0) = 0$, one has that

$$\alpha''(0) = 2((k/n)^2-1)\int_0^{2\pi} x_0\cos n\tau(x_{0\beta} - x_0\cos n\tau)d\tau,$$

where $x_{0\beta}$ is a solution (by differentiating the Hill equation) of

$$y'' + k^2y = (k^2-n^2)\cos n\tau x_0.$$

With $x_0 = \cos k\tau/\sqrt{\pi}$, one has that

$$x_{0\beta} = (n^2-k^2)/(2n\sqrt{\pi})(\cos(n+k)\tau/(n+2k) + \cos(n-k)\tau/(n-2k)).$$

From this, it is easy to prove that $\alpha''_{k/n}(0) = 3(k/n)^2(1-(k/n)^2))(4(k/n)^2-1)^{-1}$.

3. Orthogonal degree. Let $\Gamma = T^n \times A$ be an abelian group, with $|A| < \infty$ and T^n generated by $(\varphi_1,\dots,\varphi_n)$. Take V a finite dimensional representation of Γ and define $A_jx = \partial\gamma x/\partial\varphi_j|_{\gamma=Id}$, the infinitesimal generator of that part of the torus.

Definition 3.1. A continuous map $f : V \to V$ is Γ-*orthogonal* if

1. $f(\gamma x) = \gamma f(x)$,
2. $f(x) \cdot A_j x = 0$.

If $f(x)$ is non-zero on the boundary $\partial\Omega$ of an open and bounded subset Ω of V, we have defined, in [8], an orthogonal degree

$$\deg_\perp(f;\Omega) = \Sigma d_H[F_H]$$

where d_H are integers, one for each isotropy subgroup of Γ, and $[F_H]$ is an explicit generator.

This degree has all the properties of a degree theory, i.e.:

1. If $d_H \neq 0, f(x) = 0$ has a solution in V^H.
2. If f is deformable to g on $\partial\Omega$, via a Γ-orthogonal map, then $\deg_\perp(f;\Omega) = \deg_\perp(g;\Omega)$.
3. If $\Omega = \Omega_1 \cup \Omega_2$, with $\Omega_1 \cap \Omega_2 = \emptyset$ and each of the following degrees is defined, then
$$\deg_\perp(f;\Omega) = \deg_\perp(f;\Omega_1) + \deg_\perp(f;\Omega_2).$$
4. If $f(x) \neq 0$ for x in $\Omega\backslash\bar{\Omega}_1$, then
$$\deg_\perp(f;\Omega) = \deg_\perp(f;\Omega_1).$$
5. If Ω is a ball, $B(0,R)$, and $\deg_\perp(f;\Omega) = 0$, then $f|\partial\Omega$ has a Γ-orthogonal extension to B which is never 0.

One extends this degree to an infinite dimensional setting provided the maps are of the form Id - compact.

In order to compute the orthogonal degree, assume that $f(x)$ is a $C^1 - \Gamma$-orthogonal map and that $f(x_0) = 0$, with Γx_0 an isolated k-dimensional orbit with isotropy H_0. Thus, Γ/H_0 has dimension k and $A_j x_0$ are tangent to the orbit, with $A_1 x_0, ..., A_k x_0$ linearly independent. Assume that $ker Df(x_0)$ is k-dimensional and generated by the above vectors. (See [8] for the proofs of all these facts). Then, the orthogonal index of f at x_0, denoted $i_\perp(f;x_0)$, is well defined, i.e. the orthogonal degree of f with respect to any small tubular neighborhood of Γx_0.

Now $Df(x_0)$ is H_0-orthogonal and has a block diagonal structure on H_0-irreducible representations. There is a unique isotropy subgroup $\underline{H} < H_0$, minimal with the property that $|H_0/\underline{H}| < \infty$. Then, if $\underline{H} < H < H_0$, one has that $Df^H(x_0) = diag(Df^{H_0}(x_0), D_\perp f^H(x_0))$, where the second matrix is invertible. Furthermore, if $H < \underline{H}$, then $Df^H(x_0) = diag(Df^{\underline{H}}(x_0), D_\perp f^H(x_0))$, where the second matrix is invertible and complex self-adjoint and has also a diagonal structure $diag(B_1, ..., B_m)$, where B_j corresponds to all equivalent irreducible representations of H_0 on $(V^{\underline{H}})^\perp$, with isotropy K_j, hence $H_0/K_j \cong \mathbb{S}^1$. The self-adjoint matrix B_j has a complex Morse number n_j.

In these conditions, $i_\perp(f;x_0) = i_1 \times i_2$, where

$$i_1 = d_{H_0}[F_{H_0}] + \sum d_{H_i}[F_{H_i}] + \sum d_{\tilde{H}_i}[F_{\tilde{H}_i}],$$

with $d_{H_0} = (-1)^{n_{H_0}}$, where n_{H_0} is the number of negative eigenvalues of $Df^{H_0}(x_0)$, $H_i < H_0$ is such that $H_0/H_i \cong \mathbb{Z}_2$ and $d_{H_i} = d_{H_0}$ (*Sign det* $D_\perp f^{H_i}(x_0) - 1)/2$, $\tilde{H}_i < H_0$ is such that $H_0/\tilde{H}_i \cong \mathbb{Z}_2 \times ... \times \mathbb{Z}_2$ and $d_{\tilde{H}_i}$ is determined by d_{H_j} of the preceding sum. The second element of the product is

$$i_2 = [F_\Gamma] - \sum n_j[F_{K_j}] - \sum_{s=2}^{n-k}(\Pi n_j)[F_{\cap K_j}],$$

where in the second sum one has $n_{i_1}, ... n_{i_s}$ with $dim\, H_0/(K_{i_1} \cap ... \cap K_{i_s}) = s$. Finally, $[F_H] \times [F_K] = [F_{H\cap K}]$. (See Theorem 4 of [8]).

4. Application to the mechanical systems. As we have seen, both systems are $\mathbb{S}^1 \times \mathbb{S}^1$-orthogonal. Fix n and assume that the vertical line, corresponding to a fixed α, crosses the line $\alpha_{k/n}(\beta)$ at (α_0, β_0). (The points of tangency are finite). On that line the nonlinear system have the solution $\nu = \nu_0/n, x = y = 0, Z_n = a\cos(n\tau + \varphi)$, where a is proportional to β, that is a family, parametrized by β, of one dimensional orbits. Let Ω be the following tubular neighborhood of $(\nu/n, 0, 0, a_0\cos(n\tau + \varphi))$, where a_0 corresponds to β_0 :

$$\Omega = \{(\nu, x, y, a\cos(n\tau + \varphi) + \tilde{Z}) : \tilde{Z}\ L^2\text{-orthogonal to } \cos n\tau \text{ and } \sin n\tau,$$
$$|\nu - \nu_0/n| < 2\varepsilon, ||x||^2 + ||y||^2 < 2\varepsilon^2, |a - a_0| < 2\rho, ||\tilde{Z}|| < 2\varepsilon\},$$

and consider, from $\mathbb{R} \times C^2$ into $\mathbb{R} \times C^0$, the following pair

$$f_\varepsilon(\nu, X) \equiv (d^2(\nu, X) - \varepsilon^2, f(\nu, X))$$

where $d^2(\nu, X) = |\nu - \nu_0/n|^2 + ||x||^2 + ||y||^2 + ||\tilde{Z}||^2$, with $||x||$ being the L^2-norm of $x(\tau)$, is the distance to the plane $\nu = \nu_0/n, x = y = 0, Z = a\cos(n\tau + \varphi)$, which will be called the trivial solutions.

Choose ε so small that the only ν in Ω of the form ν_0/m is for $m = n$. In particular, any zero in Ω of $f_\varepsilon(\nu, X)$, for $\varepsilon \neq 0$, must have $||x||^2 + ||y||^2 \neq 0$, from the form of the equations. Furthermore, for $|a - a_0| = 2\rho$ small enough, the (x, y)-part of the linearization is invertible and the only solution, for ε small enough, will be on the plane, i.e. with $d = 0$. Thus, $f_\varepsilon(\nu, X)$ is non-zero on $\partial\Omega$ and its orthogonal degree is well defined (the extension to an infinite dimensional context is explained in [8] and requires to take in fact $-f(\nu, X)$ so that the linearizations are elliptic).

Choosing ρ and ε appropriately, one may assume that whenever $f(\nu, X) = 0$ in Ω, and $|a - a_0| > \rho/2$, then $d(\nu, X) = 0$, then one may perform the orthogonal deformation

$$(\lambda(d^2 - \varepsilon^2) + (1 - \lambda)(\rho^2 - (a - a_0)^2), f(\nu, X)).$$

Then, $\deg_\perp(f;\Omega) = i_-(f) + i_+(f)$, where $i_\pm(f)$ is the orthogonal index of $(\rho^2 - (a - a_0)^2, f)$ at $a = a_0 \pm \rho, \nu = \nu_0/n, x = y = 0, Z_n = a\cos(n\tau + \varphi)$, with isotropy subgroup $H_0 = \mathbb{Z}_n \times \mathbb{S}^1$.

For $\varphi = 0$, the linearization of the pair, at $Z_n = (a_0 \pm \rho)\cos n\tau$, will be $(\mp 2\rho\varepsilon_1, L_n X)$, where $Z = Z_n + z$ and $z = \varepsilon_1 \cos n\tau + \varepsilon_2 \sin n\tau + \tilde{Z}$. Hence, the kernel of the linearization is generated by $\mu = 0, x = y = 0, z = \varepsilon_2 \sin n\tau$, i.e. by X_0'. Thus, both indices may be computed from the results given in the preceding section.

Since $H_0 = \mathbb{Z}_n \times \mathbb{S}^1$, one has that $\underline{H} = \{e\} \times \mathbb{S}^1$ with $V^{\underline{H}}$ corresponding to $x = y = 0, z$ arbitrary, while V^{H_0} corresponds to $x = y = 0, z(2\pi/n)$-periodic, i.e. with modes which are multiples of n. Furthermore, if H_1 is such that $H_0/H_1 \cong \mathbb{Z}_2$, then n is even and V^{H_1} corresponds to $z(\tau)$ which are $4\pi/n$-periodic, i.e. modes which are multiples of $n/2$. Thus, d_{H_0} and d_{H_1} are given by the number of negative eigenvalues λ of the system

$$(\mp\varepsilon_1 - \lambda\mu, -z'' - n^2 z + bZ_n\mu - \lambda z)$$

in the space of $(2\pi/n)$ and $(4\pi/n)$-periodic functions. In both systems b is positive.

Since, for $\lambda < 0$, the second equation is non-resonant, its particular solution has to be $z = \varepsilon_1 \cos n\tau$, with $\varepsilon_1 = b\mu(a_0 \pm \rho)/\lambda = \mp\lambda\mu$. Since $b > 0$, one has a contribution, for $\mu \neq 0$, only at $a_0 - \rho$ with $\lambda = -(b(a_0 - \rho))^{1/2}$. For $\mu = 0$, one has non-trivial solutions only for $n^2 + \lambda = k^2$, hence in the spaces under consideration, only for $k = n/2$ and $\lambda = -3n^2/4$, with a 2−dimensional kernel, (one dimensional when restricting to $z(\tau)$ even in which case the linearization is invertible).

Thus, $d_{H_0}(a_0 - \rho) = -d_{H_0}(a_0 + \rho) = -1$.

On the other hand for H_1 (and n even), there is an additional two dimensional eigenvalue, hence $d_{H_1}(a_0 \pm \rho) = 0$. Thus,

$$i_1(a_0 \pm \rho) = \mp[F_{H_0}].$$

It remains to identify the irreducible representations of H_0 in $(V^{\underline{H}})^\perp$, that is for x and y only, their isotropy K_k, the operators B_k and their Morse number n_k, as well as V^{K_k}.

Lemma 4.1. *There are n different irreducible representations of H_0 in $(V^{\underline{H}})^\perp$, with $K_k = \{(l, \varphi \equiv 2\pi lk/n), l = 0, ..., n-1\}$, for $k = 0, ...n-1$. The space V^{K_k} is spanned by functions $x(\tau), y(\tau), z(\tau)$ with the property that*

$$\mathcal{R}_{2\pi k/n}\begin{pmatrix} x(\tau + 2\pi/n) \\ y(\tau + 2\pi/n) \end{pmatrix} = \begin{pmatrix} x(\tau) \\ y(\tau) \end{pmatrix}$$

and $z(\tau)$ is $(2\pi/n)$-periodic. More precisely $x(\tau) = x_1(\tau) + x_2(\tau)$, with

$$\overline{x}_2(\tau) = x_1(\tau) = \sum_{-\infty}^{\infty} x_m e^{ik\tau} e^{imn\tau}$$

and $y(\tau) = i(x_1(\tau) - x_2(\tau))$. If $(x(\tau), y(\tau))$ is in V^{K_k}, then $(x(\tau), -y(\tau))$ is in $V^{K_{n-k}}$.

Proof. The action of H_0 on $(x(\tau), y(\tau))$ is by a time shift of $2\pi l/n$, for $l = 0, ..., n-1$ and by a rotation $\mathcal{R}_\varphi$. Hence on the mode m, one has

$$\mathcal{R}_\varphi \begin{pmatrix} x_m \\ y_m \end{pmatrix} e^{2\pi ilm/n} = p^{-1}\begin{pmatrix} e^{-i\varphi} & 0 \\ 0 & e^{i\varphi} \end{pmatrix} p \begin{pmatrix} x_m \\ y_m \end{pmatrix} e^{2\pi ilm/n}$$

where

$$2p = \begin{pmatrix} 1 & -i \\ 1 & i \end{pmatrix}.$$

Hence $\begin{pmatrix} x_m \\ y_m \end{pmatrix}$ will be fixed, if either $\varphi \equiv 2\pi lm/n, [2\pi]$, and $y_m = ix_m$ or $\varphi \equiv -2\pi lm/n, [2\pi]$, and $y_m = -ix_m$. Two modes m and $\tilde{m}$ will give the same action of H_0 if $m - \tilde{m}$ is a multiple of n, with $y_m = ix_m, y_{\tilde{m}} = ix_{\tilde{m}}$, or if $m + \tilde{m}$ is a multiple of n and $y_m = ix_m, y_{\tilde{m}} = -ix_{\tilde{m}}$. Hence, if $0 \leq k < n$ is fixed, one has, for $\varphi \equiv 2\pi kl/n, l = 0, ..., n-1$, that one has the modes $k + nm$, with $y_{k+nm} = ix_{k+nm}$ and the modes $-k + nm$, with $y_{-k+nm} = -ix_{-k+nm}$. Since $x(\tau)$ must be real, one obtains that $\bar{x}_2(\tau) = x_1(\tau)$. Finally, the action of K_k on $z(\tau)$ is limited to the time shift, hence $z(\tau)$ is $(2\pi/n)$-periodic.

Note that for $k = 0$ or $n/2$, the modes in $x_1(\tau)$ and $x_2(\tau)$ are the same. For $k = 0, x(\tau)$ and $y(\tau)$ are arbitrary $(2\pi/n)$-periodic functions, while for $k = n/2$, they have to be $(2\pi/n)$-antiperiodic. Note also that for $m = 0$, the elements of V^{K_k} are $(\cos k\tau, -\sin k\tau)$ and $(\sin k\tau, \cos k\tau)$. □

In order to compute $n_k(a_0 \pm \rho)$, one has to see the eigenvalue problem

$$-(x'' + n^2(\alpha + 2\beta\cos\tau)x = \lambda x,$$

$$-(y'' + n^2(\alpha + 2\beta\cos\tau)y = \lambda y$$

with $\lambda < 0$, for (x, y) in V^{K_k} in the first system and the analogous linear system in the second system. By plugging the Fourier series of Lemma 4.1 one arrives at an infinite system of equations which can be analyzed as in Lemmas 2.1 and 2.2.

However, it is simpler to see that this Morse number is constant on the regions separated by the curves $\alpha_{k/n}(\beta)$. In particular, one may compute them for $\beta = 0$. Here the index k in V^{K_k} in the residue class, mod n, of k in k/n.

Lemma 4.2. *If $2k/n$ is not an integer, then $n_k(a_0 - \rho) = 2[2k/n]$ and $n_k(a_0 + \rho) = 2([2k/n] + 1)$ if $\beta_0\alpha'_{k/n}(\beta_0) > 0$ and exchanging ρ by $-\rho$ for $\beta_0\alpha'_{k/n}(\beta_0) < 0$. If $2k/n = k_1$, then $n_k(a_0 - \rho) = 2(k_1 - 1)$, if $\beta_0\alpha'_{k/n}(\beta_0) > 0$ and one is on the left transition curve, $n_k(a_0 - \rho) = 2k_1$ if one is on the right transition curve, and $n_k(a_0 + \rho) = n_k(a_0 - \rho) + 2$ in this case. If $\beta_0\alpha'_{k/n}(\beta_0) < 0$, then $n_k(a_0 - \rho) = 2k_1$ on the left curve, $2(k_1 + 1)$ on the right curve and $n_k(a_0 + \rho) = n_k(\alpha_0 - \rho) - 2$. In all cases $n_{n-k}(a_0 \pm \rho) = n_k(a_0 \pm \rho)$.*

Proof. At $\alpha = (k-\varepsilon)^2/n^2, \beta = 0$, the system reduces to $((mn+k)^2 - (k-\varepsilon)^2 - \lambda)x_m = 0$, i.e., if $k_m \neq 0, \lambda = (mn + \varepsilon)(mn + 2k - \varepsilon) < 0$ for $\varepsilon < |mn| < 2k - \varepsilon$. For $\varepsilon > 0$, one has to count those m with $|m|$ between 1 and $2k/n$, while for $\varepsilon < 0$, one has to add $m = 0$. The y-component will give a contribution only for the case $2k/n$ an integer. In that case, a crossing of a transition curve will increase (or decrease according to the sign of $\beta_0\alpha'_{k/n}(\beta_0)$) the Morse number by 1, but then the y-component will give another contribution by one. The contribution for $V^{K_{n-k}}$ is then clear, if $2k/n$ is not an integer. □

Since H_0 has dimension one, one has that

$$i_2(f) = [F_\Gamma] - n_k[F_k] - n_{n-k}[F_{n-k}],$$

with only one if $k=0$ or $n/2$, F_k corresponds to V^{K_k}. Thus, since $n_{n-k}=n_k$, one has:

Theorem 4.1. *The orthogonal degree* $\deg_\perp(f_\varepsilon;\Omega)=2\eta([F_k]+[F_{n-k}])$, *where* $\eta=$ *Sign* $\beta_0\alpha'_{k/n}(\beta_0)$, *and only one generator if* $2k/n$ *is an integer. From* $(\beta_0,\alpha_{k/n}(\beta_0))$ *there is a global bifurcation in* V^{K_k} *and* $V^{K_{n-k}}$ *of a branch of non-trivial solutions which is either unbounded in* $(\nu,x,y,\tilde{Z})$ *or returns to another intersection of the line* $\alpha=\alpha_{k/n}(\beta_0)$ *with the curve* $\alpha_{k/n}(\beta)$, *with its* η *of opposite sign.*

Solutions on the branch have $2k$ *zeros in* $[0,2\pi)$ *for* x *and* y, *which are not identically zero unless trivial, i.e. with* $\nu=\nu_{0/n},\beta=\alpha_{k/n}(\beta_0)$.

Proof. The argument for the global bifurcation is standard. The fact that it takes place in V^{K_k} is proved in [8]. The relation between the two branches is given by Remark 1.1 and the isomorphism $(x,y)\to(x,-y)$. The nodal properties follow from the fact that the equations for x and y are of the form $x''+f(t)x=0$. Hence, the branch cannot return to a point on a curve $\alpha_{k'/n}(\beta)$, with $k'\neq k$.

Since (x,y) is in V^{K_k}, then if $x(\tau)\equiv 0$ one has $y(\tau)\equiv 0$ unless $2k/n=k_1$. Hence, if (x,y) tends to $(0,0)$ on the branch, one goes to $(\nu=\nu_0/\tilde{n},Z=a\cos(\tilde{n}\tau+\varphi))$ and $\alpha=\alpha_{\tilde{k}/\tilde{n}}(\beta)$. In the limit, the elements on that curve shall have $2k$ zeros, hence $k=\tilde{k}$ and, from the periodicity of $Z(\tau),\tilde{n}$ should be a multiple of n. From the fact that on each curve $\tilde{\alpha}_{\tilde{k}/\tilde{n}}$ there are only two linearizations which are not invertible, corresponding to $V^{K_{\tilde{k}}}$ and $V^{K_{\tilde{n}-\tilde{k}}}$, the above argument is reversible and $n=\tilde{n}$. thus, the only $(x,y)=(0,0)$ on the branch are the bifurcation points from the trivial solution.
□

Remark 4.1. When varying α one obtains "surfaces" bifurcating from the curve $\alpha_{k/n}(\beta)$, following the arguments of Γ-epi maps of [7].

Remark 4.2. If one wishes to use the reversibility, then one may restrict the study to fixed points subspaces of R_ε, i.e. to (x,y) of the same parity and $z(\tau)$ even. This will destroy the equivariance with respect to the time shift and keep only that with respect to $\mathcal{R}_\varphi$, which is still orthogonal. In this case, $H_0=\mathbb{S}^1$, and $K=\{e\}$. The orthogonal degree reduces to one component.

In this case, $\deg_\perp(f_\varepsilon;\Omega)=4\eta[F_e]$, when crossing $\alpha_{k/n}(\beta)$, with $2k/n$ not an integer, and $2\eta[F_e]$ in the later case. Solutions on the branch conserve the parity and the nodal properties, but there is no topological argument to prevent the branch coming out of $(\beta_0,\alpha_{k/n}(\beta_0))$ to go to a point $(\beta_1,\alpha_{k/\tilde{n}}(\beta_1)=\alpha_{k/n}(\beta_0))$, for a $\tilde{n}$ different from n, since the periodicity of $Z(\tau)$ is only 2π.

Note also that taking $y\equiv 0$, one may use the usual Leray-Schauder degree on even $(x(\tau),z(\tau))$ or on $x(\tau)$ odd and $z(\tau)$ even. The one dimensional kernel will give the same bifurcation results, which will be now planar solutions. It is clear that these solutions are likely to generate the solutions given by the reversibility. However, except for $2k/n$ an integer, they are different from the ones given in V^{K_k}. Hence, one has double bifurcation from $(\beta_0,\alpha_{k/n}(\beta_0))$, if $2k/n$ is not an integer, of planar and non-planar solutions.

Note finally that V^{K_k} can be decomposed in sums of functions of the form ($x(\tau)$ even, $y(\tau)$ odd) and ($x(\tau)$ odd, $y(\tau)$ even), the first being for coefficients x_m of $x_1(\tau)$ real. Since the equations have the additional symmetry that they preserve the fact that x

and y have opposite parities and $z(\tau)$ is even, then one may also study the equations in $V^{K_k} \cap$ (even, odd, even) or $V^{K_k} \cap$ (odd, even, even), where one has a jump of one eigenvalue when crossing $\alpha_{k/n}(\beta)$ and may use the Leray-Schauder theory in that space.

From the stability in the complement of the Arnold's tongues, it seems likely that the first bifurcation will correspond to a crossing of a transition curve, i.e. with $2k/n$ an integer and a planar solution.

References

[1] H. W. BROER, G. A. LUNTER, G. VEGTER: *Equivariant singularity theory with distinguished parameters: two case studies of resonant Hamiltonian systems,* Physica D **112** (1998), 64-80

[2] R. BROUCKE, P. A. BAXA: *Periodic solutions of a spring-pendulum system,* Celestial Mechanics **8** (1973), 261-267

[3] J. T. CUSHING: *The spring-mass system revisited,* Amer. J. Phys. **52** (1984), 925-933

[4] J. DAMON: *Applications of singularity theory to the solutions of nonlinear equations,* in: Topological Nonlinear Analysis [Ed.: M. MATZEU, A. VIGNOLI], Birkhäuser, Basel 1995, p. 178-302

[5] N. GOMEZ: *Soluciones periódicas de la ecuación de Mathieu,* Thesis, Univ. Nacional Autón. México 1994

[6] J. IZE: *Topological bifurcation,* in: Topological Nonlinear Analysis [Ed.: M. MATZEU, A. VIGNOLI], Birkhäuser, Basel 1995, p. 341-463

[7] J. IZE, I. MASSABÒ, A. VIGNOLI: *Global results on continuation and bifurcation for equivariant maps,* NATO Adv. Sci. Int. Ser. **173** (1986), 75-111

[8] J. IZE, A. VIGNOLI: *Equivariant degree for abelian actions. Part III: Orthogonal maps,* Topol. Methods Nonlin. Anal. **13, 1** (1999), 105-146

[9] L. G. KHAZIN, F. KH. TSEL'MAN: *Nonlinear interactions of resonating oscillators,* Soviet Phys. Dokl. **15** (1971), 677-679

[10] H. M. LAI: *On the recurrence phenomenon of a resonant spring pendulum,* Amer. J. Phys. **52** (1984), 219-223

[11] J. MONTALDI, M. ROBERTS, I. STEWART: *Existence of nonlinear normal modes of symmetric Hamiltonian systems,* Nonlinearity **3** (1990), 695-730

[12] M. G. OLSON: *Why does a mass on a spring sometimes misbehave?* Amer. J. Phys. **44** (1976), 1211-1212

[13] M. G. RUSBRIDGE: *Motion of the spring pendulum,* Amer. J. Phys. **48** (1980), 146-151

[14] D. J. TRITTON: *Ordered and chaotic motion of a forced spherical pendulum,* Eur. J. Phys. **7** (1986), 162-169

[15] V. A. YAKUBOVICH, V. M. STARZHINSKII: *Linear Differential Equations with Periodic Coefficients,* Vol. 2, John Wiley & Sons, New York 1975

JORGE IZE, IIMAS-FENOMEC, Universidad Nacional Autónoma de México, A. P. 20-726, México 20, D. F.; `jil@uxmym1.iimas.unam.mx`

Progress in Nonlinear Differential Equations
and Their Applications, Vol. 40

On the Semilinear Dirichlet Problem for a Class of Nonlocal Operators Generating Dirichlet Forms

NIELS JACOB, VITALY MOROZ

Dedicated to Alfonso Vignoli on the occasion of his 60th birthday

Summary: We consider the semilinear boundary value problem for pseudo differential operators generating symmetric Dirichlet forms. Using a variational approach we establish the existence of solutions under some growth assumptions on the nonlinearity. We also develop a certain truncation technique based on the specific properties of Dirichlet forms. Such technique allows us to obtain results about existence of bounded positive solutions and to relax growth conditions on the nonlinearity.

Keywords: Dirichlet form, boundary value problem, nonlocal operator, positive solutions, truncation of nonlinearity, variational methods.

Classification: 35S15, 31C25, 47H07.

Acknowledgement: The first author has been supported by the DFG-grant Ja 522/7-1. The second author has been supported by a one-year scholarship of the DAAD (German Academic Exchange Service) and by the Belorussian Fund of Fundamental Research.

1. Introduction. We consider a class of pseudo-differential operators

$$p(x,D)u(x) = (2\pi)^{-(N/2)} \int_{\mathbb{R}^N} e^{ix\xi} p(x,\xi)\hat{u}(\xi)d\xi \tag{1}$$

where $p : \mathbb{R}^N \times \mathbb{R}^N \to \mathbb{R}$ is a real valued continuous symbol such that $p(x,\cdot) : \mathbb{R}^N \to \mathbb{R}$ is negative definite in the sense of I. J. Schoenberg. Under suitable conditions $p(x,D)$ extends from $C_0^\infty(\mathbb{R}^N)$ to a generator of a symmetric Dirichlet form $(B, D(B))$ with domain $D(B) \subset L_2(\mathbb{R}^N)$ and

$$B(u,v) = \int_{\mathbb{R}^N} p(x,D)u(x) \cdot v(x)\,dx \quad \text{for} \quad u,v \in C_0^\infty(\mathbb{R}^N).$$

In this paper we are interested in the semilinear boundary value problems for $p(x,D)$ on some open set $\Omega \subset \mathbb{R}^N$. The main difficulty which arise is that $p(x,D)$ is in general a non-local operator not satisfying the transmission condition. Nonlocality means that *supp* $(u) \subseteq \Omega'$ does not imply *supp* $(p(x,D)u) \subseteq \Omega'$ for all open sets $\Omega' \subseteq \mathbb{R}^N$. Several approaches to the linear boundary value problems for the operator $p(x,D)$ were considered in the papers [6,10], see also [9]. From the considerations in these papers it seems to be reasonable to give the following formulation of the semilinear Dirichlet problem for $p(x,D)$ on Ω: *given a Carathéodory function* $f:\mathbb{R}^N\times\mathbb{R}\to\mathbb{R}$, *find* $u:\mathbb{R}^N\to\mathbb{R}$ *such that*

$$\text{(2)}\qquad \begin{cases} p(x,D)u = f(x,u) & \text{a.e. in } \Omega,\\ u = 0 & \text{a.e. in } \Omega^c,\end{cases}$$

where $\Omega\subset\mathbb{R}^N$ is an open bounded set with sufficiently smooth boundary $\partial\Omega$ and complement $\Omega^c=\mathbb{R}^N\setminus\Omega$.

We will use a classical variational approach to handle a weak formulation of this problem. According to such approach the weak solutions of the problem (2) correspond to the minima of the energy functional for (2) on the certain anisotropic Sobolev space. The existence of the minima for the energy functional is provided by the usual one-sided estimates for the nonlinearity. Further we will see that the property of being a Dirichlet form enables us to develop truncation techniques for the problem (2) which seems to be new for nonlocal pseudo-differential operators. Such techniques, typical for the second-order partial differential operators allow to obtain some results about existence of bounded positive solutions for the problem (2) and to relax the growth conditions on the nonlinearity. The motivation for considering such a problem is given by the theory of superprocesses, see E. B. Dynkin [4]. We will come back to these relations in another paper.

2. A class of pseudo-differential operators. Let us recall some results from [7], see also [9]. Let $a^2:\mathbb{R}^N\mapsto\mathbb{R}$ be a real valued continuous negative definite function, that is a^2 is a continuous function such that $a^2(0)\geq 0$ and for all $t>0$ the function $\xi\mapsto e^{-ta^2(\xi)}$ is positive definite. We define for $s\geq 0$ the norm

$$\|u\|^2_{a^2,s}=\int_{\mathbb{R}^N}(1+a^2(\xi))^{2s}|\hat{u}(\xi)|^2d\xi$$

and the anisotropic Sobolev spaces

$$H^{a^2,s}(\mathbb{R}^N)=\{u\in L_2(\mathbb{R}^N):\|u\|_{a^2,s}<\infty\}.$$

The space $H^{a^2,s}(\mathbb{R}^N)$ is a real Hilbert space with the scalar product

$$(u,v)_{a^2,s}=\int_{\mathbb{R}^N}(1+a^2(\xi))^{2s}\hat{u}(\xi)\overline{\hat{v}(\xi)}d\xi$$

and $C_0^\infty(\mathbb{R}^N)$ is a dense subspace of $H^{a^2,s}(\mathbb{R}^N)$. For $a^2(\xi)=|\xi|^2$ the space $H^{a^2,s}(\mathbb{R}^N)$ coincides with the usual Sobolev space $H^{2s}(\mathbb{R}^N)$.

It is known that a real valued continuous negative definite function a^2 satisfies for some $c>0$ the estimate

$$\text{(3)}\qquad 0\leq a^2(\xi)\leq c(1+|\xi|^2).$$

Suppose also that for some $r\in(0,1]$ the function a^2 satisfies the condition

$$(a_r)\qquad a^2(\xi)\geq c|\xi|^{2r}\text{ for some } c>0 \text{ and all } \xi\in\mathbb{R}^N,\ |\xi|\geq R>0.$$

Then $H^{a^2,s}(\mathbb{R}^N)$ is continuously embedded in $H^{2sr}(\mathbb{R}^N)$. Using embedding theorems for the Sobolev scale $H^t(\mathbb{R}^N)$ we can obtain embedding results for $H^{a^2,s}(\mathbb{R}^N)$.

In the following we will always suppose that $p : \mathbb{R}^N \times \mathbb{R}^N \to \mathbb{R}$ is a real-valued continuous symbol such that for any fixed $x \in \mathbb{R}^N$ the function $p(x,\cdot) : \mathbb{R}^N \to \mathbb{R}$ is negative definite and $p(x,\xi)$ has the decomposition

$$p(x,\xi) = p_1(\xi) + p_2(x,\xi)$$

where for a suitable $m \in \mathbb{N}$

(p_1) $|p_1(\xi)| \le c(1 + a^2(\xi))$ for some $c > 0$ and all $\xi \in \mathbb{R}^N$;

(p_2) $p_2(\cdot,\xi) \in C^m(\mathbb{R}^N)$ and for all $\beta \in \mathbb{N}_0^n$, $|\beta| \le m$,

$$|\partial_x^\beta p_2(x,\xi)| \le \varphi_\beta(x)(1 + a^2(\xi))$$

holds for all $\xi \in \mathbb{R}^N$ with some $\varphi_\beta \in L_1(\mathbb{R}^N)$;

(p_3) $p_1(\xi) \ge 2\gamma_0 a^2(\xi)$ for some $\gamma_0 > 0$ and all $\xi \in \mathbb{R}^N$, $|\xi| \ge R > 0$;

(p_4) $\sum_{|\alpha|\le m} \|\varphi_\alpha\|_{L_1}$ is small w.r.t. γ_0 (in a very precise sense, see [7]).

Then the operator $p(x,D)$ as defined in (1) maps $C_0^\infty(\mathbb{R}^N)$ into the space $C(\mathbb{R}^N)$ and the bilinear form $B(u,v)$ associated with $p(x,D)$ is defined for $u,v \in C_0^\infty(\mathbb{R}^N)$. In the following we will suppose that the operator $p(x,D)$ is symmetric on $C_0^\infty(\mathbb{R}^N)$. Then $p(x,D)$ has a selfadjoint extension on $L_2(\mathbb{R}^N)$ with domain $H^{a^2,1}(\mathbb{R}^N)$. The bilinear form B extends to a continuous symmetric Dirichlet form with domain $H^{a^2,1/2}(\mathbb{R}^N)$, see [5,11] for the general theory of Dirichlet forms and their properties. In particular the form B is positive definite on $H^{a^2,1/2}(\mathbb{R}^N)$, i.e.

$$B(u,u) \ge 0 \quad \text{for all} \quad u \in H^{a^2,1/2}(\mathbb{R}^N).$$

Moreover the form B satisfies Gårding inequality

$$B(u,u) \ge \gamma_0 \|u\|^2_{a^2,1/2} - \lambda_0 \|u\|^2_{L_2}, \tag{4}$$

here γ_0 is taken from condition (p_3) and $\lambda_0 > 0$.

3. Variational setting of the problem. Let $\Omega \subseteq \mathbb{R}^N$ ($N > 2r$) be a bounded open set with smooth boundary $\partial\Omega$ and $u \in C_0^\infty(\Omega)$. We can extend u to $\mathbb{R}^N$ by setting it in Ω^c equal to zero obtaining a function in $C_0^\infty(\mathbb{R}^N)$ with support in Ω. For this reason we can identify $C_0^\infty(\Omega)$ as a subspace of $C_0^\infty(\mathbb{R}^N)$. Since $C_0^\infty(\mathbb{R}^N) \subseteq H^{a^2,s}(\mathbb{R}^N)$ we can take the closure of $C_0^\infty(\Omega)$ in $H^{a^2,1/2}(\mathbb{R}^N)$ which we denote by $H_0^{a^2,1/2}(\Omega)$. Suppose that $u \in H_0^{a^2,1/2}(\Omega)$. For any $\varphi \in C_0^\infty(\Omega^c)$ we find

$$\int_{\mathbb{R}^N} u(x)\varphi(x)\,dx = \lim_{n\to\infty} \int_{\mathbb{R}^N} u_n(x)\varphi(x)\,dx = 0$$

where $(u_n) \subset C_0^\infty(\Omega)$ converges to u in the norm $\|\cdot\|_{a^2,1/2}$. Thus we find $u = 0$ a.e. in Ω^c showing that elements in $H_0^{a^2,1/2}(\Omega)$ fulfil the "boundary" condition in a generalised sense.

It is shown in [7] that under condition (a_r) the space $H_0^{a^2,1/2}(\Omega)$ is continuously embedded into the standard Sobolev space $H_0^r(\Omega)$. Since $N > 2r$ by the Sobolev embedding theorems we obtain the sequence of continuous embedding

$$(5) \qquad H_0^{a^2,1/2}(\Omega) \subseteq L_{\frac{2N}{N-2r}}(\Omega) \subseteq L_2(\Omega) \subseteq L_{\frac{2N}{N+2r}}(\mathbb{R}^N).$$

Moreover $L_{\frac{2N}{N+2r}}(\mathbb{R}^N)$ embedded into $[H_0^{a^2,1/2}(\Omega)]^*$ in the sense that

$$l_h(u) = \int_{\mathbb{R}^N} h(x)u(x)\,dx$$

is a linear continuous functional on $H_0^{a^2,1/2}(\Omega)$ for each $h \in L_{\frac{2N}{N+2r}}(\mathbb{R}^N)$. Since Ω is bounded the embedding of $H_0^{a^2,s}(\Omega)$ into $L_2(\Omega)$ is compact and there exists the precise embedding constant $\sigma = \sigma(\Omega) > 0$ such that

$$(6) \qquad \|u\|_{L_2(\mathbb{R}^N)} \leq \sigma \|u\|_{a^2,1/2} \quad \text{for all} \quad u \in H_0^{a^2,1/2}(\Omega).$$

Combining (6) with Gårding's inequality (4) we obtain the estimate

$$(7) \qquad B(u,u) \geq (\gamma_0 - \lambda_0\sigma^2)\|u\|^2_{a^2,1/2} \quad \text{for all} \quad u \in H_0^{a^2,1/2}(\Omega),$$

that is we can assert that B is strictly positive definite for domains Ω with sufficiently small embedding constant $\sigma(\Omega)$. However the form B is always positive definite on $H_0^{a^2,1/2}(\Omega)$ since B is Dirichlet form and we can replace (7) by the estimate

$$(8) \qquad B(u,u) \geq (0 \vee (\gamma_0 - \lambda_0\sigma^2))\|u\|^2_{a^2,1/2} \quad \text{for all} \quad u \in H_0^{a^2,1/2}(\Omega).$$

Now let $u \in H_0^{a^2,1/2}(\Omega)$ such that $p(x,D)u \in L_2(\mathbb{R}^N)$ be a solution of the Dirichlet problem (2) and suppose that the function $f(x,u(x))$ is integrable. Multiplying with $\varphi \in C_0^\infty(\Omega)$ we find

$$(9) \qquad B(u,\varphi) = \int_{\mathbb{R}^N} f(x,u(x))\varphi(x)\,dx$$

for all $\varphi \in C_0^\infty(\Omega)$. Conversely, it is clear that if $u \in H_0^{a^2,1/2}(\Omega)$ satisfies (9) for all $\varphi \in C_0^\infty(\Omega)$ and $p(x,D)u \in L_2(\mathbb{R}^N)$ then u is a solution of (2). For this reason we will say that u is *a weak solution* of the Dirichlet problem (2) if (9) holds for all $\varphi \in C_0^\infty(\Omega)$.

Let us note that in general the function $f(x,\cdot)$ has a nontrivial dependence on x even on the complement of Ω. For example the class of functions $f(x,u) = g(u) + h(x)$ with *supp* $(h) = \mathbb{R}^N$ is admissible. It is easy to see that actually the solutions of Dirichlet problem (2) do not depend on the behaviour of $x \mapsto f(x,u)$ on Ω^c. It is possible to assume that $f(x,u) \equiv 0$ on Ω^c. However we prefer to consider the more general case $f : \mathbb{R}^N \times \mathbb{R} \to \mathbb{R}$ because of the probabilistic motivation of the Dirichlet problem (2), see [8] for the discussion in the linear case.

We define the *energy functional* J for problem (2) by means of formula

$$J(u) = \frac{1}{2}B(u,u) - \int_{\mathbb{R}^N} F(x,u(x))\,dx,$$

here

$$F(x,u) = \int_0^u f(x,\xi)\,d\xi$$

is the primitive of f with respect to the second variable, note that $F(x,0) \equiv 0$.
We are interested in the conditions on a nonlinearity $f(x,u)$ which ensure that the energy functional J is well defined on $H_0^{a^2,1/2}(\Omega)$ and each local minimum of J corresponds to a weak solution of the original boundary value problem (2).

Lemma 1. *Suppose that $f(x,u)$ satisfies the assumption*

(f_r) *there exists $c > 0$, $h \in L_{\frac{2N}{N+2r}}(\mathbb{R}^N)$ such that*

$$|f(x,u)| \le c|u|^{\frac{N+2r}{N-2r}} + h(x).$$

Then the functional J is defined on $H_0^{a^2,1/2}(\Omega)$.

Proof. From condition (f_r) it follows that the primitive $F(x,u)$ satisfy the estimate

$$|F(x,u)| \le c_1|u|^{\frac{2N}{N-2r}} + c_2 h(x)u.$$

Then

$$\begin{aligned}\int_{\mathbb{R}^N} F(x,u(x))\,dx &\le c_1 \int_\Omega |u(x)|^{\frac{2N}{N-2r}}\,dx + c_2 \int_{\mathbb{R}^N} h(x)u(x)\,dx \\ &< c_1(\|u\|_{L_{\frac{2N}{N-2r}}})^{\frac{2N}{N-2r}} + c_2\|h\|_{L_{\frac{2N}{N+2r}}}\|u\|_{L_{\frac{2N}{N-2r}}} < \infty\end{aligned}$$

since $u \in H_0^{a^2,1/2}(\Omega) \subseteq L_{\frac{2N}{N-2r}}(\Omega)$ and $h \in L_{\frac{2N}{N+2r}}(\mathbb{R}^N)$. □

Lemma 2. *Suppose that assumption (f_r) holds. Then the functional J is Gâteaux differentiable on the space $H_0^{a^2,1/2}(\Omega)$ and its derivative for all $\varphi \in H_0^{a^2,1/2}(\Omega)$ is given by the formula*

$$J'(u)(\varphi) = B(u,\varphi) - \int_{\mathbb{R}^N} f(x,u(x))\varphi(x)\,dx. \tag{10}$$

Moreover each local minimum $u \in H_0^{a^2,1/2}(\Omega)$ of the functional J is a weak solution of the Dirichlet problem (2).

Proof. Clearly $B(u,u)$ is differentiable on $H_0^{a^2,1/2}(\Omega)$ as a continuous bilinear form and its derivative for all $\varphi \in H_0^{a^2,1/2}(\Omega)$ is given by the formula

$$B'(u,u)(\varphi) = B(u,\varphi).$$

We check the differentiability of the nonlinear term

$$J_F(u) = \int_{\mathbb{R}^N} F(x,u(x))\,dx.$$

By the mean value theorem for each $u, \varphi \in H_0^{a^2,1/2}(\Omega)$ and $\tau \in [-1,1]$ there exists a function $\theta_\tau(x)$ such that $0 \le \theta_\tau(x) \le 1$ and

$$\frac{J_F(u+\tau\varphi) - J_F(u)}{\tau} = \int_\Omega f(x, u(x) + \tau\theta_\tau(x)\varphi(x))\varphi(x)\,dx. \tag{11}$$

It is known [1] that the function θ_τ may be chosen to be measurable so $\theta_\tau \in L_\infty(\Omega)$ and the right hand side of (11) makes sense.

We shall verify that the integral on the right hand side of (11) does exist. Recall that $H_0^{a^2,1/2}(\Omega) \subseteq L_{\frac{2N}{N-2r}}(\Omega)$ by the sequence of embeddings (5). Hence

$$u(x) + \tau\theta_\tau(x)\varphi(x) \in L_{\frac{2N}{N-2r}}(\Omega).$$

Further from assumption (f_r) it follows that

$$f(x, u(x) + \tau\theta_\tau(x)\varphi(x)) \in L_{\frac{2N}{N+2r}}(\mathbb{R}^N),$$

see e.g. [1]. Since $\varphi \in H_0^{a^2,1/2}(\Omega) \subseteq L_{\frac{2N}{N-2r}}(\Omega)$ we have

$$f(x, u(x) + \tau\theta_\tau(x)\varphi(x))\varphi(x) \in L_1(\mathbb{R}^N).$$

Hence the integral in the right hand side of (11) does exist. Let $\tau \to 0$. Clearly

$$u(x) + \tau\theta_\tau(x)\varphi(x) \to u(x)$$

in measure and form an U-bounded family of functions in $L_{\frac{2N}{N-2r}}(\Omega)$, i.e. there exists $U \in L_{\frac{2N}{N-2r}}(\Omega)$ such that

$$|u(x) + \tau\theta_\tau(x)\varphi(x)| \le U(x) \quad \text{for all} \quad \tau \in [-1,1].$$

Hence for $|\tau| \le 1$ and $\tau \to 0$

$$f(x, u(x) + \tau\theta_\tau(x)\varphi(x)) \to f(x, u(x))$$

in measure and makes an U-bounded family of functions in $L_{\frac{2N}{N+2r}}(\mathbb{R}^N)$. So the Lebesgue dominated convergence theorem can be applied to (11) and we have

$$J_F'(u)(\varphi) = \frac{d}{d\tau} J_F(u+\tau\varphi)\,|_{\tau=0} = \lim_{\tau\to 0} \frac{J_F(u+\tau\varphi) - J_F(u)}{\tau}$$

$$= \lim_{\tau\to 0} \int_\Omega f(x, u(x) + \tau\theta_\tau(x)\varphi(x))\varphi(x)\,dx = \int_\Omega f(x, u(x))\varphi(x)\,dx.$$

Since $f(x, u(x)) \in L_{\frac{2N}{N+2r}}(\mathbb{R}^N)$ it follows that $J_F'(u,\cdot)$ is a linear continuous functional on $H_0^{a^2,1/2}(\Omega)$ and therefore J_F is Gâteaux differentiable on $H_0^{a^2,1/2}(\Omega)$.

Finally let $u \in H_0^{a^2,1/2}(\Omega)$ be a minimum for J. Clearly (f_r) implies that the function $f(x, u(x))$ is integrable. By the classical Euler-Fermat principle we have

$$J'(u)(\varphi) = B(u,\varphi) - \int_{\mathbb{R}^N} f(x, u(x))\varphi(x)\,dx = 0$$

for all $\varphi \in H_0^{a^2,1/2}(\Omega)$. Hence (9) holds for all $\varphi \in C_0^\infty(\Omega)$ and therefore is u is a weak solution for the Dirichlet problem (2). □

Remark 1. By a standard arguments we can also prove that actually under assumption (f_r) the functional J is continuously differentiable on $H_0^{a^2,1/2}(\Omega)$. This follows by the standard arguments from the continuity of the embedding $H_0^{a^2,1/2}(\Omega) \subseteq L_{\frac{2N}{N-2r}}$. However we do not need it in the further consideration.

4. Existence of a minimum. In this section we are interested in investigating the conditions which lead to the existence of a minimum for J. According to the classical Weierstrass principle (see e.g. [13]) it suffices to verify that J is coercive and (sequentially) lower semicontinuous with respect to the weak topology on $H_0^{a^2,1/2}(\Omega)$, that is

$$J(u) \to +\infty \quad \text{as} \quad \|u\|_{a^2,1/2} \to \infty.$$

Let F satisfy the usual one-sided *coercivity condition*

(F) there exist $\alpha < 0 \vee (\frac{\gamma_0}{\sigma^2} - \lambda_0)$ and functions $h \in L_{\frac{2N}{N+2r}}(\mathbb{R}^N)$, $g \in L_1(\mathbb{R}^N)$ such that

$$F(x,u) \leq \frac{\alpha}{2}u^2 + h(x)u + g(x),$$

here γ_0, λ_0 is a taken from Gårding inequality (4) and σ is the embedding constant from (6). We assert that under the condition (F) the functional J attains its infimum.

Lemma 3. *Suppose that assumptions $(f_r), (F)$ hold. Then the functional J is bounded from below and coercive.*

Proof. From (F) and estimates (6), (8) it follows that

$$J(u) \geq \frac{1}{2}B(u,u) - \int_{\mathbb{R}^N} \left\{ \frac{\alpha}{2}u^2(x) + h(x)u(x) + g(x) \right\} dx$$

$$= \frac{1}{2}\left(B(u,u) - \alpha\|u\|_{L_2}^2\right) - \int_{\mathbb{R}^N} h(x)u(x)\,dx - \int_{\mathbb{R}^N} g(x)\,dx$$

$$\geq \frac{1}{2}[(0 \vee (\gamma_0 - \lambda_0\sigma^2)) - \alpha\sigma^2]\|u\|_{a^2,1/2}^2 - \|h\|_{a^2,-1/2}\|u\|_{a^2,1/2} - \|g\|_{L_1} \to +\infty$$

as $\|u\|_{a^2,1/2} \to +\infty$ since by assumption $\alpha < 0 \vee (\frac{\gamma_0}{\sigma^2} - \lambda_0)$. □

Lemma 4. *Suppose that assumptions $(f_r), (F)$ holds. Then the functional J is lower semi-continuous in the weak topology of $H_0^{a^2,1/2}(\Omega)$.*

Proof. The energy functional J can be rewritten in the form:

$$J(u) = \frac{1}{2}B(u,u) - \int_{\mathbb{R}^N} F(x,u(x))\,dx$$

$$= \frac{1}{2}B(u,u) - \int_{\mathbb{R}^N} \frac{\alpha}{2}u^2(x) + h(x)u(x) + g(x)\,dx$$

$$+\int_{\mathbb{R}^N}\left\{\frac{\alpha}{2}u^2(x)+h(x)u(x)+g(x)\right\}-F(x,u(x))\,dx$$
$$=\left\{\frac{1}{2}B(u,u)-\frac{\alpha}{2}\|u\|^2_{L_2}\right\}-\int_{\mathbb{R}^N}h(x)u(x)\,dx-\int_{\mathbb{R}^N}g(x)\,dx$$
$$+\int_{\mathbb{R}^N}\left\{\frac{\alpha}{2}u^2(x)+h(x)u(x)+g(x)\right\}-F(x,u(x))\,dx.$$

Let us consider each term separately. We have by (F) and (6),(8)

$$\frac{1}{2}B(u,u)-\frac{\alpha}{2}\|u\|^2_{L_2}\geq\frac{1}{2}(0\vee(\gamma_0-\lambda_0\sigma^2)-\alpha\sigma^2)\|u\|^2_{a^2,1/2}\geq 0$$

for all $u\in H_0^{a^2,1/2}(\Omega)$. Hence the quadratic term is (sequentially) weakly lower semicontinuous as a positive definite form on $H_0^{a^2,1/2}(\Omega)$ (see e.g. [14]). Also the linear term generated by $h\in L_{\frac{2N}{N+2r}}(\mathbb{R}^N)$ is continuous and hence weakly continuous on $H_0^{a^2,1/2}(\Omega)$.

Now let us consider the last term

$$J_F(u)=\int_{\mathbb{R}^N}\left\{\frac{\alpha}{2}u^2(x)+h(x)u(x)+g(x)\right\}-F(x,u(x))\,dx.$$

Let $(u_n)\subset H_0^{a^2,1/2}(\Omega)$ be a sequence weakly converging to u_0. Then (u_n) is bounded in $H_0^{a^2,1/2}(\Omega)$. Since the embedding of $H_0^{a^2,1/2}(\Omega)$ into $L_2(\Omega)$ is compact, (u_n) contains a subsequence converging in $L_2(\Omega)$. It is easy to see that weak convergence in $H_0^{a^2,1/2}(\Omega)$ and convergence in $L_2(\Omega)$ are consistent in the sense that if the sequence converges in both topologies then limits coincide. We conclude that (u_n) converges to u_0 in $L_2(\Omega)$ and hence converges to u_0 in measure. Then the sequence

$$v_n(x)=\frac{\alpha}{2}u_n^2(x)+h(x)u_n(x)+g(x)-F(x,u_n(x))$$

also converges to

$$v_0(x)=\frac{\alpha}{2}u_0^2(x)+h(x)u_0(x)+g(x)-F(x,u_0(x))$$

in measure. From (F) it follows that the sequence (v_n) is nonnegative. Now applying the Fatou-Lemma we obtain

$$\int_{\mathbb{R}^N}v_0(x)\,dx\leq\liminf_{n\to\infty}\int_{\mathbb{R}^N}v_n(x)\,dx,$$

which means that J_F is (sequentially) weakly lower semicontinuous. □

Theorem 1. *Suppose that assumptions $(f_r),(F)$ hold. Then J is bounded from below and has a point of minimum on $H_0^{a^2,1/2}(\Omega)$. Moreover if the primitive $-F(x,u)$ is strictly convex in u then the minimum point of J in $H_0^{a^2,1/2}(\Omega)$ is unique.*
The proof of the theorem follows immediately from Lemmas 3 and 4 (see [13,14]).

Remark 2. The results of this section remain true without any growth assumptions on $|f(x,u)|$ that is without condition (f_r). Of course in this case J may be not defined on the whole space $H_0^{a^2,1/2}(\Omega)$. However under one-sided condition (F) the minimum still exists on $Dom(J) \subseteq H_0^{a^2,1/2}(\Omega)$, see [12] for a close consideration.

5. A basic solvability result. The basic existence results for the Dirichlet problem (2) follow immediately from Theorem 1 and Lemma 2.

Theorem 2. *Suppose that assumptions* $(f_r),(F)$ *hold. Then problem* (2) *has at least one weak solution in the space* $H_0^{a^2,1/2}(\Omega)$. *Moreover if the primitive* $-F(x,u)$ *is strictly convex in* u *then the solution of* (2) *in* $H_0^{a^2,1/2}(\Omega)$ *is unique.*

In the previous sections we discussed the Dirichlet problem (2) using the Dirichlet space $H_0^{a^2,1/2}(\Omega)$ generated by $p(x,D)$. Actually we did not further use the Dirichlet space structure. In the next sections we need this Dirichlet structure in order to develop the truncation technique for problem (2). This technique should allow to avoid the growth restriction (f_r) on the nonlinearity $f(x,u)$ and to obtain some additional information on the properties of the solutions of (2).

Before doing this, let us consider as an example the following problem

$$(12)\qquad \begin{cases} p(x,D)u + u|u|^{\rho-1} = h(x) & \text{a.e. in } \Omega, \\ u = 0 & \text{a.e. in } \Omega^c. \end{cases}$$

Corollary 1. *For each* $\rho \in (0, \frac{N-2r}{N+2r})$ *and* $h \in L_{\frac{2N}{N+2r}}(\mathbb{R}^N)$ *the problem* (12) *has a unique weak solution in the space* $H_0^{a^2,1/2}(\Omega)$. *Moreover if the function* h *is nonnegative on* $\mathbb{R}^N$ *then the corresponding solution is nonnegative on* Ω.

Proof. The existence and uniqueness of the solution follows immediately from Theorem 2. Let the function h be nonnegative on $\mathbb{R}^N$ and let $\tilde{u} \in H_0^{a^2,1/2}(\Omega)$ be the corresponding solution to (12). Then $\tilde{u}$ is a minimum of J on $H_0^{a^2,1/2}(\Omega)$. Since $H_0^{a^2,1/2}(\Omega)$ is a Dirichlet space it follows that $|\tilde{u}| \in H_0^{a^2,1/2}(\Omega)$ and

$$B(|\tilde{u}|,|\tilde{u}|) \leq B(\tilde{u},\tilde{u}),$$

see e.g. [11, p. 35/36]. Therefore we have

$$J(|\tilde{u}|) = \frac{1}{2}B(|\tilde{u}|,|\tilde{u}|) + \frac{1}{\rho+1}\int_\Omega |\tilde{u}|^{\rho+1}\,dx - \int_{\mathbb{R}^N} h(x)|\tilde{u}|\,dx \leq J(\tilde{u}).$$

This means that $|\tilde{u}|$ is also a minimum of J on $H_0^{a^2,1/2}(\Omega)$. Since the minimum point of J is unique it follows that $\tilde{u} = |\tilde{u}| \geq 0$. □

6. Truncation of the nonlinearity. In this section we will always suppose that the form B is strictly positive definite, that is $\gamma_0 - \lambda_0\sigma^2 > 0$. Given a continuous function $f : \mathbb{R} \to \mathbb{R}$, let us consider the problem

$$(13)\qquad \begin{cases} p(x,D)u = f(u) + h(x) & \text{a.e. in } \Omega, \\ u = 0 & \text{a.e. in } \Omega^c. \end{cases}$$

Theorem 3. *Suppose that the function $f(u)$ satisfies*

$$\inf_{u\geq 0} f(u) = -\infty, \qquad \sup_{u\leq 0} f(u) = +\infty. \tag{14}$$

Then for each $h \in L_\infty(\mathbb{R}^N)$ the problem (13) has at least one weak solution $\tilde{u} \in H_0^{a^2,1/2}(\Omega) \cap L_\infty(\Omega)$. Moreover if the function h is nonnegative on $\mathbb{R}^N$ then the solution $\tilde{u}$ is nonnegative on Ω.

Proof. We will follow the lines of the proof in [3, Proposition 1] where the semilinear Dirichlet problem for the Laplacian $-\Delta$ was considered. By (14) there exist constants $a \leq 0 \leq b$ such that

$$f(a) \geq \max\{-\inf_{\mathbb{R}^N} h, 0\} \quad \text{and} \quad f(b) \leq \min\{-\sup_{\mathbb{R}^N} h, 0\}.$$

We define a truncation of the nonlinearity $f(u)$ by means of formula

$$\tilde{f}(u) = \begin{cases} f(a), & \text{if} \quad u < a, \\ f(u), & \text{if} \quad a \leq u \leq b, \\ f(b), & \text{if} \quad u > b. \end{cases}$$

The truncation $\tilde{f}$ is a bounded continuous function. Hence $\tilde{f}$ satisfies the growth condition (f_r) and the primitive

$$\tilde{F}(u) = \int_0^u \tilde{f}(\xi)d\xi$$

satisfies the coercivity condition (F) since $\gamma_0 - \lambda_0\sigma^2 > 0$. Therefore by Theorem 1 the truncated problem

$$\begin{cases} p(x,D)u &= \tilde{f}(u) + h(x) \qquad \text{a.e. in } \Omega, \\ u &= 0 \qquad \text{a.e. in } \Omega^c. \end{cases}$$

has at least one weak solution $\tilde{u} \in H_0^{a^2,1/2}(\Omega)$. We will prove that

$$a \leq \tilde{u}(x) \leq b \quad \text{a.e. in } \Omega.$$

Then $\tilde{u}$ is a (bounded) solution of original problem (13).

Let $(\tilde{u} - b)^+ \geq 0$ be a test function. Since $H_0^{a^2,1/2}(\Omega)$ is a Dirichlet space $(\tilde{u} - b)^+ \in H_0^{a^2,1/2}(\Omega)$. Then we have

$$\begin{aligned} B(\tilde{u}, (\tilde{u} - b)^+) &= \int_{\mathbb{R}^N} (\tilde{f}(\tilde{u}(x)) + h(x))(\tilde{u} - b)^+(x)\,dx \\ &= \int_{supp\,((\tilde{u}-b)^+)} (\tilde{f}(\tilde{u}(x)) + h(x))(\tilde{u} - b)^+(x)\,dx \leq 0 \end{aligned} \tag{15}$$

since

$$\tilde{f}(\tilde{u}(x)) + h(x) = f(b) + h(x) \leq 0 \quad \text{on} \quad supp\,((\tilde{u} - b)^+)$$

by the definition of $\tilde{f}$. Further

$$0 \leq B((\tilde{u} - b)^+, (\tilde{u} - b)^+) = B(\tilde{u}, (\tilde{u} - b)^+) - B(\tilde{u} \wedge b, (\tilde{u} - b)^+) \leq 0$$

by (15) and since for all $b \geq 0$

$$B(\tilde{u} \wedge b, (\tilde{u} - b)^+) \geq 0$$

by the property of Dirichlet forms (see e.g. [11, p.32]). Therefore $(\tilde{u} - b)^+ = 0$ and $\tilde{u} \leq b$. In the same way taking as a test function $(\tilde{u} + a)^- \geq 0$ we can show that $\tilde{u} \geq a$. In particular if the function h is nonnegative we have $a = 0$. This means that $\tilde{u} \geq 0$. □

Corollary 2. *For each $\rho > 0$ and $h \in L_\infty(\mathbb{R}^N)$ the problem* (12) *has at least one weak solution $\tilde{u} \in H_0^{a^2,1/2}(\Omega) \cap L_\infty(\Omega)$. Moreover if the function h is nonnegative on $\mathbb{R}^N$ then the solution $\tilde{u}$ is nonnegative on Ω.*

Now, consider the problem

$$\left\{ \begin{array}{rcll} p(x,D)u & = & \lambda f(u) & \text{a.e. in } \Omega, \\ u & = & 0 & \text{a.e. in } \Omega^c, \end{array} \right. \tag{16}$$

here $f : \mathbb{R} \to \mathbb{R}$ is a continuous function such that $f(0) = 0$ and $\lambda > 0$ is a real parameter. Clearly $u = 0$ is a trivial solution of (16). We are interesting in the existence of nontrivial solutions.

Theorem 4. *Suppose that the function $f(u)$ satisfies the assumption*

$$\liminf_{u \to +\infty} f(u) < 0, \tag{17}$$

and there exists $\xi > 0$ such that $f(\xi) > 0$. Then for each $\lambda > 0$ sufficiently large the problem (16) *has at least one nontrivial nonnegative weak solution $\tilde{u} \in H_0^{a^2,1/2}(\Omega) \cap L_\infty(\Omega)$.*

Proof. The arguments in the case of the Dirichlet problem for the Laplacian $-\Delta$ is well-known, see [2]. By (17) and the continuity of f there exist a constant $b > \xi$ such that $f(b) = 0$. We define a truncation of the nonlinearity $f(u)$ by means of formula

$$\tilde{f}(u) = \begin{cases} 0, & \text{if } \ u < 0, \\ f(u), & \text{if } \ 0 \leq u \leq b, \\ 0, & \text{if } \ u > b. \end{cases}$$

The truncation $\tilde{f}$ is a bounded continuous function. Let us note that for all $\lambda > 0$ the nonlinearity $\lambda \tilde{f}$ has the same "zeros" as $\tilde{f}$. Therefore $\lambda \tilde{f}$ satisfies the growth condition (f_r), the primitive $\lambda \tilde{F}$ satisfies the coercivity condition (F), and by Theorem 1 the truncated functional

$$\tilde{J}_\lambda(u) = \frac{1}{2} B(u,u) - \lambda \int_\Omega \tilde{F}(u(x))\, dx$$

has a point of minimum $\tilde{u}_\lambda \in H_0^{a^2,1/2}(\Omega)$ for each $\lambda > 0$. Clearly $\tilde{u}_\lambda$ is a weak solution of the "truncated" problem

$$\left\{ \begin{array}{rcll} p(x,D)u & = & \lambda \tilde{f}(u) & \text{a.e. in } \Omega, \\ u & = & 0 & \text{a.e. in } \Omega^c. \end{array} \right.$$

We will prove that

$$0 \le \tilde{u}_\lambda(x) \le b \quad \text{a.e. in } \Omega.$$

Then $\tilde{u}_\lambda$ is a (bounded) solution of original problem (16).
Let $(\tilde{u}_\lambda - b)^+ \in H_0^{a^2,1/2}(\Omega)$ be a test function. Then we have

$$(18) \qquad \begin{aligned} B(\tilde{u}_\lambda, (\tilde{u}_\lambda - b)^+) &= \lambda \int_{\mathbb{R}^N} \tilde{f}(\tilde{u}_\lambda(x))(\tilde{u}_\lambda - b)^+(x)\,dx \\ &= \lambda \int_{supp\,((\tilde{u}_\lambda - b)^+)} \tilde{f}(\tilde{u}_\lambda(x))(\tilde{u}_\lambda - b)^+(x)\,dx = 0 \end{aligned}$$

since

$$\tilde{f}(\tilde{u}_\lambda(x)) = f(b) = 0 \quad \text{on} \quad supp\,((\tilde{u}_\lambda - b)^+)$$

by the definition of $\tilde{f}$. Further

$$0 \le B((\tilde{u}_\lambda - b)^+, (\tilde{u}_\lambda - b)^+) = B(\tilde{u}_\lambda, (\tilde{u}_\lambda - b)^+) - B(\tilde{u}_\lambda \wedge b, (\tilde{u}_\lambda - b)^+) \le 0$$

by (18) and since

$$B(\tilde{u}_\lambda \wedge b, (\tilde{u}_\lambda - b)^+) \ge 0$$

for $b \ge 0$ by the property of Dirichlet forms. Therefore $(\tilde{u}_\lambda - b)^+ = 0$ and $\tilde{u}_\lambda \le b$.
Similarly taking as a test function $(\tilde{u}_\lambda)^- \in H_0^{a^2,1/2}(\Omega)$ we obtain

$$(19) \qquad B(\tilde{u}_\lambda, (\tilde{u}_\lambda)^-) = \lambda \int_{\mathbb{R}^N} \tilde{f}(\tilde{u}_\lambda(x))(\tilde{u}_\lambda)^-(x)\,dx = 0$$

and

$$0 \le B((\tilde{u}_\lambda)^-, (\tilde{u}_\lambda)^-) = B((\tilde{u}_\lambda)^+, (\tilde{u}_\lambda)^-) - B(\tilde{u}_\lambda, (\tilde{u}_\lambda)^-) \le 0$$

by (19) and since

$$B((\tilde{u}_\lambda)^+, (\tilde{u}_\lambda)^-) \le 0$$

by the property of Dirichlet forms (see e.g. [11, p.33]. Hence $(\tilde{u}_\lambda)^- = 0$ and $\tilde{u}_\lambda \ge 0$.
Finally, we will show that $\tilde{u}_\lambda \ne 0$ for $\lambda > 0$ sufficiently large. Let us note that by condition (f_r) the functional

$$J_F(u) = \int_{\mathbb{R}^N} F(u(x))\,dx$$

is defined and continuous on the space $L_{\frac{2N}{N-2r}}(\Omega)$, see the proof of Lemma 1 and [1]. Let $u_0(x) \equiv \xi$ on Ω. Clearly $u_0 \in L_{\frac{2N}{N-2r}}(\Omega)$ and $J_F(u_0) > 0$. Further the space $H_0^{a^2,1/2}(\Omega)$ is densely embedded into $L_{\frac{2N}{N-2r}}(\Omega)$. By continuity arguments for ε small enough we can take an element $u_\varepsilon \in H_0^{a^2,1/2}(\Omega)$ such that

$$J_F(u_\varepsilon) \ge J_F(u_0) - \varepsilon > 0.$$

Then for $u_\varepsilon \in H_0^{a^2,1/2}(\Omega)$ we obtain

$$J_\lambda(u_\varepsilon) = \frac{1}{2} B(u_\varepsilon, u_\varepsilon) - \lambda J_F(u_\varepsilon) < \frac{1}{2} B(u_\varepsilon, u_\varepsilon) - \lambda (J_F(u_0) - \varepsilon) < 0$$

for $\lambda > 0$ sufficiently large. Hence

$$\min_{H_0^{a^2,1/2}(\Omega)} J_\lambda < 0.$$

Since $J_\lambda(0) = 0$ by definition of the primitive $\tilde{F}$ and $\tilde{u}_\lambda$ is a minimum of J_λ it follows that $\tilde{u}_\lambda \neq 0$ for all $\lambda > 0$ sufficiently large. □

Remark 3. In this section we considered just simple examples showing that the typical truncation technique known for second-order elliptic partial differential operators can be applied to the nonlocal Dirichlet problem (2). It seems that much more delicate and involved results could be obtained by a combination of sub- and super-solution techniques with topological methods of critical points theory, see [13] for the case of local problems.

References

[1] J. Appell, P. P. Zabrejko: *Nonlinear Superposition Operators,* Cambridge Univ. Press, Cambridge 1990

[2] H. Brézis, L. Nirenberg: *Remarks on finding critical points,* Comm. Pure. Appl. Math. **44** (1991), 939-963

[3] D. de Figueiredo, J.-P. Gossez: *A semilinear elliptic problem without growth condition,* C. R. Acad. Sci. Paris **308** (1989), 277-280

[4] E. B. Dynkin: *An Introduction to Branching Measure-Valued Processes,* Amer. Math. Soc., Providence RI 1994

[5] M. Fukushima, Y. Oshima, H. Takeda: *Dirichlet Forms and Symmetric Markov Processes,* Walter de Gruyter, Berlin 1994

[6] W. Hoh, N. Jacob: *On the Dirichlet problem for pseudodifferential operators generating Feller semigroups,* J. Funct. Anal. **137** (1996), 19-48

[7] N. Jacob: *A class of Fellers semigroups generated by pseudo-differential operators,* Math. Zeitschr. **215** (1994), 151-166

[8] N. Jacob: *Various approaches to the Dirichlet problem for non-local operators satisfying the positive maximum principle,* in: Proc. Intern. Conf. Potential Theory [Ed.: J. Kral et al.], Walter de Gruyter, Berlin 1996, p.367-375

[9] N. Jacob: *Pseudo-Differential Operators and Markov Processes,* Akademie-Verlag, Berlin 1996

[10] N. Jacob, R. Schilling: *Somc Dirichlet spaces obtained by subordinate reflected diffusion,* Rev. Mat. Iberoamericana, to appear

[11] Z.-M. Ma, M. Röckner: *Introduction to the Theory of Non-Symmetric Dirichlet Forms,* Springer, Berlin 1992

[12] V. MOROZ, P. P. ZABREJKO: *On the Hammerstein equations with natural growth conditions,* Zeitschr. Anal. Anw. **18, 3** (1999), 625-638

[13] M. STRUWE: *Variational Methods,* Springer, Berlin 1990

[14] M. M. VAINBERG: *Variational Methods in the Study of Nonlinear Operators* [in Russian], Gostekhizdat, Moscow 1956; Engl. transl.: Holden Day, San-Francisco 1964

NIELS JACOB, Institut für Theoretische Informatik und Mathematik, Universität der Bundeswehr München, Werner-Heisenberg-Weg 39, D-85577 Neubiberg, Germany; jacob@informatik.unibw-muenchen.de

VITALY MOROZ, Belgosuniversitet, Matematicheskij Fakul'tet, pr. Skariny 4, BY-220050 Minsk, Belorussia; moroz@mmf.bsu.unibel.by

Progress in Nonlinear Differential Equations
and Their Applications, Vol. 40

Bifurcation for One-Parameter Families of Scalar Maps: A Geometric Viewpoint

MARIO MARTELLI, GAREGUIN MIKAELIAN, SUZANNE SINDI

Ad Alfonso Vignoli, con l'affetto e la stima di sempre

Summary: Using some elementary geometric ideas we provide (sufficient) transversality conditions for bifurcation of one-parameter families of scalar maps. The approach has two advantages: first, it shows how the conditions arise and, second, it provides very simple proofs of sufficiency.

Keywords: bifurcation, transversality, multiplicity, stability.

Classification: 47H15, 58E07, 26B10.

1. Introduction. The four typical bifurcation phenomena arising in discrete dynamical systems governed by one-parameter families of scalar maps are fold, transcritical, pitchfork and period-doubling. Almost all recent books on discrete dynamical systems, designed either for teachers, researchers, or both, include a discussion about sufficient (transversality) conditions for each one of the four types of bifurcation. However, the origin of the sufficient conditions is never discussed, and, possibly as a consequence of this omission, two not so desirable consequences frequently arise. First, now and then some incorrect conditions are included and inaccurate results are derived [3]. Second, even when the conditions and the results are correct, the road to arrive at the desired conclusion is not as simple as it could be [1]. In addition to this, the exchange of stability between different branches of fixed points or periodic orbits is frequently omitted, despite its obvious relevance to the processes modeled by the dynamical systems.

The situation described above motivates the present paper, which illustrates how the necessary and the sufficient conditions arise and when and how the exchange of stability is taking place. Notice that we are not talking about necessary and sufficient conditions. The proofs that the listed sufficient conditions do indeed guarantee the respective type of bifurcation will appear in a forthcoming paper, together with some results on bifurcation of one-parameter families of vector functions.

The following elementary geometric ideas will be used repeatedly throughout the paper. Let $G : \mathbb{R}^2 \to \mathbb{R}$ be continuous together with its partial derivatives up to the order that will be needed. Consider the level curve of level 0 of G, i.e., the set $C = \{(x, y) : G(x, y) = 0\}$, and let $P = (x_0, y_0)$ be a non isolated point of C. Let L be the family of lines passing through P. The family depends on a parameter m, representing the slope of the line corresponding to it: $L = \{y = y_0 + m(x - x_0) : m \in \mathbb{R}\}$. To study the intersections between any line of L and the curve C, consider the zeros of the function $g(x) = G(x, y_0 + m(x - x_0))$. As expected, x_0 is one of them, namely $g(x_0) = 0$, since

P belongs to C and to every line of the family L. The line tangent to C at P is the one corresponding to the value of m for which x_0 is a double root of the equation $g(x) = 0$. The root x_0 is double for $g(x) = 0$ when, in addition to $g(x_0) = 0$, we have $g'(x_0) = 0$. Since

(1) $$g'(x) = G_x(x, y) + mG_y(x, y),$$

we obtain

(2) $$m = -\frac{G_x(P)}{G_y(P)}, \quad P = (x_0, y_0).$$

The tangent line to C at P is the vertical line $x = x_0$ if and only if $G_y(P) = 0$ and $G_x(P) \neq 0$.

It may happen that the curve C "passes through" $P = (x_0, y_0)$ twice. In this case x_0 is a double root of $g(x) = 0$ regardless of the value of m. For this to happen, (1) must have infinitely many solutions when the partial derivatives are evaluated at P. Thus,

(3) $$G_x(P) = G_y(P) = 0.$$

There are now two tangent lines to the curve, one for each of the two branches of C passing through P. Their slopes are found by searching those values of m for which $g''(x_0) = 0$ (triple root). This requires that

(4) $$G_{xx}(P) + 2mG_{xy}(P) + m^2G_{yy}(P) = 0.$$

The solutions of the quadratic equation (4) are the slopes of the two tangent lines. It may happen that (4) is a perfect square, in which case the two branches have the same tangent at P. When one of the tangent lines is vertical (one solution has to be $m = \infty$), the quadratic term $m^2G_{yy}(P)$ must disappear, i.e.,

(5) $$G_{yy}(P) = 0.$$

Both tangent lines could be vertical. Then we must have

(6) $$G_{yy}(P) = G_{xy}(P) = 0.$$

In any case, provided that P is only a double point for the curve C, (4) cannot be an identity since this would imply that P is at least a triple point. Hence, one (or more) of the three coefficients of (4) has to be different from 0.

These simple ideas, valid for algebraic curves and adjustable to transcendental ones, will be used to derive the results on transversality and exchange of stability. The paper is organized as follows. Section 2 contains definitions and notations to be used throughout the paper. In Section 3 we present the results on transversality conditions and exchange of stability. Section 4 contains some concluding remarks.

2. Notations and definitions. Let I and J be two non empty open intervals and let $F : I \times J \to \mathbb{R}; (a, x) \to F(a, x)$, be differentiable with continuous partial derivatives up to the order 3. Consider the one-parameter family of scalar maps $\{F(a, \cdot) : a \in I\}$ and assume that for every value of the parameter a there is an interval J_a which is invariant under the action of $F(a, \cdot)$.

Given $x_0 \in J_a$, the orbit of x_0, denoted by $O(x_0)$, is the sequence

$$x_0,\ x_1 = F(a, x_0),\ x_2 = F(a, x_1) = F(a, F(a, x_0)), \dots .$$

We shall use the symbols F^2, F^3 etc. to denote the second, third etc. iterates of $F(a,\cdot)$. Consequently, $x_2 = F^2(a, x_0)$, and, in general, $x_n = F^n(a, x_0)$. We say that x_0 is a fixed point if $x_1 = F(a, x_0) = x_0$ and is a periodic point of period p if $x_p = F^p(a, x_0) = x_0$, but $x_i \neq x_0$ for every $i = 1, 2, ..., p-1$. Fixed points are usually denoted with the symbol x_s, where the subscript s stands for stationary. A fixed point x_s is a *sink* if there exists $r > 0$ such that $|x_0 - x_s| \leq r$ implies that $O(x_0)$ converges to x_s [5]. A periodic point of period p is a sink if all its points are sinks for the pth iterate of $F(a,\cdot)$.

For notational convenience we shall use $F_x(P)$ to denote the partial derivative of F with respect to x evaluated at the point P. Similar symbols are used for the partial derivative of F with respect to a and for higher-order partial derivatives.

It can be easily proved that x_s is a sink if either one of the following two conditions is verified:

(i) F_x is continuous at (a_s, x_s) and $|F_x(a_s, x_s)| < 1$.

(ii) There is $r > 0$ such that $|F_x(a_s, x)| < 1$ for every x such that $0 < |x - x_s| \leq r$.

Using the chain rule, the above conditions can be easily rephrased for periodic orbits. In particular, a periodic orbit of period 2: $\{p = F(a, q),\ q = F(a, p),\ q \neq p\}$ is a sink if $|F_x(a, p)F_x(a, q)| < 1$.

A fixed point x_s is a *source* if there exists $r > 0$ such that $O(x)$ intersects $(-\infty, x_s - r) \cup (x_s + r, \infty)$ for every $x \in [x_s - r, x_s + r]$, $x \neq x_s$. It can be easily seen that x_s is a source if either one of the following two conditions is verified:

(i) F_x is continuous at (a_s, x_s) and $|F_x(a_s, x_s)| > 1$.

(ii) There is $r > 0$ such that $|F_x(a_s, x)| > 1$ for every x such that $0 < |x - x_s| \leq r$.

Using the chain rule, the above conditions can be easily rephrased for periodic orbits. In particular, a periodic orbit of period 2: $\{q = F(a, p),\ p = F(a, q),\ q \neq p\}$ is a source if $|F_x(a, p)F_x(a, q)| > 1$.

A branch of fixed points of the one-parameter family is the graph of a C^2 function $x_s(a)$, defined in some open interval $I_1 \subset I$, such that $x_s(a) = F(a, x_s(a))$ for every $a \in I_1$. Branches of periodic points of period p are defined in a similar way.

A point $a_0 \in J$ is a bifurcation point for the one-parameter family if there are $q \geq 2$ branches $x_i(a), i = 1, 2, \dots, q$ of fixed or periodic points defined in intervals $J_i \subset J, a_0 \in J_i$, such that

$$x_i(a_0) = x_k(a_0),\ x_i(a) \neq x_k(a) \text{ for } a \in J_i \cap J_k,\ a \neq a_0, \tag{7}$$

$(1 \leq i < k \leq q)$. Each J_i is of the form (b, c), $[a_0, b)$, or $(b, a_0]$, $b \neq a_0$. If there are only two branches then we require $J_1 = J_2$. The period $p(\geq 1)$ of all points of all branches may be the same, or the points of some branches may have period p and the points of the remaining branches period $2p$.

As mentioned in the introduction, there are four fundamental types of bifurcation: fold, transcritical, pitchfork, and period-doubling. In the fold case two branches exist in the same interval, which is of the form $[a_0, b)$ (supercritical fold) or $(b, a_0]$ (subcritical fold).

In the transcritical case two branches exists in the same interval, which is of the form (b,c). In the pitchfork there are three branches, one existing in an interval of the form (b,c) and the other two in an interval of the form $[a_0,b)$ (supercritical case) or $(b,a_0]$ (subcritical case). In all these three cases the points of all branches are periodic of the same period p. In the last case there is a branch of periodic points of period p which exists in an interval of the form (b,c) and two additional branches of periodic points of period $2p$ which exist in an interval of the form $[a_0,b)$ (supercritical case) or $(b,a_0]$ (subcritical case).

We say that at $a=a_0$ there is an exchange of stability if the points of one branch are sinks for $a<a_0(a>a_0)$ and sources for $a>a_0(a<a_0)$, and the points of the remaining branches are sinks for $a>a_0(a<a_0)$. We shall see that in the transcritical bifurcation there is always an exchange of stability, while in the pitchfork and period-doubling case we have exchange of stability only if certain conditions are satisfied. In the fold case there is no exchange of stability. However, for a sufficiently close to a_0, the points of one branch are sinks and the points of the other branch are sources.

3. Results. This section is divided into four parts. In the first we establish two fundamental equalities which play a major role in the rest of the paper. In the second we derive the necessary and the sufficient conditions for the different types of bifurcation. In the third part we discuss stability and exchange of stability. Finally, in the last part, we prove that the sufficient conditions do the job expected from them. Bifurcation can take place also when they are not verified, since the conditions are simply sufficient, but bifurcation will always happen when the conditions are fulfilled.

3.1. Two fundamental equalities. We start with two elementary theorems for scalar functions. Both results can be extended to higher-dimensional spaces.

Theorem 1. *Let I and J be two open intervals and $F: I\times J\to\mathbb{R}$ be of class C^2. Assume that F_{xxx} is continuous and $\phi: I\to J$ is of class C^2 and such that $F(a,\phi(a))=\phi(a)$ for all $a\in I$. Then there exists $T: I\times J\to\mathbb{R}$ of class C^2 such that $x-F(a,x)=(x-\phi(a))T(a,x)$.*

Proof. Define $g:[0,1]\to\mathbb{R}$ by $g(t)=tx-F(a,tx+(1-t)\phi(a))$. Apply the fundamental theorem of calculus to g'. □

Theorem 2. *Let I and J be two open intervals and $F: I\times J\to\mathbb{R}$ be of class C^2. Assume that $\psi: J\to I$ is of class C^2 and such that $F(\psi(x),x)=x$ for all $x\in J$. Then there exists $T: I\times J\to\mathbb{R}$ of class C^1 with T_{xx} and T_{xa} continuous, such that $x-F(a,x)=(a-\psi(x))T(a,x)$.*

Proof. Define $g:[0,1]\to\mathbb{R}$ by $g(t)=x-F(ta+(1-t)\psi(x),x)$. Apply the fundamental theorem of calculus to g'. □

3.2. Transversality conditions for bifurcation. We now use the geometric ideas presented in the introduction in combination with Theorems 1 and 2 to detect the transversality conditions for each one of the four types of bifurcation. We start with the case of fold bifurcation. The function G is $G(a,x)=x-F(a,x)$. The point P has coordinates $P=(a_0,x_0)$ with $x_0=x_1(a_0)=x_2(a_0)$, where $(a,x_1(a))$ and $(a,x_2(a))$ are the two branches of fixed points starting (supercritical case) or ending at $a = a_0$. The tangent line at $P = (a_0, x_0) = (a_0, x_1(a_0)) = (a_0, x_2(a_0))$ is vertical and the point P is not a double point of the curve $G(a,x) = 0$. Using x as the independent variable, the generic line through P has the form $a = a_0 + m(x - x_0)$ (the horizontal axis is the a-axis). According to (1), the slope m of the tangent line must satisfy

$$(8) \qquad 0 = g'(x_0) = 1 - F_x(a_0, x_0) + mF_a(a_0, x_0).$$

The equation of the tangent line is $a = a_0$. Hence, $m = 0$ is a solution of (8). Moreover, it is the only one since P is a simple point for the curve. Thus necessary conditions for fold bifurcation are

$$(9) \qquad 1 - F_x(a_0, x_0) = 0, \ F_a(a_0, x_0) \neq 0.$$

We can go one step further. Assume that the fold bifurcation is supercritical and let us regard a as function of x, namely $a = a(x)$, along the curve $G(a,x) = 0$. The tangent line at P is $a = a_0$. Hence, $a'(x_0) = 0$. The fold is guaranteed by $a''(x_0) \neq 0$ and it is supercritical if $a''(x_0) > 0$. Differentiating twice the equality $x - F(a(x), x) = 0$, and evaluating at P, we obtain

$$(10) \qquad F_{xx}(P) + a''(x_0)F_a(P) = 0.$$

Hence,

$$(11) \qquad F_{xx}(P) \neq 0, \ F_{xx}(P)F_a(P) < 0$$

are the transversality conditions for supercritical fold and

$$F_{xx}(P) \neq 0, \ F_{xx}(P)F_a(P) > 0$$

are the ones for subcritical fold.

We now examine the transcritical and pitchfork cases. The point $P = (a_0, x_0)$ is a double point of the curve $G(a,x) = 0$. Hence, the two partial derivatives of G have to be 0 at P, namely

$$(12) \qquad F_a(P) = 0, \ F_x(P) = 1.$$

The above conditions are necessary for both cases. Moreover, when the bifurcation is transcritical, we can assume that neither one of the two tangent lines is vertical, while in the case of pitchfork one tangent line is vertical and the other one is not vertical. Hence, according to (5), a transversality condition for transcritical bifurcation is

$$(13) \qquad F_{xx}(P) \neq 0,$$

while transversality conditions for pitchfork bifurcation are

$$(14) \qquad F_{xx}(P) = 0, \ F_{xa}(P) \neq 0.$$

From (14) we derive that in the pitchfork case the two tangent lines are distinct. We require the same condition for transcritical bifurcation. Therefore, we assume that

$$(15) \qquad [F_{xa}(P)]^2 - F_{xx}(P)F_{aa}(P) > 0$$

in both cases. The analysis of the pitchfork can go one step further. Recall that a branch $x_1(a)$ of fixed points exists in an open interval J, $a_0 \in J$ (see Section 2). Using Theorem 1 write

$$x - F(a,x) = G(a,x) = (x - x_1(a))(x - H(a,x)). \tag{16}$$

Differentiating (16) with respect to x, and evaluating at P, we obtain $x_0 = H(P)$. Differentiating twice with respect to x, and evaluating again at P, gives $1 - H_x(P) = 0$. Differentiating (16) with respect to x and with respect to a, and evaluating at P, we obtain $H_a(P) = F_{xa}(P) \neq 0$. Clearly, a_0 is a natural candidate for fold bifurcation for the map H. The only conditions to be included is $H_{xx}(P) \neq 0$. Differentiating (16) three times with respect to x, and evaluating at P, we obtain

$$F_{xxx}(P) = 3H_{xx}(P). \tag{17}$$

Therefore, transversality conditions for transcritical bifurcation are

$$F_{xx}(P) \neq 0, [F_{xa}(P)]^2 - F_{xx}(P)F_{aa}(P) > 0 \tag{18}$$

and for pitchfork

$$F_{xx}(P) = 0, F_{xa}(P) \neq 0, F_{xxx}(P) \neq 0. \tag{19}$$

The pitchfork bifurcation is supercritical when

$$F_{ax}(P)F_{xxx}(P) = 3H_a(P)H_{xx}(P) < 0.$$

When a_0 is a period-doubling bifurcation, we can repeat the considerations made above for the pitchfork case by replacing $F(a,x)$ with $F(a, F(a,x))$. However, first let us notice that $P = (a_0, x_0)$ is not a double point of the curve $G(a,x) = x - F(a,x) = 0$. Hence, at least one of the partial derivatives of G has to be different from 0 at P. Thus,

$$[1 - F_x(P)]^2 + [F_a(P)]^2 > 0. \tag{20}$$

At the same time, P is at least a double point for the curve $K(a,x) = x - F(a, F(a,x)) = 0$. Hence, both partial derivatives of K are 0 at P, i.e.,

$$0 = 1 - (F_x(P))^2, \; 0 = F_a(P)[1 + F_x(P)]. \tag{21}$$

To satisfy (20) and (21) we need $F_x(P) = -1$. In fact, if $F_x(P) = 1$, then the second equality of (21) requires that $F_a(P) = 0$. Both conditions imply that (20) is violated. Thus $F_x(P) = -1$ is necessary for period-doubling bifurcation.

We recognize that the tangent line to the branch of fixed points $x = x_1(a)$ passing through P is not vertical since its slope m is given by $m = F_a(P)/2$. Assuming that P is a double point for the curve $K(a,x) = x - F(a, F(a,x)) = 0$, the slopes of the tangent lines to $K(a,x) = 0$ at the point P are the solutions of the equation

$$K_{aa}(P) + 2mK_{xa}(P) + m^2 K_{xx}(P) = 0. \tag{22}$$

Since

$$K_{xx}(P) = -F_{xx}(P)[F_x(P)]^2 - F_x(P)F_{xx}(P) \tag{23}$$

and $F_x(P) = -1$, we obtain $K_{xx}(P) = 0$. Hence, one of the tangent lines is vertical, i.e., its equation is $a = a_0$. Since the other one is not we must have $K_{xa}(P) \neq 0$. This implies that

$$r = F_{xx}(P)F_a(P) + 2F_{xa}(P) \neq 0. \tag{24}$$

To complete the picture of what is taking place at P, let us write $x - F(a, F(a,x)) = K(a,x)$ in the form

$$x - F(a, F(a,x)) = K(a,x) = (x - x_1(a))(x - H(a,x)). \tag{25}$$

where $x_1(a)$ is the branch of fixed points of $F(a,x)$ passing through P. We can now repeat the procedure previously used for the pitchfork bifurcation. We obtain $x_0 - H(P) = 0$, $1 - H_x(P) = 0$, and $H_a(P) = -F_{xx}(P)F_a(P) - 2F_{xa}(P) \neq 0$. Hence a_0 is a natural candidate for fold bifurcation for the function H. The only condition which remains to be required is $H_{xx}(P) \neq 0$. Differentiating (25) three times with respect to x and evaluating at P, gives

$$s = 2F_{xxx}(P) + 3[F_{xx}(P)]^2 = -3H_{xx}(P). \tag{26}$$

Summarizing, the transversality conditions for period-doubling bifurcation are

$$\begin{aligned} r &= F_{xx}(P)F_a(P) + 2F_{ax}(P) \neq 0; \\ s &= 2F_{xxx}(P) + 3[F_{xx}(P)]^2 \neq 0, \end{aligned} \tag{27}$$

with $rs < 0$ giving the supercritical case.

Before concluding this second part of Section 3 we would like to mention that Theorem 4.3 of [3] replaces condition (24) with $F_{ax}(P) \neq 0$. The following example shows that period-doubling bifurcation may not take place.

Example 3.1. Let $F(a,x) = a - x - ax + x^2 + x^3$. Notice that $F(0,0) = F_x(0,0) + 1 = 0$, $F_{ax}(0,0) = -1$, and $3[F_{xx}(0,0)]^2 + 2F_{xxx}(0,0) = 24$. Hence, all conditions of Theorem 4.3 are satisfied. Assume that $F(a,p) = q$ and $F(a,q) = p$ with a, p, and q close to 0 and $p \neq q$. From $q - p = F(a,p) - F(a,q)$ and $q^2 - p^2 = qF(a,p) - pF(a,q)$ we obtain

$$p^2 + q^2 = p^2 q + q^2 p$$

an equality which is obviously false since $|p^2 q| < p^2$ and $|q^2 p| < q^2$.

Assuming that $F(a,0) = 0$, for all $a \in J$, where J is an open interval containing 0, the conditions provided in Theorem 4.3 are correct. In fact, in this case, $F_a(0,0) = 0$ and (24) reduces to $F_{ax}(0,0) \neq 0$.

Example 3.1 can be used to show that the necessary condition $F_x(P) = -1$ does not guarantee that period-doubling bifurcation is taking place. Likewise, the conditi- on $F_x(P) = 1$ is necessary for fold, transcritical and pitchfork, but, as the following example shows, does not guarantee any one of them.

Example 3.2. Let $F(a,x) = a + x + x^3$. Since $F_x(P) = 1$ requires $x = 0$, and $F(a,0) = 0$ implies $a = 0$ we see that 0 is the only possible choice for bifurcation. However, the equation $F(a,x) = x$ has one and only one solution for every $a \in \mathbb{R}$. Hence, $a = 0$

is not a bifurcation point. Notice that $F_a(0,0) = 1$ which implies that 0 cannot be a transcritical or pitchfork bifurcation, and $F_{xx}(0,0) = 0$ which does not satisfy one of the transversality conditions for fold bifurcation.

We summarize the results obtained so far in the following table, where $P = (a_0, x_0)$.

	$F_x(P)$	$F_a(P)$	$F_{xx}(P)$	*Other*
Fold	$= 1$	$\neq 0$	$\neq 0$	$F_{xx}(P)F_a(P) < 0$ supercritical
Transcritical	$= 1$	$= 0$	$\neq 0$	$[F_{xa}(P)]^2 - F_{xx}(P)F_{aa}(P) > 0$
Pitchfork	$= 1$	$= 0$	$= 0$	$F_{xa}(P) \neq 0,\ F_{xxx}(P) \neq 0$ $F_{xa}(P)F_{xxx}(P) < 0$ supercritical
Period-doubling	$= -1$			$r = F_{xx}(P)F_a(P) + 2F_{ax} \neq 0$ $s = 2F_{xxx}(P) + 3(F_{xx}(P))^2 \neq 0$ $rs < 0$ supercritical

Before concluding this section we would like to point out that the conditions listed in the first two columns of the above table are necessary for the corresponding type of bifurcation. The conditions of the third and fourth column are then sufficient, if we assume that in the transcritical, pitchfork and period-doubling case there is a branch of fixed points $(a, x(a))$ that exists for $a \in J$, where J in an open interval containing a_0. This result will be established in the fourth part of this section. Bifurcation may take place when one or more of the sufficient conditions are not satisfied, as the following example shows.

Example 3.3. Let $F(a,x) = a^6 + x - x^2$. Then $a_0 = 0$ is a transcritical bifurcation point. The two branches of fixed points are $x_1(a) = a^3$ and $x_2(a) = -a^3$. The necessary conditions are satisfied since $F_x(0,0) = 1$, $F_a(0,0) = 0$. Moreover, $F_{xx}(0,0) \neq 0$. However, the tangent lines to the two branches coincide. Hence, the condition $[F_{xa}(P)]^2 - F_{xx}(P)F_{aa}(P) > 0$ is not verified.

3.3. Bifurcation: exchange of stability. Let us fist examine the situation at a fold bifurcation. To understand what is happening we consider

$$\frac{d}{dx}F_x(a(x), x) = F_{xa}(a(x), x)a'(x) + F_{xx}(a(x), x). \tag{28}$$

Computing (28) at (a_0, x_0) we obtain $F_{xx}(P)$. Since this quantity is different from zero we know that $F_x(a(x), x)$ is either increasing or decreasing at x_0. Consequently, one of the branches is made of sources, and the other of sinks, at least for a sufficiently close to a_0.

A similar analysis can be performed in the case of transcritical bifurcation. This time we differentiate with respect to a.

$$\frac{d}{da}F_x(a, x_1(a)) = F_{xa}(a, x_1(a)) + F_{xx}(a, x_1(a))x_1'(a). \tag{29}$$

The assumption $[F_{xa}(P)]^2 - F_{xx}(P)F_{aa}(P) > 0$ implies that the above derivative is non zero at P. Hence, for $a < a_0$, the points of the branch $(a, x_1(a))$ are sinks if $F_{xa}(P) + F_{xx}(P)x_1'(a) > 0$ and sources if $F_{xa}(P) + F_{xx}(P)x_1'(a) < 0$. Moreover, since

$$(F_{xa}(P) + F_{xx}(P)x_1'(a)) \times (F_{xa}(P) + F_{xx}(P)x_2'(a)) = -([F_{xa}(P)]^2 - F_{xx}(P)F_{aa}(P)) < 0, \tag{30}$$

there is always an exchange of stability at P.

For the pitchfork bifurcation the situation is a bit more complicated. Assume first that $F_{xa}(P) > 0$ and the bifurcation is supercritical, namely $F_{xxx}(P) < 0$. Then, the points of the branch $x_1(a)$ are sinks for $a < a_0$ and sources for $a > a_0$. For the points of the branch $a(x)$ we have

$$\frac{d}{dx}F_x(a(x), x) = 0 \text{ at } P \tag{31}$$

Differentiating one more time with respect to x and computing at P we obtain

$$\frac{d^2}{dx^2}F_x(a(x), x)[x = x_0] = F_{xa}(P)a''(x_0) + F_{xxx}(P). \tag{32}$$

This implies

$$\frac{d^2}{dx^2}F_x(a(x), x)[x = x_0] = \frac{2}{3}F_{xxx}(P) < 0. \tag{33}$$

Thus, $F_x(a(x), x)$ has a maximum point at x_0 and the points of the branch $a(x)$ are sinks for $a > a_0$. We can easily see that if $F_{xa}(P) < 0$, and the bifurcation is still supercritical, then the points of $a(x)$ are sources while the points of $x_1(a)$ are sinks for $a > a_0$. The situation for the subcritical case is similar. In summary:

- *Supercritical case* $(F_{xa}(P)F_{xxx}(P) < 0)$.
  If the fixed points of the branch $x_1(a)$ are sinks before and sources after a_0 $(F_{xa}(P) > 0)$ the points of the branch $a(x)$ are sinks since $F_{xxx}(P) < 0$. There is an exchange of stability at a_0.
  If the fixed point of the branch $x_1(a)$ are sources before and sinks after a_0 $(F_{xa}(P) < 0)$ the points of the branch $a(x)$ are sources since $F_{xxx}(P) > 0$. There is no exchange of stability at a_0.

- *Subcritical case* $(F_{xa}(P)F_{xxx}(P) > 0)$.
  The conclusions of the two statements above should be reversed.

For the period-doubling bifurcation replace $F(a, x)$ with $F(a, F(a, x))$. Let $r = 2F_{xa}(P) + F_{xx}(P)F_a(P)$, $s = 2F_{xxx}(P) + 3[F_{xx}(P)]^2$. Using the strategy just outlined for the pitchfork case we obtain the following.

- *Supercritical period-doubling* $(rs < 0)$.
  If the fixed points of the branch $x_1(a)$ are sinks before and sources after a_0 $(r < 0)$, the periodic orbit is a sink, since $s > 0$. There is an exchange of stability at a_0.
  If the fixed point of the branch $x_1(a)$ are sources before and sinks after a_0 $(r > 0)$, the periodic orbit is a source, since $s < 0$. There is no exchange of stability at a_0.

- *Subcritical period-doubling* $(rs > 0)$.
  The conclusions of the two statements above should be reversed.

3.4. Bifurcation: the given transversality conditions are sufficient. In 3.2 we have shown that $F_x(P) = 1$ is necessary for fold, transcritical, and pitchfork bifurcation, while $F_x(P) = -1$ is necessary for period-doubling. Both results can easily be derived from the implicit function theorem. We shall always assume that $P = (a_0, x_0)$.

Theorem 3. *Fold, transcritical, or pitchfork bifurcation can take place at $P = (a_0, x_0)$ only if $F_x(P) = 1$. A period-doubling bifurcation can take place only if $F_x(P) = -1$.*

In the three theorems below we shall always assume that the condition $F_x(P) = 1$ is verified. The proofs will appear in a forthcoming paper.

Theorem 4. *P is a fold bifurcation provided that $F_a(P)F_{xx}(P) \neq 0$.*

The setting for Theorems 5, 6, and 7 is the following. We assume the existence of a C^2 function $x_1(a)$ defined in some open interval J containing a_0 such that $x_1(a) - F(a, x_1(a)) = 0$ for every $a \in J$.

Theorem 5. *P is a transcritical bifurcation provided that $F_a(P) = 0$, $F_{xx}(P) \neq 0$ and $[F_{xa}(P)]^2 - F_{xx}(P)F_{aa}(P) > 0$.*

Theorem 6. *P is a pitchfork bifurcation provided that $F_a(P) = 0$, $F_{xx}(P) = 0$, $F_{xa}(P) \neq 0$ and $F_{xxx}(P) \neq 0$.*

For the next theorem we assume that $F_x(P) = -1$.

Theorem 7. *P is a period-doubling bifurcation provided that $F_{xx}(P)F_a(P) + 2F_{ax}(P) \neq 0$, and $2F_{xxx}(P) + 3[F_{xx}(P)]^2 \neq 0$.*

Bifurcation phenomena are so important in nonlinear analysis that there is consensus among the experts regarding the precise meaning of a bifurcation point [2] . In the case of one-parameter families of scalar maps the consensus translates into the following setting. Let U be an open subset of $\mathbb{R}^2$ and $G : U \to \mathbb{R}$ be a C^3 map. Let $S = \{\mathbf{x} \in \mathbb{R}^2 : G(\mathbf{x}) = 0\}$ and assume that M is a one-dimensional manifold contained in S. M is called the *trivial set* of solutions of $G(\mathbf{x}) = 0$. Let $\mathbf{x_0} \in M$ be such that M is locally parametrized at $\mathbf{x_0} = (a_0, x_0)$ by a smooth simple curve $\mathbf{a} : J \to \mathbb{R}^2$, $\mathbf{a}(t) = (a(t), x(t))$, where J is an open interval containing 0, and $\mathbf{a}(0) = \mathbf{x_0}$. The point $\mathbf{x_0}$ is a bifurcation point for the equation $G(\mathbf{x}) = G(a, x) = 0$ relative to the manifold M if for every $d > 0$ the ball $B(\mathbf{x_0}, d)$ contains points of S outside of M.

It is clear that fold bifurcation does not qualify as a bifurcation point in this setting, while period-doubling does for the second iterate of F. In the new approach we can prove a basic result, which incorporates both transcritical and pitchfork bifurcation. Without loss of generality we may assume that $\mathbf{x_0} = \mathbf{0}$.

Theorem 8. *Let G be such that $G_x(\mathbf{0}) = 0$ and assume that either one of the following two sets of conditions is verified:*

(i) *$a'(0) \neq 0$ and $\frac{d}{dt}[G_x(a(t), x(t))] \neq 0$ at $t = 0$;*

(ii) *$a'(0) = 0$, $a''(0) \neq 0$, $\frac{d}{dt}[G_x(a(t), x(t))] = 0$ and $\frac{d^2}{dt^2}[G_x(a(t), x(t))] \neq 0$ at $t = 0$.*

Then $\mathbf{x_0} = (0, 0)$ is a bifurcation point relative to the manifold M.

The two theorems below can be regarded as the multi-dimensional version of Theorem 8 (i) and (ii), respectively.

Theorem 9. *Let $F : J \times \mathbb{R}^q \to \mathbb{R}^q$ and assume that there is a branch of fixed points $\mathbf{x} = \mathbf{x}(a)$ which exists in an open interval $J_1 \subset J$. Assume that* $\det(I - F_{\mathbf{x}}(a, \mathbf{x}(a)))$ *changes sign at $a = a_0 \in J_1$. Then $P = (a_0, \mathbf{x}(a_0))$ is a bifurcation point.*

Theorem 10. *Let $F : J \times \mathbb{R}^q \to \mathbb{R}^q$ and assume that there is a surface of fixed points $a = a(\mathbf{x})$ which exists in an open ball $B(\mathbf{0}, d)$, for some $d > 0$ and $a'(\mathbf{0}) = \mathbf{0}$. Assume that $F_a(0, \mathbf{0}) = \mathbf{0}$, and* $\det(F_{a\mathbf{x}}(0, \mathbf{0})) \neq 0$. *Then $(0, \mathbf{0})$ is a bifurcation point.*

4. Conclusion. The analysis of the bifurcation phenomena in the scalar case provided in the second and third part of the previous section has the advantage of being based on simple geometrical ideas and the chain rule. The inverse function theorem plays a key role in part 4 of Section 3. Overall, the presentation given in the paper shows why the sufficient conditions look the way they do. At a first glance, they appear counter-intuitive, but once we go through the motion of searching for their origin, we discover why they are the way they are, and we avoid the risk of providing incorrect listing. The approach used in the one-dimensional setting benefits the investigation of bifurcation phenomena in higher dimensional spaces, since it permits an easy recognition and proof of some important cases.

References

[1] R. L. Devaney: *An Introduction to Chaotic Dynamical Systems*, Addison-Wesley, Reading MA 1989

[2] M. Furi, M. Martelli, M. P. Pera: *General bifurcation theory: local results and applications*, in: Differential Equations and Applications to Biology and to Industry [Ed.: M. Martelli et al.], World Scientific, Singapore 1995, p. 101-116

[3] Yu. A. Kuznetsov: *Elements of Applied Bifurcation Theory*, Springer, Berlin 1994

[4] M. Martelli: *Introduction to Discrete Dynamical Systems and Chaos*, John Wiley & Sons, New York 1999

[5] M. Martelli, D. Marshall: *Stability and attractivity in discrete dynamical systems,* Math. Biosci. **128** (1995), 347-355

Mario Martelli, California State University, Fullerton, Mathematics Department, Fullerton, CA 92834, USA; mmartelli@fullerton.edu

Gareguin Mikaelian, California State University, Fullerton, Mathematics Department, Fullerton, CA 92834, USA

Suzanne Sindi, California State University, Fullerton, Mathematics Department, Fullerton, CA 92834, USA

Progress in Nonlinear Differential Equations
and Their Applications, Vol. 40

Mountain Pass and Linking Type Solutions for Semilinear Dirichlet Forms

MICHELE MATZEU

Dedicated to Alfonso Vignoli on the occasion of his 60th birthday

Summary: Using mountain pass and linking type techniques, we obtain some existence results of (possibly positive) solutions for certain semilinear problems associated with general Dirichlet forms.

Keywords: Dirichlet form, semilinear problem, mountain pass theorem, linking theorem.

Classification: 31C25, 35J60, 58E05.

Acknowledgement: The author wishes to thank Umberto Mosco for very useful and stimulating discussions on this subject and also for having brought the paper by Falconer [12] to his attention. This work is supported by MURST, Project "Metodi Variazionali ed Equazioni Differenziali Non Lineari".

0. Introduction. This paper is motivated by the recent increasing development of the theory of Dirichlet forms, in order to approach the study of some semilinear equations of the form

$$(P) \qquad \Delta u(x) + f(x, u(x)) = 0 \qquad (x \in \Omega),$$

to be interpreted in this abstract framework, by using some basic results of critical point theory.

Some papers also appeared on this subject. In [1] the theory of Dirichlet forms is used by referring to the Laplace operator for variational fractals and a problem as (P) with Dirichlet boundary condition is considered.

In [6] the framework is very general, the "Laplace" operator is replaced by the generator of a Dirichlet form with suitable properties, and no further boundary condition appears in the possible "concrete" differential example (P) (i.e. $\Omega = \mathbb{R}^N$ in (P)).

Both papers [6,12] deal with the problem of finding solutions of (P) as critical points of a suitable functional. It is well known that this kind of approach is based on the test of the *Palais-Smale condition* whose validity relies on the use of some appropriate (continuous and compact) embedding results of Sobolev spaces into L^p-spaces.

Actually, a deep investigation of the structural properties of the domains suitably associated with Dirichlet forms and their related generators has been developed in many recent papers by Biroli and Mosco (see [3,4,19] for the case of variational fractals). This type of analysis has allowed them to get some important functional embedding results, analogous to the above mentioned theorems in the Sobolev space setting, in the more general context of Dirichlet forms.

In [1] the continuous embedding properties shown in [3,4,19], as well as a *compact* embedding property proved in [5] (see also [10]) have been used in order to guarantee that the Palais-Smale condition actually holds. On the other hand, [6] assumes a suitable coercivity condition on the "potential" related to the problem which indeed guarantees this compactness property. In both papers [6] and [12] the solutions of the respective problems are found by a natural mountain pass technique.

In the present paper the purpose is twofold. Firstly, we state the same kind of results obtained in [12] (in particular assuming Dirichlet boundary conditions), but in the very general abstract framework presented in [6] for the "free" problem (i.e. $\Omega = \mathbb{R}^N$). Secondly, we consider a generalization of the problem, due to the presence of a "linear perturbation" of the operator A associated with the Dirichlet form, say a term of the type $\lambda a(x)u(x)$, where $u(x)$ represents the solution of the problem. In connection with the interaction of the parameter λ with the eigenvalues of the operators $A/a(x)$, another existence result is obtained by a linking type technique, which generalizes a well known result due to Rabinowitz [22] in case that $-A$ is the usual Laplacian in Euclidean space.

Actually, various examples of "differential" problems can be considered as possible applications of the results stated in the present paper: the case, as mentioned above, considered in [12], but also other cases as, for example, those related to Dirichlet forms associated with weighted Sobolev spaces (see [7,8,21]), as well as those related to Dirichlet forms associated with the Kohn-Heisenberg Laplacian, see [2,9,15,17,23] for some results on similar problems for this operator. Moreover, some more general classes of uniformly subelliptic operators can be included in this context.

In any case, it has to be remarked that the common problem is essentially to prove the above mentioned compact embedding property. For example, in the case of weighted Sobolev spaces it is guaranteed when a suitable relation between the growth exponent at infinity of the nonlinear term, the integrability exponent of the "weight function", and the Euclidean dimension of the space holds.

We point out that, when the present paper was finished, the author became aware of the preprint [13] where a deep investigation of this kind of problems is developed in case of an odd nonlinearity and in the framework of Dirichlet forms associated with the Laplace operator related to the Sierpiński Gasket.

1. The linear framework. Let $a(\cdot,\cdot)$ be a given *strongly local irreducible Dirichlet form*, with domain $\mathcal{D}[a]$ in $L^2(X,m)$, where X is a *locally compact separable Hausdorff* space, whose topology is endowed with a *pseudodistance* d, with respect to which X is supposed to be *complete* and m is a *positive Radon measure* on X with supp $[m] = X$. Denoting, for any $(u,v) \in \mathcal{D}[a] \times \mathcal{D}[a]$, by $\alpha(u,v)$ the *signed Radon measure* on X such that

$$a(u,v) = \int_X d\alpha(u,v),$$

we suppose that the following Poincaré property holds:

$$(P)\quad \begin{cases} \forall R_0 > 0\ \exists c_0 = c_0(R_0) > 0\ ,\ k_1 = k_1(R_0) \geq 1 \text{ such that } \quad \forall r \leq R_0 \\ \displaystyle\int_{B(x,r)} |u - \overline{u}_{x,r}|^2 m(dx) \leq c_0 \int_{B(x,k_1 r)} d\alpha(u,u) \forall u \in \mathcal{D}_2[a, B(x,k_1 r)], \end{cases}$$

where

- $\forall \rho > 0$, $B(x,\rho)$ denotes the *ρ-ball* centered at $x \in X$, w.r.t. the pseudodistance d,
- $\overline{u}_{x,\rho} = \dfrac{1}{m(B(x,\rho))} \displaystyle\int_{B(x,\rho)} u\, m(dx)$,
- for any *open bounded* subset Ω of X,

$$\mathcal{D}_2[a,\Omega] = \left\{ u \in \mathcal{D}_{\text{loc}}[a,\Omega] : \int_\Omega d\alpha(u,u) + \int_\Omega u^2 m(dx) < +\infty \right\},$$

 where $\mathcal{D}_{\text{loc}}[a,\Omega]$ is the set of all *measurable* functions on X coinciding m-almost everywhere with some function of $\mathcal{D}[a]$ on every compact subset of Ω.

Furthermore, we suppose that the so called "duplication" property holds:

$$(D)\quad \begin{cases} \forall R_0 > 0\ \exists c_1 = c_1(R_0) \text{ such that, for } r \leq R/2 < R \leq R_0 \\ \text{one has } m(B(x,r)) \geq c_1 \left(\frac{r}{R}\right)^\nu m(B(x,R)), \end{cases}$$

where ν is a *positive real number independent* of r, R, and R_0.
At this point we fix an open bounded subset Ω of X such that the capacity of $X \backslash \Omega$ is non zero, and consider the space

$$\mathcal{D}_0(\Omega) = \overline{\mathcal{D}(\Omega) \cap C_0(\Omega)}$$

where $\mathcal{D}(\Omega)$ denotes the set of the restrictions to Ω of the functions in $\mathcal{D}[a]$, $C_0(\Omega)$ is the set of *continuous* functions with *compact support* in Ω, and the *closure* is intended with respect to the "intrinsic" norm

$$\|u\|_{a,\Omega} = [a_\Omega(u,u) + \|u\|^2_{L^2(X,m)}]^{1/2},$$

where, for any u, $v \in \mathcal{D}(\Omega) \cap C_0(\Omega)$,

$$a_\Omega(u,v) = a(\tilde{u},\tilde{v}) = \int_\Omega d\alpha(\tilde{u},\tilde{v}) \text{ with } \tilde{u},\tilde{v} \in \mathcal{D}[a] \text{ such that } \tilde{u}|_\Omega = u \text{ and } \tilde{v}|_\Omega = v.$$

One verifies that a_Ω is a *strongly local regular Dirichlet form* with domain $\mathcal{D}[a_\Omega] = \mathcal{D}_0(\Omega)$. On the other hand it follows, as a consequence of (P), that the norm $\|u\|_{a,\Omega}$ is *equivalent* to the norm

$$\|u\| = a_\Omega(u,u)^{1/2}.$$

From now on, we shall consider the space $\mathcal{D}_0 = \mathcal{D}_0(\Omega)$ equipped with this norm. Finally, we define A_0 as the *linear selfadjoint* operator, defined on a domain $\mathcal{D}_{A_0} \subset \mathcal{D}_0(\Omega)$ with values in $L^2(\Omega)$ in such a way that

$$a_\Omega(u,v) = \int_\Omega (A_0 u)v m(dx) \quad \forall\, u \in \mathcal{D}_{A_0}\ \forall\, v \in \mathcal{D}_0.$$

Examples. A large class of Dirichlet forms satisfying all assumptions listed above can be exhibited. Here we mention some main cases which have been considered in the recent literature, namely

- forms associated with degenerate elliptic operators with a weight in the Muckenhoupt's class (see [4] for the proof of (D) and (P));
- forms associated with subelliptic Laplace operators generated by vector fields satisfying a Hörmander condition in the case that the length of the vector corresponding to the commutators spanning $\mathbb{R}^N$ is bounded from above and from below by strictly positive constants (see [16] for the proof of (D) and (P));
- energy forms on nested fractals as considered in [18];
- forms associated with elliptic operators on C^∞-Riemannian manifolds with Ricci curvature bounded from below.

In the following we shall use some well known results concerning the embeddings of $\mathcal{D}_0$ into the L^p-spaces as well as into the space C^0. Let us summarize with the following

Lemma 1. *The space $\mathcal{D}_0$ is*
(a) *continuously embedded in $L^{\frac{2\nu}{\nu-2}}$, if $\nu > 2$;*
(b) *continously embedded in $L^p(\Omega)$ for any $p \geq 2$, if $\nu = 2$;*
(c) *continuously embedded in $C^0(\overline{\Omega})$, if $\nu < 2$;*
(d) *compactly embedded in $L^2(\Omega)$ for any $\nu > 0$.*
For the proof of (a), (b), (c) see [3,4] and, for variational fractals, [19]; for the proof of (d) see [5] and Remark 4.3 of [10].

2. The nonlinear framework. Fix a function $f : \overline{\Omega} \times \mathbb{R} \to \mathbb{R}$ such that

$$(f_1) \qquad f \in C^0(\overline{\Omega} \times \mathbb{R}),$$

and suppose that

$$(f_2) \qquad |f(x,t)| \leq a_1 + a_2 |t|^s\ \forall\,(x,t) \in \overline{\Omega} \times \mathbb{R}\ , \text{ for some } a_1, a_2 > 0,$$

and, for some $s \in \left(1, \frac{\nu+2}{\nu-2}\right)$ if $\nu > 2$, or for some $s > 1$ if $\nu = 2$,

$$(f_3) \qquad f(x,t) = o(|t|) \text{ as } t \to 0 \text{ uniformly w.r.t. } x \in \overline{\Omega},$$

and

$$(f_4) \qquad \begin{cases} \exists\,\mu > 2,\ r > 0 \quad \text{such that} \\ 0 < \mu F(x,t) \le t f(x,t) \quad \forall x \in \overline{\Omega}\,, \quad \forall t \ge r, \end{cases}$$

where

$$F(x,t) = \int_0^t f(x,\tau)d\tau \quad \forall x \in \overline{\Omega}\,, \quad \forall t \in \mathbb{R}.$$

Remark 1. No condition is assumed in (f_2) in case $\nu < 2$.

Remark 2. Observe that condition (f_4) implies

$$(F_4) \qquad F(x,t) \ge \text{ const. } t^\mu \quad \forall t \ge r.$$

Remark 3. In the second of the two problems we are going to deal with, assumption (f_4) will be reinforced by the following one:

$$(f_5) \qquad \begin{cases} \exists\,\mu > 2\,, \quad r > 0 \quad \text{such that} \\ 0 < \mu F(x,t) \le t f(x,t) \quad \forall x \in \overline{\Omega}\,, \quad \forall t \in \mathbb{R} \text{ with } |t| \ge r. \end{cases}$$

In this case, (F_4) will be consequently reinforced by

$$(F_5) \qquad F(x,t) \ge \text{ const. } |t|^\mu \quad \forall t \in \mathbb{R} \text{ with } |t| \ge r.$$

In the first problem we are going to deal with, we shall consider the "truncated" function

$$\overline{f}(x,t) = \begin{cases} 0 \quad \forall x \in \overline{\Omega},\ t < 0, \\ f(x,t) \quad \forall x \in \overline{\Omega},\ t \ge 0 \end{cases}$$

and put

$$\overline{F}(x,t) = \int_0^t \overline{f}(x,t)d\tau \quad \forall x \in \overline{\Omega}\,, \quad \forall t \in \mathbb{R}.$$

It is easy to check that $\overline{f}$ verifies all the analogous properties $(\overline{f}_1)$, $(\overline{f}_2)$, $(\overline{f}_3)$, $(\overline{f}_4)$, $(\overline{F}_4)$ obtained by replacing f with $\overline{f}$ in (f_1), (f_2), (f_3), (f_4), (F_4), respectively.

3. First problem. We are interested in the following problem:

$$(\mathcal{P}) \qquad \begin{cases} \text{Find } \ u \in \mathcal{D}_{A_0} \text{ such that } u \ge 0,\ u \not\equiv 0 \ m\text{-a.e. on } \Omega, \\ A_0 u(x) = f(x, u(x)) \quad m\text{-a.e. on } \Omega. \end{cases}$$

Obviously, by the definition of $\overline{f}$, problem $(\mathcal{P})$ is equivalent to the problem

$$(\overline{\mathcal{P}}) \qquad \begin{cases} \text{Find } \ u \in \mathcal{D}_{A_0} \text{ such that } u \ge 0,\ u \not\equiv 0 \quad m\text{-a.e. on } \Omega, \\ A_0 u(x) = \overline{f}(x, u(x)) \quad m\text{-a.e. on } \Omega. \end{cases}$$

A "weak" formulation of problem $(\overline{\mathcal{P}})$ may be stated as follows:

$$(\overline{\mathcal{P}}_w) \qquad \begin{cases} \text{Find } u \in \mathcal{D}_{A_0} \text{ such that } u \geq 0,\ u \not\equiv 0\ m\text{-a.e. on } \Omega, \\ a_\Omega(u,v) - \int_\Omega \overline{f}(x,u(x))v(x)m(dx) = 0 \quad \forall v \in \mathcal{D}_0. \end{cases}$$

The definition of $\mathcal{D}_0$, A_0, $\mathcal{D}_{A_0}$ enables us to verify that the problems $(\overline{\mathcal{P}})$ and $(\overline{\mathcal{P}}_w)$ are indeed equivalent. So we are reduced to deal with solutions of $(\overline{\mathcal{P}}_w)$. On the other hand, any solution u of $(\overline{\mathcal{P}}_w)$ can be viewed as a non-negative non-trivial critical point of the functional

$$I(v) = \tfrac{1}{2} a_\Omega(v,v) - \int_\Omega \overline{F}(x,v(x))m(dx)$$

$$= \frac{1}{2}\|v\|^2 - \int_\Omega \overline{F}(x,v(x))m(dx) \quad \forall v \in \mathcal{D}_0,$$

that is, an element u such that

$$(I') \qquad u \in \mathcal{D}_0\,, \quad \langle I'(u), v\rangle = 0 \quad \forall v \in \mathcal{D}_0,$$

where $\langle\cdot,\cdot\rangle$ denotes the pairing between $\mathcal{D}_0$ and its dual space $\mathcal{D}_0'$.

Indeed, properties $(\overline{f}_1)$, $(\overline{f}_2)$ and Lemma 1 easily imply that I is actually a C^1-functional on $\mathcal{D}_0$. If u is a solution of $(\overline{\mathcal{P}}_w)$, then the third relation in $(\overline{\mathcal{P}}_w)$ implies that I is critical. Conversely, if u is critical for I, then (I') yields the third relation in $(\overline{\mathcal{P}}_w)$ which actually implies the further fact that u belongs to $\mathcal{D}_{A_0}$, as $g(x) = f(x,u(x))$ belongs to $L^2(\Omega)$, as a consequence of $(\overline{f}_1)$, $(\overline{f}_2)$.

Therefore, from now on, we look for non-negative, non-trivial critical point of the functional I. Indeed, one can prove the following

Theorem 1. *Let f verify $(f_1),\ldots,(f_4)$. Then there exists at least one solution of problem $(\mathcal{P})$.*

Proof. We proceed by steps.

Step 1. Reduction to the problem

$$(I'_+) \qquad u \in \mathcal{D}_0\backslash\{0\}\,, \quad u \geq 0: \langle I'(u), v\rangle = 0 \quad \forall v \in \mathcal{D}_0.$$

Step 2. The functional I verifies the Palais-Smale condition, that is

$$(PS) \qquad \begin{cases} \text{Any } \{v_k\} \subset \mathcal{D}_0 \text{ such that } |I(v_k)| \leq \text{ const., and } \{I'(v_k)\} \to 0 \text{ in } \mathcal{D}_0' \\ \text{admits a subsequence strongly converging in } \mathcal{D}_0'. \end{cases}$$

Proof of Step 1 and 2. First of all, let us prove that $\{v_k\}$ is bounded in $\mathcal{D}_0$. Indeed, using the fact that $\{I'(v_k)\} \to 0$, hence in particular

$$|\langle I'(v_k), v_k\rangle| \leq \mu\|v_k\|$$

for k large enough, and the boundedness of $\{I(v_k)\}$, one gets

$$\begin{aligned} \text{const.} + \|v_k\| &\geq I(v_k) - \frac{1}{\mu}\langle I'(v_k), v_k\rangle \\ (1) \qquad &= \frac{1}{2}\|v_k\|^2 - \int_\Omega \overline{F}(v_k)m(dx) - \frac{1}{\mu}\left(\|v_k\|^2 - \int_\Omega \overline{f}(v_k)\cdot v_k m(dx)\right) \\ &= \left(\frac{1}{2} - \frac{1}{\mu}\right)\|v_k\|^2 - \int_\Omega \left(\overline{F}(v_k) - \frac{1}{\mu} f(v_k)v_k\right) m(dx) \quad \forall k. \end{aligned}$$

Observing that $\overline{f}$ verifies $(\overline{f}_4)$ not only for $t > r$, but trivially for $t \leq 0$, too, one has

$$(2) \qquad \int_\Omega \left(\overline{F}(v_k) - \frac{1}{\mu}\overline{f}(v_k)v_k\right) m(dx) \leq 0 \qquad \forall k.$$

Thus, by (1) and (2),

$$\text{const.} + \|v_k\| \geq \left(\frac{1}{2} - \frac{1}{\mu}\right)\|v_k\|^2 \qquad \forall k$$

which yields the boundedness of $\{v_k\}$, as $\mu > 2$. Therefore, by Lemma 1 one gets, respectively,

$$(3) \qquad v_k \rightharpoonup v_0 \text{ in } \mathcal{D}_0 \quad \text{and} \quad v_k \to v_0 \text{ in } L^2(\Omega)$$

for a suitable subsequence of $\{v_k\}$, still denoted by $\{v_k\}$.
At this point, in order to prove the *strong* convergence of $\{v_k\}$ in $\mathcal{D}_0$, one can argue as in [6]. For the sake of completeness, we report all the details here.
First, let us prove that

$$(4) \qquad \int_\Omega |\overline{f}(v_k)(v_k - v_0)|m(dx) \to 0 \quad \text{as} \quad k \to \infty.$$

It is convenient to separate the two possible cases $\nu < 2$ and $\nu \geq 2$.
Case $\nu < 2$. Since, by Lemma 1, $\|v_k\|_{L^\infty} \leq$ const., from $(\overline{f}_1)$ it follows

$$(5) \qquad |\overline{f}(x, v_k(x))| \leq \text{ const. } \forall x \in \overline{\Omega}\,,$$

so (5) and the strong convergence of $\{v_k\}$ in $L^2(\Omega)$ yield (4).
Case $\nu \geq 2$. In this case $(\overline{f}_2)$ yields

$$(6) \qquad \begin{cases} \overline{f}(x,t) = o(|t|^{\beta-1}) \quad \text{as } |t| \to \infty\,, \text{ uniformly w.r.t. } x \in \overline{\Omega} \\ \text{with } \beta = \frac{2\nu}{\nu-2} > 2 \text{ if } \nu > 2 \text{ and } \beta = s' + 1\,,\ \ s' > s, \text{ if } \nu = 2. \end{cases}$$

By (6), $(\overline{f}_1)$, $(\overline{f}_2)$, $(\overline{f}_3)$, one easily checks that

$$(7) \qquad \forall \varepsilon > 0\ \exists C_\varepsilon > 0 : |\overline{f}(x,t)| \leq \varepsilon(|t| + |t|^{\beta-1}) + C_\varepsilon|t| \quad \forall t \in \mathbb{R}\,, \quad \forall x \in \overline{\Omega};$$

therefore,

$$
\begin{aligned}
&\int_\Omega |\overline{f}(v_k)(v_k - v_0)|m(dx) \\
(8) \qquad &\leq \varepsilon \int_\Omega (|v_k|^2 + |v_k|^\beta + |v_0|^2 + |v_0|^\beta)|m)dx \\
&+C_\varepsilon \int_\Omega |v_k|\ |v_k - v_0|m(dx).
\end{aligned}
$$

At this point, since ε is an *arbitrary* positive number, (4) follows from (8), Lemma 1, and property (d). Let us prove that (4) gives

$$
(9) \qquad \lim_{k\to\infty} (\|v_k\|^2 - (v_k, v_0)_{\mathcal{D}_0}) = 0,
$$

where $(\cdot,\cdot)_{\mathcal{D}_0}$ denotes the inner product in $\mathcal{D}_0$, which, together with the first relation in (3), yields the strong convergence of $\{v_k\}$ to v_0 in $\mathcal{D}_0$, due to the fact that $\mathcal{D}_0$ is a Hilbert space.

Actually, the definition of I and the evaluation of I' yield

$$
\begin{aligned}
&\|v_k\|^2 - (v_k, v_0)_{\mathcal{D}_0} = a_\Omega(v_k, v_k) - a_\Omega(v_k, v_0) \\
(10) \qquad &= \langle I'(v_k), v_k\rangle + \int_\Omega \overline{f}(v_k)(v_k - v_0)m(dx) + + \int_\Omega \overline{f}(v_k)(v_k - v_0)m(dx) \\
&-\langle I'(v_k), v_0\rangle - \int_\Omega \overline{f}(v_k)v_0 m(dx).
\end{aligned}
$$

Therefore, by the boundedness of $\{v_k\}$ in $\mathcal{D}_0$ and the fact that $\{I'(v_k)\} \to 0$, one deduces from (10) that

$$
(11) \qquad \lim_{k\to+\infty} (\|v_k\|^2 - (v_k, v_0)_{\mathcal{D}_0}) = \lim_{k\to+\infty} \int_\Omega \overline{f}(v_k)(v_k - v_0)m(dx).
$$

Finally, (5) and (11) imply (9), so the assertion of Step 2 follows.

Step 3. There exist some ρ, $\gamma > 0$ such that $I(v) > \gamma$ for any $v \in \mathcal{D}_0$ with $\|v\| = \rho$.

Proof of Step 3. As in the proof of Step 2, we distinguish the two possible cases for ν.

Case $\nu < 2$. For any $\varepsilon > 0$, $(\overline{f}_3)$ yields

$$
\overline{F}(x,t) < \varepsilon|t|^2 \quad \forall x \in \overline{\Omega}\,, \quad \forall t \in \mathbb{R} \text{ with } |t| < \delta = \delta_\varepsilon;
$$

therefore, by the continuous embedding of $\mathcal{D}_0$ in both $L^2(\Omega)$ and $L^\infty(\Omega)$, one deduces

$$
(12) \qquad \int_\Omega \overline{F}(x, v(x))m(dx) \leq \varepsilon \text{ const. } \|v\|^2
$$

if $\|v\| = \rho = \rho(\delta_\varepsilon) = \rho_\varepsilon$ sufficiently small. Thus (12) and the definition of I yield the statement for $\nu < 2$.

Case $\nu \geq 2$. First of all, formulating (7) for the primitive $\overline{F}$ gives

$$
(13) \qquad \forall \varepsilon > 0\ \exists C_\varepsilon > 0 : \overline{F}(x,t) \leq \varepsilon(|t|^2 + |t|^\beta) + C_\varepsilon|t|^\beta \quad \forall x \in \overline{\Omega}\,, \forall t \in \mathbb{R}
$$

with $\beta = \frac{2\nu}{\nu-2} > 2$ if $\nu > 2$, and $\beta = s' + 1$, $s' > s$, if $\nu = 2$. By taking into account the continuous embedding of $\mathcal{D}_0$ in $L^2(\Omega)$ and in $L^\beta(\Omega)$, one deduces from (13) that

$$\int_\Omega \overline{F}(x, v(x))m(dx) \leq \text{const. } (\varepsilon\|v\|^2 + (\varepsilon + C_\varepsilon)\|v\|^\beta) \quad \forall v \in \mathcal{D}_0 ,$$

hence

$$I(v) \geq \left(\frac{1}{2} - \varepsilon \text{ const.}\right) \|v\|^2 - (\varepsilon + C_\varepsilon)\|v\|^\beta \quad \forall v \in \mathcal{D}_0.$$

So the assertion of Step 3 follows, as $\beta > 2$.

Step 4. There exists $v \in \mathcal{D}_0$ such that $I(v_-) < 0$.

Proof of Step 4. Let $\overline{v} \in \mathcal{D}_0$ with $\overline{v}(x) \geq c > 0$ for all $x \in \overline{\Omega}$ (for example, one can choose a strictly positive continuous function of $\mathcal{D}_0$) and consider the functional I on the half-line $\{t\overline{v} : t > 0\}$. One has

$$I(t\overline{v}) = \frac{1}{2}t^2 a_\Omega(\overline{v}, \overline{v}) - \int_\Omega \overline{F}(x, t\overline{v}(x))m(dx) \quad \forall t > 0.$$

Then, by $(\overline{F}_4)$ one gets for $t > r/c$

$$I(t\overline{v}) \leq 1/2t^2 a_\Omega(\overline{v}, \overline{v}) - \overline{c}t^\beta, \ \overline{c} > 0; \tag{14}$$

so the assertion follows by taking $v_- = t\overline{v}$, with sufficiently large $t > r/c$.

Step 5. There exists a non-trivial critical point u of I.

One can apply the mountain pass theorem by Ambrosetti and Rabinowitz [1] which, thanks to the statements of Step 2, 3, 4, yields the existence of a critical point u of I such that

$$I(u) \geq \gamma > 0,$$

hence $u \not\equiv 0$, as $I(0) = 0$.

Step 6. Conclusion: the critical point u of I in Step 5 is such that

$$u(x) \geq 0 \quad m\text{-a.e. in } \Omega.$$

Proof of Step 6. Let us put $u_- = \min(u, 0)$. From the properties of a it follows that u_- belongs to $\mathcal{D}_0$. Moreover, since u is critical we have

$$a_\Omega(u, u_-) = \int_\Omega \overline{f}(x, u(x))u_-(x)m(dx) = \int_\Omega \overline{f}(x, 0)u_-(x)m(dx) = 0. \tag{15}$$

On the other hand, the Markovian property of the Dirichlet form a, which is preserved by its restriction a_Ω, yields

$$a_\Omega(u - u_-, u_-) = 0. \tag{16}$$

Thus (15) and (16) imply

$$a_\Omega(u_-, u_-) = 0$$

which, by the regularity of a_Ω, gives $u_- = 0$, so $u(x) \geq 0$ m-a.e. on Ω. □

4. Second problem. Let us fix now an arbitrary number $\lambda \neq 0$, a real function $a \in \mathrm{Lip}\,(\overline{\Omega})$ with $a(x) > 0$ for all $x \in \overline{\Omega}$, and consider the following problem:

$$(\mathcal{P}^\lambda) \qquad \begin{cases} \text{Find } u \in \mathcal{D}_{A_0} \text{ such that } u \not\equiv 0 \quad m\text{-a.e. on } \Omega, \\ A_0 u(x) = \lambda a(x) u(x) + f(x, u(x)) \quad m\text{-a.e. on } \Omega. \end{cases}$$

Obviously, for $\lambda = 0$ one obtains problem $(\mathcal{P})$, apart from the non-negativity of u. As in the case of problem $(\mathcal{P})$, one can give an equivalent "weak" formulation of problem $(\mathcal{P}^\lambda)$ as follows:

$$(\mathcal{P}^\lambda_w) \qquad \begin{cases} \text{Find } u \in \mathcal{D}_{A_0} \text{ such that } u \not\equiv 0 \quad m\text{-a.e. on } \Omega, \\ a_\Omega(u, v) - \lambda \int_\Omega a(x) u(x) v(x) m(dx) \\ - \int_\Omega f(x, u(x)) v(x) m(dx) = 0 \ \forall v \in \mathcal{D}_0, \end{cases}$$

and then one can find any solution of $(\mathcal{P}^\lambda)$ as a non-trivial critical point of the functional

$$I^\lambda(v) = \frac{1}{2} a_\Omega(v, v) - \frac{\lambda}{2} \int_\Omega a(x) v(x)^2 m(dx) \\ - \int_\Omega F(x, v(x)) m(dx) \quad \forall v \in \mathcal{D}_0.$$

Note that, under the previous assumptions (f_1), (f_2) on f, as well as for $\lambda = 0$, in this case I^λ is also a C^1-functional on $\mathcal{D}_0$, whose non-trivial critical points yield solutions of $(\mathcal{P}^\lambda_w)$, thus of $(\mathcal{P}^\lambda)$. Actually, the approach to the study of critical points of I^λ is *different* in case that either $\lambda < \lambda_1$ or $\lambda \geq \lambda_1$, where λ_1 is the first eigenvalue of the operator $A_0/a(x)$. Indeed, the problem can be solved in a quite analogous way as problem $(\mathcal{P})$ (i.e. for $\lambda = 0$) in case that $\lambda < \lambda_1$ and one finds a solution which is moreover non-negative, exactly under the same assumptions $(f_1), \ldots, (f_4)$ of Theorem 1.

On the other hand, in case $\lambda \geq \lambda_1$ one can use a different critical point technique which does not yield the non-negativity of the solution, so one cannot use the "truncation argument", where f is replaced by $\overline{f}$; so one has to replace assumption (f_4) with the stronger assumption (f_5) (see Remark 3). Moreover, one considers the sequence $\{\lambda_k\}$ of eigenvalues of $A_0/a(x)$, with $0 < \lambda_1 \leq \lambda_2 \leq \ldots \leq \lambda_k \leq \lambda_{k+1} \leq \ldots$, counting multiplicities. (We refer to [20] for general spectral properties of fractal structures, and to [14] for the particular case of the Sierpiński Gasket.) Assume the following further condition

$$(F_6) \qquad \int_\Omega F(x, v(x)) m(dx) \geq 0 \ \forall\, v \in V_k \text{ if } \lambda \in [\lambda_k, \lambda_{k+1}) \text{ with } \lambda_k < \lambda_{k+1},$$

where V_k is the k-dimensional space spanned by the eigenfunctions related to the eigenvalues $\{\lambda_1, \ldots, \lambda_k\}$. Then one can state the following

Theorem 2. *Let $\lambda < \lambda_1$ and let f verify $(f_1), \ldots, (f_4)$. Then there exists at least one non-negative solution u of $(\mathcal{P}^\lambda)$. Let $\lambda \geq \lambda_1$, more precisely $\lambda \in [\lambda_k, \lambda_{k+1})$ with*

$\lambda_{k+1} > \lambda_k$, and let f verify (f_1), (f_2), (f_3), (f_5); moreover, let (F_6) hold. Then there exists at least one solution of $(\mathcal{P}^\lambda)$.

The proof of the second statement of Theorem 2 (i.e. for $\lambda \geq \lambda_1$) relies on the use of the following "Linking theorem" due to Rabinowitz [22].

Lemma 2. *Let E be a real Banach space with $E = V \oplus V_0$, where V_0 is finite dimensional. Let us suppose that $J \in C^1(E)$ satisfies* (PS) *and the further conditions*

$$(J_1) \qquad \exists\, \rho, \alpha > 0 : J(v) \geq \alpha \ \forall v \in V : \|v\| = \rho$$

and

$$(J_2) \qquad \begin{cases} \exists\, e \in V,\ \|e\| = 1 \text{ and } R > \rho \text{ such that } J(v) \leq 0 \ \forall v \in \partial Q, \\ \text{where } Q = \{w \in E : w = v_0 + re, v_0 \in V_0, \|v_0\| \leq R,\ r \in [0, R]\}. \end{cases}$$

Then J possesses a critical point u such that $J(u) \geq \alpha$.

Proof of Theorem 2. As for the first statement, that is in case $\lambda < \lambda_1$, the variational characterization of λ_1, i.e.

$$\lambda_1 = \inf_{v \in \mathcal{D}_0 \setminus \{0\}} \frac{a_\Omega(v, v)}{\int_\Omega a(x) v(x)^2 m(dx)}$$

enables us to state that the square root of the quadratic form

$$a_\Omega^\lambda(v, v) = a_\Omega(v, v) - \lambda \int_\Omega a(x) v(x)^2 m(dx)$$

is a norm on $\mathcal{D}_0$ which is equivalent to the usual norm given by $a_\Omega(v, v)^{1/2}$. Therefore, for $\lambda < \lambda_1$, one can argue exactly in the same way as in the proof of Theorem 1 in order to find a critical point of mountain pass type u for the functional I^λ associated to a_Ω^λ, so a solution of $(\mathcal{P}^\lambda)$. Moreover, as in the case $\lambda = 0$, one has the property

$$a_\Omega^\lambda(v - v_-, v_-) = 0 \quad \forall v \in \mathcal{D}_0,\ v_- = \min(v, 0);$$

so the non-negativity of u follows by the same argument used in Step 6 of the proof of Theorem 1.

Now let us consider the case $\lambda \geq \lambda_1$, more precisely let $\lambda \in [\lambda_k, \lambda_{k+1})$, with $\lambda_k < \lambda_{k+1}$. Choose the decomposition $E = \mathcal{D}_0 = V \oplus V_0$ of E, where $V_0 = V_k$ is the k-dimensional space spanned by the eigenfunctions related to $\{\lambda_1, \ldots, \lambda_k\}$, and $V = (V_k)^\perp$.

We claim that the functional $J = I^\lambda$ satisfy (J_1) and (J_2). As for (J_1), one notes that the variational characterization of $1_{\lambda+1}$, i.e.

$$\lambda_{k+1} = \inf_{v \in (V_k)^\perp \setminus \{0\}} \frac{a_\Omega(v, v)}{\int_\Omega a(x) v(x)^2 m(dx)} \tag{17}$$

yields

$$a_\Omega^\lambda(v, v) \geq \left(1 - \frac{\lambda}{\lambda_{k+1}}\right) \|v\|^2 = c_\lambda \|v\|^2, \text{ with } c_\lambda > 0, \text{ for any } v \in V_k^\perp, \tag{18}$$

while, arguing as in Step 3 of the proof of Theorem 1, one gets

$$\int_\Omega F(x,v(x))m(dx) = o(\|v\|^2) \text{ as } \|v\| \to 0 \text{ in } \mathcal{D}_0. \tag{19}$$

Therefore $J = I^\lambda$ satisfies (J_1) as a consequence of (18) and (19).
As for the proof of (J_2), first of all note that, as observed also in [22], Remark 5.5, it is enough to prove the two properties

$$I^\lambda(v) \le 0 \qquad \forall v \in V_k \tag{J_2'}$$

and

$$\begin{cases} \exists e \in V_k^\perp,\ \|e\| = 1 \text{ and } \overline{R} > 0 \text{ such that} \\ I^\lambda(v) \le 0 \quad \forall v \in V_k \oplus \text{span } \{e\} \text{ with } \|v\| \ge \overline{R}. \end{cases} \tag{J_2''}$$

Let us verify (J_2'). For this purpose, let $\{v_1, \dots, v_k\}$ be an orthonormal basis of V_k where v_i is an eigenfunction related to λ_i for $i = 1, \dots, k$. Any $v \in V_k$ is then given by

$$v = \sum_{i=1}^k t_i v_i$$

for some $t_1, \dots, t_k \in \mathbb{R}$. So, for any such an element v in V_k one has

$$\begin{aligned} a_\Omega^\lambda(v,v) &= a_\Omega(v,v) - \lambda \int_\Omega a(x) v(x)^2 m(dx) \\ &= \sum_{i=1}^k (\lambda_i - \lambda) t_i^2 \int_\Omega a(x) v_i(x)^2 m(dx) \le 0. \end{aligned} \tag{20}$$

Therefore (F_6) and (20) imply (J_2').
As for the proof of (J_2''), let us fix $e = v_{k+1}$, a normalized eigenfunction related to λ_{k+1}. Let $v \in V_k \oplus \text{span } \{v_{k+1}\}$, so

$$v = t v_{k+1} + \sum_{i=1}^k t_i v_i \qquad \text{for some } t, t_1 \dots t_k \in \mathbb{R}.$$

Then

$$\begin{aligned} a_\Omega^\lambda(v,v) &= t^2(\lambda_{k+1} - \lambda) \int_\Omega a(x)(v_{k+1}(x))^2 m(dx) \\ &\quad + \sum_{i=1}^k (\lambda_i - \lambda) t_i^2 \int_\Omega a(x)(v_i(x))^2 m(dx) \\ &\le (\lambda_{k+1} - \lambda)\|a\|_{L^\infty(\Omega)} \|v\|_{L^2(\Omega)} \le \text{ const. } \|v\|^2_{L^\mu(\Omega)}, \end{aligned} \tag{21}$$

while (f_5) implies

$$\int_\Omega F(x,v(x)) \ge C_1 \int_\Omega |v|^\mu - C_2, \qquad C_1, C_2 > 0. \tag{22}$$

Therefore (21) and (22) yield, as $\mu > 2$,

$$I^\lambda(v) \to -\infty \text{ as } \|v\|_{L^\mu(\Omega)} \to +\infty \text{ in } V_k \oplus \text{ span } \{v_{k+1}\}. \tag{23}$$

So one gets, by (23) and the fact that V_k is finite dimensional,

$$I^\lambda(v) \to -\infty \quad \text{as} \quad \|v\| \to +\infty \text{ in } V_k \oplus \text{ span } \{v_{k+1}\}$$

and (J_2'') follows.

Finally one has to verify the Palais-Smale condition. Thus, let $\{v_k\} \subset \mathcal{D}_0$ with $|I^\lambda(v_k)| \leq$ const. and $(I^\lambda)'(v_k) \to 0$ in $\mathcal{D}_0'$. By choosing $\beta \in (1/\mu, 1/2)$ and taking into account (f_5) (and its consequence (F_5)), one gets

$$\begin{aligned} &\text{const. } + \beta\|v_k\| \geq \\ &(\tfrac{1}{2} - \beta)\|v_k\|^2 - \lambda(\tfrac{1}{2} - \tfrac{1}{\beta})\|a\|_{L^\infty(\Omega)}\|v_k\|_{L^2(\Omega)} \\ &+(\beta\mu - 1) \text{ const. } \|v_k\|^\mu_{L^\mu(\Omega)}. \end{aligned} \tag{24}$$

At this point, by the continuous embedding of $\mathcal{D}_0$ into $L^\mu(\Omega)$ and the fact that $\mu > 2$, (24) easily implies the boundedness of $\{v_k\}$ in $\mathcal{D}_0$. Finally, by the same argument as given in Step 2 of the proof of Theorem 1, the boundedness of $\{v_k\}$ implies its strong convergence in $\mathcal{D}_0$. So (PS) is proved and one can apply Lemma 2 for the proof of the second statement of Theorem 2. □

Remark 4. An obvious condition implying (F_5) is the following

$$tf(t) \geq 0 \qquad \forall t \in \mathbb{R}. \tag{25}$$

Actually, in case that one assumes (25) one cannot expect to find a solution with constant sign if $\lambda \geq \lambda_1$, unless some further condition are imposed on f. Indeed, as also observed in [22] (see Remark 5.19), if v_1 is a positive eigenfunction corresponding to λ_1, and u solves $(\mathcal{P}^\lambda)$, one has

$$\begin{aligned} &\int_\Omega [\lambda a(x)u(x) + f(x, u(x))]v_1(x)m(dx) = a_\Omega(u, v_1) \\ &= a_\Omega(v_1, u) = \lambda_1 \int_\Omega a(x)v_1(x)u(x)m(dx). \end{aligned}$$

Thus,

$$(\lambda_1 - \lambda)\int_\Omega a(x)u(x)v_1(x)m(dx) = \int_\Omega f(x, u(x))v_1(x)m(dx). \tag{26}$$

If u is non-negative, the left-hand side in (26) is non-positive for $\lambda \geq \lambda_1$ while the right-hand side is non-negative in case that (25) holds. So it should be zero; but this is possible, by (25), if and only if $f(x, u(x)) \equiv 0$.

References

[1] A. Ambrosetti, P. H. Rabinowitz: *Dual variational methods in critical point theory and applications*, J. Funct. Anal. **14** (1973), 349-381

[2] I. Birindelli, I. Capuzzo Dolcetta, A. Cutri: *Liouville theorems for semilinear equations on the Heisenberg group*, Ann. Inst. H. Poincaré, Analyse Non-linéaire **14, 3** (1997), 295-308

[3] M. Biroli, U. Mosco: *Sobolev and isoperimetric inequalities for Dirichlet forms on homogeneous spaces*, Rend. Accad. Naz. Lincei **9** (1995), 37-44

[4] M. Biroli, U. Mosco: *Sobolev inequalities on homogenous spaces*, Potential Anal. **4** (1995), 311-324

[5] M. Biroli, N. Tchou: *Asymptotic behaviour of relaxed Dirichlet problems involving a Dirichlet-Poincaré form*, J. Anal. Appl. **16, 2** (1997), 281-309

[6] M. Biroli, S. Tersian: On the existence of nontrivial solutions to a semilinear equation relative to a Dirichlet form, to appear

[7] P. Caldiroli, R. Musina: *On a variational degenerate elliptic problem*, to appear

[8] P. Caldiroli, R. Musina: *On the existence of extremal functions for a weighted Sobolev embedding with critical exponent*, to appear

[9] G. Citti: *Semilinear Dirichlet problem involving critical exponent for the Kohn Laplacian*, Annali Mat. Pura Appl. **169** (1995), 375-392

[10] G. Dalmaso, V. Di Cicco, L. Notarantonio, N. Tchou: *Limits of variational problems of Dirichlet forms in varying domains*, J. Math. Pure Appl. **77** (1998), 89-116

[11] J. E. Fabes, C. Kenig, R. Serapioni: *The local regularity of solutions of degenerate elliptic equations*, Comm. Part. Diff. Equ. **7** (1982), 77-116

[12] K. J. Falconer: *Semilinear partial differential equations on self-similar fractals*, to appear

[13] K. J. Falconer, J. Hu: *Multiple solutions for a class of nonlinear elliptic equations on the Sierpiński Gasket*, to appear

[14] M. Fukushima, T. Shima: *On a spectral analysis for the Sierpiński Gasket*, Potential Anal. **1** (1992), 1-35

[15] N. Garofalo, E. Lanconelli: *Existence and non existence results for semilinear equations on the Heisenberg group*, Indiana Univ. Math. J. **41** (1992), 71-98

[16] D. Jerison: *The Poincaré inequality for vector fields satisfying the Hörmander condition*, Duke Math. J. **53** (1986), 503-523

[17] E. LANCONELLI, E. UGUZZONI: *Asymptotic behaviour and non existence theorem for semilinear Dirichlet problems involving critical exponent on unbounded domains of the Heisenberg group*, Boll. Unione Mat. Ital. **B-8** (1998), 139-168

[18] U. MOSCO: *Variational metrics on selfsimilar fractals*, C. R. Acad. Sci. Paris **321** (1995), 715-720

[19] U. MOSCO: *Dirichlet forms and self-similarity*, AMS/IP Studies Adv. Math. **8** (1998), 117-155

[20] U. MOSCO: *Spectral properties of fractal structures*, Rend. Accad. Naz. Lincei, to appear

[21] D. PASSASEO: *Some concentration phenomena in degenerate semilinear elliptic problems*, Nonlin. Anal. TMA **24, 7** (1995), 1011-1025

[22] P. H. RABINOWITZ: *Minimax Methods in Critical Point Theory with Applications to Differential Equations*, CBMS Regional Conf. Series Math. **65**, Amer. Math. Soc., Providence RI 1986

[23] F. UGUZZONI: *A non existence theorem for a semilinear Dirichlet problem involving critical exponential on half-spaces of the Heisenberg group*, Nonlin. Diff. Equ. Appl., to appear

MICHELE MATZEU, Dipartimento di Matematica, Università di Roma "Tor Vergata", Via della Ricerca Scientifica, I-00133 Roma, Italy; matzeu@mat.uniroma2.it

Progress in Nonlinear Differential Equations
and Their Applications, Vol. 40

Self-Similar Measures in Quasi-Metric Spaces

UMBERTO MOSCO

Ad Alfonso con stima ed affetto per il suo 60mo compleanno

Summary: We prove the existence and uniqueness of invariant selfsimilar measures for given families of contractions in locally compact complete quasi-metric spaces. We also show that the invariant measure is an attractor in the space of measures, both for the weak topology as well as for a suitable metric on measures, defined in terms of Hölder continuous functions. Our results apply to complete quasi-metric spaces of homogeneous type.

Keywords: Quasi-distances, quasi-metric spaces, Hölder continuous functions, spaces of homogeneous type, self-similarity, invariant self-similar measures, fractals, multifractals.

Classification: 28A80, 43A85, 28A33.

Acknowledgement: The author wishes to thank the Alexander-von-Humboldt Stiftung for assisting this research with a Humboldt Research Award and the Mathematisches Institut der Universität Bonn for hospitality while this work was carried on.

1. Introduction. In the theory of fractals and multifractals an important role is played by measures on a space X, which are invariant with respect to a finite family of mappings $S_1, \ldots, S_N$ of X into itself. Namely, (positive) measures μ with compact support in X of total mass 1 that satisfy the identity

$$\mu = \sum_{i=1}^{N} m_i \, \mu \circ S_i^{-1} \tag{1}$$

where $m_1, \ldots, m_N$ are given positive constants with $\sum_{i=1}^{N} m_i = 1$.

If $S_1, \ldots, S_N$ are contractions in a complete metric space, then the existence and uniqueness of μ has been proved by Hutchinson, [5]. He proved, in addition, that μ is an attractor in the weak convergence of measures, namely, given an initial measure μ_0 the sequence of measures

$$\mu_n = \sum_{i=1}^{N} m_i \, \mu_{n-1} \circ S_i^{-1}$$

$n = 1, 2, \ldots$, converges weakly to μ as $n \to \infty$.

The basic tool of Hutchinson's proof is a suitable metric $L(\mu, \nu)$ in the space $\mathcal{M}^1$ of Borel regular measures on X with bounded support and unit mass. The metric $L(\mu, \nu)$ is defined as follows

$$L(\mu, \nu) = \sup \{ | \int \Phi \, d\mu - \int \Phi \, d\nu | : Lip \, \Phi \leq 1 \}$$

where Φ is a (real valued) function on X and $Lip\,\Phi$ is the smallest constant $c > 0$ such that $|\Phi(x) - \Phi(y)| \leq c\,d(x,y)$ for every $x, y \in X$. For arbitrary given constants $m_1, \ldots, m_N$ as before, the map $\mu \to \mathcal{T}\mu$:

(2) $$\mathcal{T}\mu = \sum_{i=1}^{N} m_i \; \mu \circ S_i^{-1}$$

is a contraction in $\mathcal{M}^1$ with respect to the metric L. This yields the desired properties.

The aim of this paper is to extend Hutchinson's result to *quasi-metric spaces*. These are topological spaces whose topology is given by a *quasi-distance*, instead than a distance. A quasi-distance on a set X enjoies the properties of a distance, except for the triangle inequality, which is satisfied by a quasi-distance only up to a multiplicative constant $c_T \geq 1$ at the right hand side. A typical quasi-distance d on $\mathbb{R}^D$ is for example

$$d(x,y) = \sum_{h=1}^{D} |x_h - y_h|^{\alpha_h},$$

where α_h are positive constants.

Quasi-distances occur naturally on homogeneous groups and in more general spaces of homogeneous type. While a homogeneous group is built on an underlying differentiable structure, a space of homogeneous type is built on a topological space endowed with a quasi-metric compatible with the topology. Examples of non-differentiable spaces of homogeneous type arise in the theory of self-similar fractals in Euclidean spaces $\mathbb{R}^D$. In this case "effective" quasi-distances have been considered, which are suitable power of the Euclidean distance, the power exponent being related to the scaling property of the intrinsic energy functional on the fractal, [9-11]. Various classes of self-affine fractals in $\mathbb{R}$ can also be treated as self-similar fractals, after a change of the Euclidean metric of $\mathbb{R}^D$, [1]. Non isotropic quasi-distances of the type mentioned before occur also in the theory of stratified Lie groups, like the Heisenberg group, [4,12,14]. Self-similar sets and fractals in the Heisenberg group have been constructed by Strichartz, [15] and Marchi, [7].

In dealing with quasi-distances, the main problem arises from their lack of regularity. In fact, a distance is Lipschitz continuous, because the inequality $|d(x,z) - d(y,z)| \leq d(x,y)$ holds for every $x, y, z \in X$. This inequality fails for quasi-distances. As a consequence, the class of Lipschitz test functions occurring in the Hutchinson metric L may be empty, making the definition of L meaningless. In order to introduce a metric in the space of measures $\mathcal{M}^1$ we need a new class of test functions.

We shall develop our theory by replacing Lipschitz functions with *Hölder continuous* functions. Indeed, as shown by Macias-Segovia, [6], every quasi-distance d on a set X is *equivalent* to a quasi-distance $\tilde{d}$, which is Hölder continuous on bounded sets.

In our theory, however, we cannot replace d with $\tilde{d}$ from the beginning. In fact, the property of S_i of being a contraction with respect to d is not invariant under a change to an equivalent quasi-distance. We have to keep both d and $\tilde{d}$ in our picture.

An additional technical difficulty arises from the fact that we cannot estimate the Hölder seminorm of $\tilde{d}$ globally on X - as in the case of a Lipschitz distance - but only locally on bounded subsets of X.

Our metric on measures is:

$$H(\mu,\nu) = \sup\{|\int \Phi\, d\mu - \int \Phi\, d\nu| : \Phi \text{ bounded on } X, |\Phi|_{0,\gamma} \le 1\}$$

where $\gamma \in (0,1]$ is a regularity parameter of the space X, depending only on the constant c_T of X, and $|\Phi|_{0,\gamma}$ is the smallest constant $c > 0$ such that

$$|\Phi(x) - \Phi(y)| \le c\, d(x,y)^\gamma \text{ for every } x, y \in X.$$

Our main results are the following ones:

Theorem 1.1. *Let $S_1, \ldots, S_N$ be contractions in a locally compact, complete quasi-metric space X. Then:*

(i) *For every set of positive constants $m_1, \ldots, m_N$ with $m_1 + \ldots + m_N = 1$, there exists, and is unique, a Radon measure μ with compact support in X and unit mass satisfying the identity* (1).

(ii) *For every given Borel regular measure μ_0 with bounded support in X and unit mass, the sequence of measures $\mathcal{T}^n\mu_0$, $n = 1, 2, \ldots$, where $\mathcal{T}$ is the map* (2), *converges weakly and in the metric H to μ, as $n \to \infty$.*

(iii) support $\mu = K$, *where K is the uniquely determined nonempty compact subset of X such that $K = \bigcup_{i=1}^N S_i(K)$.*

The mappings $S_1, \ldots, S_N$ of X into itself satisfy the *open set condition* if there exists a nonempty bounded open set $\mathcal{O} \subset X$ such that $S_i(\mathcal{O}) \subset \mathcal{O}$ for every $i = 1, \ldots, N$ and $S_i(\mathcal{O}) \cap S_j(\mathcal{O}) = \emptyset$ if $i \ne j$. They are *contractive similitudes* in X if $d(S_i(x), S_i(y)) \equiv \ell_i\, d(x,y)$ for every $x, y \in X$, with $\ell_i \in (0,1)$, $i = 1, \ldots, N$.

Theorem 1.2. *Let $S_1, \ldots, S_N$ be contractions in a complete space of homogeneous type X. Then:*

(j) *The conclusions* (i),(ii),(iii) *of the previous theorem hold.*

(jj) *If, in addition, the maps $S_1, \ldots, S_N$ are contractive similitudes satisfying the open set condition in X and $m_i = \ell_i^{d_f}$ for every $i = 1, \ldots, N$, where ℓ_i is the contraction factor of S_i and $d_f > 0$ is uniquely given by the identity $\sum_{i=1}^N \ell_i^{d_f} = 1$, then*

$$\mu = \mathcal{H}^{d_f}(K)^{-1}\mathcal{H}^{d_f}|K$$

where $\mathcal{H}^{d_f}$ is the d_f-dimensional Hausdorff measure in the space X and $H^{d_f}|K$ its restriction to K.

In Section 2 below we prove the relevant properties of Hölder functions. In Section 3 we introduce the metric on the space of measures. Properties of contractions are described in Section 4. The proofs of the theorems above are given in Section 5.

2. Hölder functions on quasi-metric spaces. A *quasi-distance* on a set X is a function $d : X \times X \to \mathbb{R}$ such that:

(i) $0 \le d(x,y) = d(y,x)$ for every $x, y \in X$;

(ii) $d(x,y)=0$ if and only if $x=y$;
(iii) there exists a constant $c_T \geq 1$ such that

$$d(x,y) \leq c_T(d(x,z)+d(y,z)) \tag{3}$$

for every $x,y,z \in X$.
For every $x \in X$ and every $r \geq 0$, we put

$$B(x,r)=\{y \in X : d(x,y)<r\}, \quad C(x,r)=\{y \in X : d(x,y) \leq r\},$$

and call $B(x,r)$ the *balls* of X.

Definition 2.1. A *quasi-metric space* X is a topological space X on which a quasi-distance d is defined, such that the balls $B(x,r)$, $r>0$, of d form a basis of neighborhoods of x for the topology of X.

A quasi-distance d on a set X uniquely defines a uniform structure, hence a topology, on X. This topology is called the *d-topology*, or *uniform topology* of X. Moreover, the balls $B(x,r)$ of d form a basis of neighborhoods for the d-topology. Therefore, the topology of a quasi-metric space X coincides with the uniform topology of X. Since the topology induced by a uniform structure is metrizable, the topology of a quasi-metric space is metrizable.
Two quasi-distances d and $\tilde{d}$ on the same set X are *equivalent*, if there exist two constants $c_1>0$ and c_2 such that

$$c_1\, d(x,y) \leq \tilde{d}(x,y) \leq c_2\, d(x,y) \tag{4}$$

for every $x,y \in X$. Uniformities and topologies defined by equivalent quasi-distances coincide.
Let $X \equiv (X,d)$ be a quasi-metric space. X is *complete* if for any sequence $x_1,x_2,\dots$ in X with $d(x_n,x_m) \to 0$ as $n,m \to \infty$ there exists $x \in X$ such that $d(x_n,x) \to 0$ as $n \to \infty$. A set $A \subset X$ is *bounded* if it is contained in some ball of d, A is *totally bounded* if for every $\varepsilon>0$ there exists a finite number of balls $B(x_i,\varepsilon)$ centered in A covering A. As in the usual metric case, if X is complete, then any totally bounded $A \subset X$ is *pre-compact* in X, that is, every sequence $y_1,y_2,\dots$ in A possesses a subsequence $y_{n_1},y_{n_2},\dots$ such that $d(y_{n_k},y) \to 0$ for some $y \in X$, as $k \to \infty$. The closure $\overline{A}$ in X of a totally bounded subset A of X is (sequentially) compact.
The following result is due to Macias-Segovia, [6] Theorem 2, see also [14], Ch.I, § 8,8.10:

Lemma 2.1. *Let d be a quasi-distance on a set X. Then there exists a quasi-distance $\tilde{d}$ on X and fixed finite constants $c_1>0$, c_2, C, $0<\gamma\leq 1$, such that* (4) *holds, moreover*

$$|\tilde{d}(x,z)-\tilde{d}(y,z)| \leq C(\tilde{d}(x,z)+\tilde{d}(y,z))^{1-\gamma}\,\tilde{d}(x,y)^{\gamma}$$

for every x,y,z. The constants γ, C, c_1, c_2 can be chosen to depend only on the constant c_T of d.

The quasi-distance $\tilde{d}$ is obtained as a suitable power of the distance associated with the (metrizable) uniform topology of X.

In the following, given a quasi-metric space $X \equiv (X, d)$, by $\tilde{d}$ and by γ, C, c_1, c_2 we always denote the equivalent quasi-distance and the fixed constants occurring in the Lemma above and we occasionally refer to γ as to the regularity exponent of d (or X). The balls of $\tilde{d}$ will be denoted by $\tilde{B}(x,r)$, *i.e.*,

$$\tilde{B}(x,r) = \{y \in X : \tilde{d}(x,y) < r\}, \quad \tilde{C}(x,r) = \{y \in X : \tilde{d}(x,y) \leq r\}.$$

Since $\tilde{d}(x,\cdot)$ is a continuous function on X, we have $\overline{\tilde{B}(x,r)} \subset \tilde{C}(x,r)$ for every $x \in X$ and every $r > 0$. Therefore,

$$\overline{B(x,r)} \subset C(x,(c_2/c_1)r) \subset B(x,R)$$

for every $x \in X$, $r > 0$ and $R > c_2/c_1$.

Lemma 2.2. *Let d be a quasi-distance on a set X. Let x_0 be a point of X and $R_0 > 0$ a constant such that $B(x_0,R_0) \neq \overline{B(x_0,R_0)}$. Then, for every $x, y \in X$ with $d(x_0,x) < R_0 \leq d(x_0,y))$, there exists $z \in X$, such that*

$$R_0 \leq d(x_0,z) \leq kR_0,$$

$$d(x,z) \leq d(x,y),$$

where k is a fixed constant depending only on c_T.

Proof. We have $\overline{B(x_0,R_0)} \subset C(x_0,(c_2/c_1)R_0)$. Now let $x \in B(x_0,R_0)$. Then $C(x_0,(c_2/c_1)R_0) \subset B(x,k_0R_0)$ where $k_0 = c_T(1 + c_2/c_1)$. Moreover, $B(x,k_0R_0) \subset B(x_0,kR_0)$ where $k = c_T(1 + k_0)$. Thus,

$$\overline{B(x_0,R_0)} \subset B(x,k_0R_0) \subset B(x_0,kR_0).$$

If $d(x_0,y) \leq kR_0$, then $z = y$ has the required properties. Now let $d(x_0,y) > kR_0$. We then choose z to be any point in $\overline{B(x_0,R_0)} - B(x_0,R_0)$. Then, $d(x_0,z) \geq R_0$. Moreover, by the preceding inclusion of balls, $d(x_0,z) < kR_0$. By our assumption, y is not in $B(x_0,kR_0)$, hence not in $B(x,k_0R_0)$, therefore $d(x,y) \geq k_0R_0$. On the other hand, $\overline{B(x_0,R_0)} \subset B(x,k_0R_0)$, therefore $d(x,z) < k_0R_0$. Thus, $d(x,z) \leq d(x,y)$ and this concludes the proof. □

Let $X \equiv (X,d)$ be a quasi metric-space, $\gamma \in (0,1]$ the regularity parameter of d. By $C_b^{0,\gamma}$ we denote the set of all functions $\Phi : X \to \mathbb{R}$ such that

$$||\Phi||_\infty = \sup\{|\Phi(x)| : x \in X\}$$

and

$$|\Phi|_{0,\gamma} = \sup\left\{\frac{|\Phi(x) - \Phi(y)|}{d(x,y)^\gamma} : x, y \in X, x \neq y\right\}$$

are both finite.

If $f_1 : X \to \mathbb{R}$, $f_2 : X \to \mathbb{R}$, we have

$$||f_1f_2||_\infty \leq ||f_1||_\infty||f_2||_\infty, \quad |f_1f_2|_{0,\gamma} \leq ||f_1||_\infty|f_1|_{0,\gamma} + ||f_2||_\infty|f_2|_{0,\gamma};$$

therefore $C_b^{0,\gamma}$ is an algebra and $1 \in C_b^{0,\gamma}$. On the other hand, the subset of all $\Phi \in C_b^{0,\gamma}$ with $|\Phi|_{0,\gamma} \leq 1$ is not an algebra.

We shall prove below that $C_b^{0,\gamma}$ contains also non-constant functions if X is not a singleton, actually, the functions in $C_b^{0,\gamma}$ separate the points of X:

Lemma 2.3. *Let d be a quasi-distance on a set X. Let $x_0, y_0 \in X$, $x_0 \neq y_0$. Then there exists a function $\Phi \in C_b^{0,\gamma}$, such that $\Phi(x_0) \neq \Phi(y_0)$.*

Proof. Let X be endowed with the uniform topology of d. Let x_0 be an isolated point of X. We choose $R_0 > 0$ such that x_0 is the only point of X in the ball $B(x_0, R_0)$. We define Φ to be zero at x_0 and $\Phi(x) = R_0$ at every $x \neq x_0$. Then Φ has the required properties.

Now let x_0 be not an isolated point of X. Let R_1 be such that $0 < R_1 \leq \tilde{d}(x_0, y_0)$. Then there exists R_0, $0 < R_0 \leq R_1$, such that $\tilde{B}(x_0, R_0) \neq \overline{\tilde{B}(x_0, R_0)}$. Otherwise, $\tilde{B}(x_0, r) \neq X$, $0 < r \leq R_1$, would be a basis of open and closed neighborhoods of x_0 in X, hence x_0 would be an isolated point of X.

Let R_0 be chosen as above, such that $\tilde{B}(x_0, R_0) \neq \overline{\tilde{B}(x_0, R_0)}$. Let $\alpha : \mathbb{R} \to \mathbb{R}$ be the function $\alpha(t) \equiv 0$ if $t < 0$, $\alpha(t) \equiv t$ if $0 \leq t < R_0$, $\alpha(t) \equiv R_0$ if $t \geq R_0$. Then $|\alpha(t) - \alpha(s)| \leq |t - s|$ for every $s.t \in \mathbb{R}$. Let Φ be the function

$$\Phi(x) = \alpha(\tilde{d}(x_0, x))$$

for every $x \in X$. Then, $|\Phi(x) - \Phi(y)| \leq |\tilde{d}(x_0, x) - \tilde{d}(x_0, y)|$ for every $x, y \in X$. Moreover, if $\tilde{d}(x_0, x) \leq R_0$, $\tilde{d}(x_0, y) \leq R_0$, then, by Lemma 2.1,

$$|\Phi(x) - \Phi(y)| = |\tilde{d}(x_0, x) - \tilde{d}(x_0, y)|$$

$$\leq C(\tilde{d}(x_0, x) + \tilde{d}(x_0, y))^{1-\gamma} \tilde{d}(x, y)^\gamma \leq C\, 2^{1-\gamma} R_0^{1-\gamma} c_2^\gamma\, d(x, y)^\gamma.$$

If $\tilde{d}(x_0, x) \geq R_0$ and $\tilde{d}(x_0, y) \geq R_0$, then $|\Phi(x) - \Phi(y)| = 0$. Now let $x, y \in X$ be such that $\tilde{d}(x_0, x) < R_0 < \tilde{d}(x_0, y)$. Let z be chosen as in Lemma 2.2, applied now to the quasi-distance $\tilde{d}$. Note that $\tilde{d}$ satisfies the quasi-triangle inequality with a constant $\tilde{c}_T = (c_2/c_1)c_T$ which depends only on c_T. Since $\tilde{d}(x_0, z) \geq R_0$, then $|\Phi(x) - \Phi(y)| = |\Phi(x) - \Phi(z)|$. Moreover, $\tilde{d}(x_0, z) \leq kR_0$, where k is the constant in Lemma 2.2, and $\tilde{d}(x, z) \leq \tilde{d}(x, y)$. Therefore,

$$|\Phi(x) - \Phi(y)| \leq |\tilde{d}(x_0, x) - \tilde{d}(x_0, z)|$$

$$\leq C(\tilde{d}(x_0, x) + \tilde{d}(x_0, z))^{1-\gamma} \tilde{d}(x, y)^\gamma$$

$$\leq C(R_0 + kR_0)^{1-\gamma} \tilde{d}(x, y)^\gamma \leq C c_2^\gamma (1 + k)^{1-\gamma} R_0^{1-\gamma}\, d(x, z)^\gamma.$$

Thus,

$$|\Phi(x) - \Phi(y)| \leq c R_0^{1-\gamma} d(x, y)^\gamma$$

for every $x, y \in X$, where the constant c depends only on c_T. Therefore, $\Phi \in C_b^{0,\gamma}$. Moreover $\Phi(x_0) = 0$ and $\Phi(y_0) > 0$. This concludes the proof. □

Lemma 2.4. *Let $X \equiv (X, d)$ be a quasi-metric space. Let T be a compact subset of X. Then, the restrictions to T of the functions in $C_b^{0,\gamma}$ are dense in $C(T)$ in the uniform norm.*

Proof. Let
$$\mathcal{A} = \{f : T \to \mathbb{R} | f(x) \equiv \Phi(x) \text{ if } x \in T, \Phi \in C_b^{0,\gamma}\}.$$
Clearly, $\mathcal{A}$ inherits the property of $C_b^{0,\gamma}$ of being an algebra, and $1 \in \mathcal{A}$. Moreover, by Lemma 2.3, for every $x, y \in T$, $x \neq y$ there exists a function $f \in \mathcal{A}$ such that $f(x) \neq f(y)$. By Stone-Weierstrass theorem, $\mathcal{A}$ is dense in $C(T)$ in the uniform norm. □

3. A metric on measures in quasi-metric spaces. Let $X \equiv (X, d)$ be a complete quasi-metric space. By $\mathcal{M}$ we shall denote the family of Borel regular measures having bounded support in X and finite mass. By $\mathcal{M}_c$ the subset of all measures in $\mathcal{M}$ with compact support. Moreover,
$$\mathcal{M}^1 = \{\mu \in \mathcal{M} : \mu(X) = 1\},$$
$$\mathcal{M}_c^1 = \{\mu \in \mathcal{M}_c : \mu(X) = 1\}.$$
Every measure $\mu \in \mathcal{M}_c$ is a Radon measure. This follows from the following approximation property of Borel regular measures μ of finite mass: if E is a Borel set of X, then

(i) $\mu(E) = \sup\{\mu(C) : C \text{ closed} \subset E\}$,

(ii) $\mu(E) = \inf\{\mu(V) : V \text{ open} \supset E\}$,

see e.g. [3], 3,2.2.2, [8]. In particular, μ can be identified with the functional $f \to \int f\, d\mu$ defined for $f \in C_0(X)$, where $C_0(X)$ denotes the space of continuous functions with compact support in X.

For arbitrary $\mu, \nu \in \mathcal{M}^1$ we define
$$H(\mu, \nu) = \sup\{|\int \phi\, d\mu - \int \phi\, d\nu| : \phi \in C_b^{0,\gamma}, |\phi|_{0,\gamma} \leq 1\}.$$
We have $H(\mu, \nu) < \infty$. In fact, let $\phi \in C_b^{0,\gamma}$ with $|\phi|_{0,\gamma} \leq 1$. Let $B = B(x, R)$ be a ball containing both support μ and support ν. If c is any constant, then $\int c\, d\mu = \int c\, d\nu = c$, therefore
$$\int \phi\, d\mu - \int \phi\, d\nu = \int (\phi - \phi(x))\, d\mu - \int (\phi - \phi(x))\, d\nu$$
$$\leq |\phi|_{0,\gamma} R^\gamma \mu(X) + |\phi|_{0,\gamma} R^\gamma \nu(X) < 2R^\gamma.$$
Obviously $0 \leq H(\mu, \nu) = H(\nu, \mu)$ and $H(\mu, \nu) = 0$ if $\mu = \nu$. The triangle inequality $H(\mu, \nu) \leq H(\mu, \chi) + H(\chi, \nu)$ for $\mu, \nu, \chi \in \mathcal{M}^1$ is also immediately verified. Therefore, H is a pseudo-distance in $\mathcal{M}^1$. We observe, incidentally, that $H(\mu, \nu) = +\infty$ if $\mu, \nu \in \mathcal{M}$ with $\mu(X) \neq \nu(X)$.

Now we prove that H is a distance on $\mathcal{M}_c^1$, that is, in addition, if $\mu, \nu, \in \mathcal{M}_c^1$ and $H(\mu, \nu) = 0$ then $\mu = \nu$. We prove that $\int f\, d\mu = \int f\, d\nu$ for every $f \in C_0(X)$. Let T be a compact subset of X containing both support μ and support ν. If f is a constant on T then the equality above is obvious, therefore we consider the case that f is not identically constant on T. Let $\varepsilon > 0$. By Lemma 2.4, there exists $\Phi \in C_b^{0,\gamma}$ such that

$$||f - \Phi||_{\infty,T} = \sup\{|f(x) - \Phi(x)| : x \in T\} < \varepsilon. \tag{5}$$

By taking ε sufficiently small, we can assume that Φ is not a constant, hence $|\Phi|_{0,\gamma} > 0$. Let $\phi = \Phi/|\Phi|_{0,\gamma}$. Then $\phi \in C_b^{0,\gamma}$ and $|\phi|_{0,\gamma} \leq 1$. Therefore,

$$\begin{aligned}
|\int f\,d\mu - \int f\,d\nu| &= |\int_T f\,d\mu - \int_T f\,d\nu| \\
&\leq 2\|f - \Phi\|_{\infty,T} + |\Phi|_{0,\gamma}|\int_T \phi\,d\mu - \int_T \phi\,d\nu| \\
&\leq 2\|f - \Phi\|_{\infty,T} + |\Phi|_{0,\gamma}|\int \phi\,d\mu - \int \phi\,d\nu| \\
&\leq 2\varepsilon + |\Phi|_{0,\gamma}\, H(\mu,\nu) = 2\varepsilon.
\end{aligned}$$

By the arbitrariness of ε this implies that $\int f\,d\mu = \int f\,d\nu$.
We summarize the preceding remarks in the following

Lemma 3.1. *H is a pseudo-distance in $\mathcal{M}^1$, it is a distance in $\mathcal{M}^1_c$.*

If $\mu_1, \mu_2, \ldots$ is a sequence of measures in $\mathcal{M}^1$ and $\mu \in \mathcal{M}^1$, then by

$$\mu_n \rightharpoonup \mu$$

we mean that $\int f\,d\mu_n \to \int f\,d\mu$ for every $f \in C_0(X)$ as $n \to 0$, where $C_0(X)$ denotes the set of continuous functions with compact support in X.

Lemma 3.2. *Let $\mu_1, \mu_2, \ldots$ and μ be measures in $\mathcal{M}^1_c$ with* support$\mu_n \subset T$ *for every n, where T is a compact subset of X. Then:*
(i) *$H(\mu_n, \mu) \to 0$ as $n \to 0$ implies $\mu_n \rightharpoonup \mu$ as $n \to \infty$.*
(ii) *If in addition X is locally compact, then the opposite implication also holds.*

Proof. (i) Let $H(\mu_n, \mu) \to 0$ as $n \to \infty$. Let $f \in C_0(X)$. We must prove that $\int f\,d\mu_n \to \int f\,d\mu$ as $n \to \infty$. We can assume that T contains also the support of μ. If f is a constant on T the limit is obvious, therefore we can suppose that f is not identically constant on T. Let $\varepsilon > 0$. By Lemma 2.4, there exists $\Phi \in C_b^{0,\gamma}$ such that (5) holds. By taking ε sufficiently small, we can assume that Φ is not a constant, hence $|\Phi|_{0,\gamma} > 0$. Let $\phi = \Phi/|\Phi|_{0,\gamma}$. Then $\phi \in C_b^{0,\gamma}$ and $|\phi|_{0,\gamma} \leq 1$. Therefore,

$$\begin{aligned}
|\int f\,d\mu_n - \int f\,d\mu| &= |\int_T f\,d\mu_n - \int_T f\,d\mu| \\
&\leq 2\|f - \Phi\|_{\infty,T} + |\Phi|_{0,\gamma}\,|\int_T \phi\,d\mu_n - \int_T \phi\,d\mu| \\
&\leq 2\varepsilon + |\Phi|_{0,\gamma}\, H(\mu_n,\mu).
\end{aligned}$$

As $n \to \infty$, we get

$$\lim |\int f\,d\mu_n - \int f\,d\mu| \leq \varepsilon$$

By the arbitrariness of ε the conclusion follows.

(ii) Let $\mu_n \rightharpoonup \mu$. Suppose that $H(\mu_n, \mu)$ does not tend to zero as $n \to 0$. By possibly passing to a subsequence, we can then find $\varepsilon_0 > 0$ and $\phi_n \in C_b^{0,\gamma}$ with $|\phi_n|_{0,\gamma} \leq 1$, such that,

$$|\int \phi_n\,d\mu_n - \int \phi_n\,d\mu| \geq \varepsilon_0 \quad (n = 1, 2, \ldots).$$

Since X is locally compact, we can find an open subset V with compact closure in X such that $T \subset V$. Let $x_0 \in T$ and $R > 0$ be such that $\bar{V} \subset B(x_0, R)$. Let $\Phi_n = \phi_n - \phi(x_0)$ for every n. We have $|\Phi_n|_{0,\gamma} \leq 1$ and $|\Phi_n(x)| \leq d(x_0, x)^\gamma$ for every n. By Urysohn's theorem, there exists a continuous function g such that $g = 1$ on T, $g = 0$ on $X - V$, $0 \leq g \leq 1$ everywhere. Let $f_n = \Phi_n g$ for every n. We have

$$||f_n||_{\infty,\bar{V}} \leq ||\Phi_n||_{\infty,\bar{V}}||g||_{\infty,\bar{V}} \leq R^\gamma,$$

$$|f_n|_{0,\gamma,\bar{V}} \leq ||\Phi_n||_{\infty,\bar{V}}|g|_{0,\gamma,\bar{V}} + ||g||_{\infty,\bar{V}}|\Phi_n|_{0,\gamma,\bar{V}} \leq R^\gamma|g|_{0,\gamma} + 1$$

By the Ascoli-Arzelà theorem, there exists $f \in C(\bar{V})$ such that, passing to a subsequence, f_n converges unformy to f on $\bar{V}$. Since $f_n = 0$ on $X - V$ for every n, then $f = 0$ on $X - \bar{V}$, therefore by extending f to be zero outside $\bar{V}$ we have $f \in C_0(X)$. Let $0 < 2\varepsilon < \varepsilon_0$. Let n_ε be such that for $n \geq n_\varepsilon$ we have

$$||f - f_n||_{\infty,T} = \sup\{|f(x) - f_n(x)| : x \in T\} \leq \varepsilon.$$

Then

$$\begin{aligned}|\int \phi_n\,d\mu_n - \int \phi_n\,d\mu| &= |\int \Phi_n\,d\mu_n - \int \Phi_n\,d\mu| = |\int_T f_n\,d\mu_n - \int_T f_n\,d\mu| \\ &\leq |\int_T f_n\,d\mu_n - \int_T f\,d\mu_n| + |\int_T f\,d\mu_n - \int_T f\,d\mu| + |\int_T f\,d\mu - \int_T f_n\,d\mu| \\ &\leq 2\varepsilon + |\int_T f\,d\mu_n - \int_T f\,d\mu| \leq 2\varepsilon + |\int f\,d\mu_n - \int f\,d\mu|.\end{aligned}$$

By letting $n \to \infty$, we get

$$\lim_{n\to\infty} |\int \phi_n\,d\mu_n - \int \phi_n\,d\mu| \leq 2\varepsilon < \varepsilon_0,$$

hence a contradiction. This concludes the proof of the Lemma. □

Lemma 3.3. *Let X be a locally compact, complete quasi-metric space. Then for every compact subset $T \subset X$ the space of all measures of $\mathcal{M}_c^1$ with support in T is complete under the metric H.*

Proof. Let $\mu_1, \mu_2, \ldots \in \mathcal{M}_c^1$, with support $\mu_n \subset T$ for every n, be such that $H(\mu_n, \mu_m) \to 0$ as $n, m \to \infty$. We construct a measure $\mu \in \mathcal{M}_c^1$ such that $H(\mu_n, \mu) \to 0$ as $n \to \infty$. Let $f \in C_0(X)$, f not a constant on T. Given $\varepsilon > 0$ small enough we take $\Phi \in C_b^{0,\gamma}$, Φ not a constant, such that (5) is satisfied. Let $\phi = |\Phi|_{0,\gamma}^{-1}\Phi$. Then

$$\begin{aligned}|\int f\,d\mu_m - \int f\,d\mu_n| &= |\int_T f\,d\mu_m - \int_T f\,d\mu_n| \\ &\leq 2||f - \Phi||_{\infty,T} + |\int_T \Phi\,d\mu_m - \int_T \Phi\,d\mu_n| \\ &\leq 2\varepsilon + |\Phi|_{0,\gamma}|\int_T \phi\,d\mu_m - \int_T \phi\,d\mu_n| \leq 2\varepsilon + |\Phi|_{0,\gamma}H(\mu_m, \mu_n)\end{aligned}$$

and in the limit as $n, m \to \infty$ we obtain

$$\lim |\int f\,d\mu_m - \int f\,d\mu_n| \leq \varepsilon.$$

Therefore $\int f\,d\mu_n$ converges to some $L(f) \in \mathbb{R}$ as $n \to \infty$ and this also holds, with $L(f) = c$, if $f = c$ on T. $L(f)$ is a linear functional of $f \in C_0(X)$. Since $|\int f\,d\mu_m| \leq ||f||_\infty$ for every n, then $|L(f)| \leq ||f||_\infty$ for every $f \in C_0(X)$. By Riesz's theorem, there exists a Radon measure μ on X, such that

$$L(f) = \int f\,d\mu$$

for every $f \in C_0X$. Moreover,

$$|\int f\,d\mu_n - \int f\,d\mu| \to 0$$

as $n \to \infty$. Since, for every n, $\int f\,d\mu_n = 0$ whenever $f \equiv 0$ on $X - T$, then support$\mu \subset T$. Moreover, by choosing $f \in C_0(X)$, $f = 1$ on T, we have

$$\mu(X) = \int_T f\,d\mu = \int f\,d\mu = \lim_{n\to\infty} \int f\,d\mu_n = \lim_{n\to\infty} \int_T f\,d\mu_n = \mu_n(X) = 1.$$

Therefore, $\mu \in \mathcal{M}_c^1$. By Lemma 3.2, $H(\mu_n, \mu) \to 0$ as $n \to \infty$ and this concludes the proof. □

4. Contractions in quasi-metric spaces. Let X be a quasi-metric space, A, B nonempty bounded subsets of X. We put

$$\operatorname{diam} A = \sup\{d(x,y) : x, y \in A\},$$

$$\operatorname{dist}(A, B) = \inf_{a\in A, b\in B} d(a,b),$$

and for every $x \in X$ and every $r > 0$

$$\operatorname{dist}(x, A) = \inf_{a\in A} d(x,a),$$

$$I_r A = \{x \in X : \operatorname{dist}(x, A) < r\}.$$

We have

$$B \subset I_{c_T(\operatorname{diam} B + \operatorname{dist}(A,B))} A,$$

$$c_T^{-1} \operatorname{dist}(x, A) \leq \operatorname{dist}(x, \bar{A}) \leq \operatorname{dist}(x, A);$$

moreover

$$B \subset I_r A \text{ implies } \bar{B} \subset I_{rc_T} A.$$

The Hausdorff distance $h(A, B)$ of two nonempty bounded subsets A, B of X is defined, as in the usual metric case, to be

$$h(A, B) = \max\{\sup_{a\in A} \operatorname{dist}(a, B), \sup_{b\in B} \operatorname{dist}(b, A)\}.$$

We have $h(A,B) < r$ if and only if $A \subset I_r B$ and $B \subset I_r A$. Moreover

$$c_T^{-1}\, h(A,B) \le h(\bar{A},\bar{B}) \le c_T\, h(A,B).$$

As shown in [7], h is a quasi-distance on the space $\mathcal{B}$ of all nonempty bounded closed subsets of X, with same constant c_T as the one of X. If X is complete, $\mathcal{B} \equiv (\mathcal{B}, h)$ is a complete quasi-metric space. Moreover, the space $\mathcal{C}$ of all nonempty compact subsets of X is a closed, hence complete, subset of $\mathcal{B}$.

Definition 4.1. A *contraction* in a quasi-metric space X is a map S of X into itself such that

$$d(S(x), S(y)) \le \ell d(x,y)$$

for every $x, y \in X$ and for some fixed constant $\ell \in (0,1)$.

It is also proved in [7] that a contraction in a complete quasi-metric space possesses a unique fixed point.

If S is a contraction in X, then

$$S(I_r A) \subset I_{\ell r} S(A)$$

$$\overline{S(I_r A)} \subset I_{\ell c_T r} S(A)$$

Let $S_1, \dots, S_N$ be contractions in X with contraction factors $\ell_1, \dots, \ell_N$. We define $\mathcal{S}$ to be the map

$$\mathcal{S}(A) = \bigcup_{i=1}^{N} S_i(A).$$

For every positive integer n, $\mathcal{S}^n$ is the n -times iterated map $\mathcal{S} \circ \cdots \circ \mathcal{S}$. We have

$$\mathcal{S}^n(A) = \bigcup_{i_1,\dots,i_n=1}^{N} S_{i_1,\dots,i_n}(A)$$

where for every set of indeces $i_1, \dots, i_n \in \{1, \dots, N\}$,

$$S_{i_1,\dots,i_n} = S_{i_1} \circ \cdots \circ S_{i_n}.$$

For every positive integer n we define $\overline{\mathcal{S}^n}$ to be the map

$$\overline{\mathcal{S}^n}(A) = \overline{\mathcal{S}^n(A)} = \bigcup_{i=1}^{N} \overline{S_{i_1,\dots,i_n}(A)}.$$

If A is compact, then $S_{i_1,\dots,i_n}(A)$ is closed, hence $\overline{\mathcal{S}^n}(A) = \mathcal{S}^n(A)$.

Lemma 4.1. *Let A be a nonempty bounded subset of X, such that $\mathcal{S}(A) \subset A$. Let B be an arbitrary nonempty bounded subset of X. Then for every $\varepsilon > 0$ there exists $\bar{n}$ such that for all $n \ge \bar{n}$ we have $\mathcal{S}^n(B) \subset I_\varepsilon A$. Moreover, $\overline{\mathcal{S}^n}(B) \subset I_{\varepsilon c_T} A$.*

Proof. Let $R_0 = c_T(\operatorname{diam} B + \operatorname{dist}(A,B))$. Then $B \subset I_{R_0} A$. For every positive integer n and every set of indices $i_1, \dots, i_n \in \{1, \dots, N\}$, we have

$$S_{i_1,\dots,i_n}(I_{R_0} A) \subset I_{\ell_{i_1} \cdots \ell_{i_n} R_0} S_{i_1,\dots,i_n}(A)$$

therefore

$$S_{i_1,\dots,i_n}(B) \subset I_{\ell_{i_1}\cdots\ell_{i_n}R_0} S_{i_1,\dots,i_n}(A) \subset I_{\ell_{i_1}\cdots\ell_{i_n}R_0} A.$$

hence

$$\mathcal{S}^n(B) \subset I_{\max\{\ell_1,\dots,\ell_N\}^n R_0} A$$

and the conclusion of the first part of the lemma follows by taking n large enough. On the other hand, if $\mathcal{S}^n(B) \subset I_\varepsilon A$ then

$$\overline{\mathcal{S}^n(B)} \subset I_{\varepsilon c_T} A$$

and this concludes the proof. □

From the preceding lemma we get immediately

Lemma 4.2. *Let A, B be nonempty bounded closed subsets of X, such that $\mathcal{S}(A) = A$, $\mathcal{S}(B) = B$. Then $A = B$.*

Proof. Since $\mathcal{S}(B) = B$, for every integer n we have $\mathcal{S}^n(B) = B$. Let $\varepsilon > 0$ and let $\bar{n}$ be as in Lemma 4.1. Then, $B = \mathcal{S}^{\bar{n}}(B) \subset I_{\varepsilon c_T} A$. Similarly, $A \subset I_{\varepsilon c_T} B$. Therefore, $h(A, B) = 0$, hence $A = B$. □

The following result is due to Marchi, [7]:

Lemma 4.3. *Let $S_1, \dots, S_N$ be contractions in a complete quasi-metric space X. Then there exists nonempty compact subset K of X such that*

$$K = \mathcal{S}(K) = \bigcup_{i=1}^{N} S_i(K).$$

Moreover, such a K is unique among all nonempty bounded closed subset B satisfying $B = \mathcal{S}(B)$.

The existence of K is obtained by showing that $\mathcal{S}$ is a contraction in the space $\mathcal{C}$ for the quasi-distance h, indeed

$$h(\mathcal{S}(A), \mathcal{S}(B)) \leq \max\{\ell_1, \dots, \ell_N\}\, h(A, B)$$

for every $A, B \in \mathcal{C}$ and by applying then the contraction principle for quasi-metric spaces mentioned before. The uniqueness of K follows from Lemma 4.2.

5. Invariant measures. In this section we assume that X is a locally compact, complete quasi-metric space. We also assume that $S_1, \dots, S_N$ are given contractions in X, with contraction factors $\ell_1, \dots, \ell_N$ and we put $\ell = \max\{\ell_1, \dots, \ell_N\}$.

We further assume that $m_1, \dots, m_N$ are given positive constants such that $m_1 + \dots + m_N = 1$.

Let $\mathcal{M}^1, \mathcal{M}^1_c$ be the spaces of measures and $H(\mu, \nu)$, $\mu, \nu \in \mathcal{M}^1$, the pseudo-distance introduced in Section 3.

If $f : X \to X$ is continuous and sends bounded sets into bounded sets, in particular if f is a contraction in X, then for every $\mu \in \mathcal{M}^1$, we have $\mu \circ f^{-1} \in \mathcal{M}^1$ and support$(\mu \circ f^{-1}) = \overline{f(\text{support}\mu)}$. Moreover support$(\mu \circ f^{-1}) = f(\text{support}\mu)$ and $\mu \circ f^{-1} \in \mathcal{M}_c^1$, if $\mu \in \mathcal{M}_c^1$.

For every $\mu \in \mathcal{M}^1$, we put

$$\mathcal{T}\mu = \sum_{i=1}^{N} m_i\, \mu \circ S_i^{-1}.$$

$\mathcal{T}$ is a map of the space $\mathcal{M}^1$ into itself and support$(\mathcal{T}\mu) = \overline{\mathcal{S}}(\text{support}\mu)$ for every $\mu \in \mathcal{M}^1$. Moreover, $\mathcal{T}$ carries $\mathcal{M}_c^1$ into itself and support$(\mathcal{T}\mu) = \mathcal{S}(\text{support}\mu)$ for every $\mu \in \mathcal{M}_c^1$.

By $\mathcal{T}^n$ we denote the n-time iterated map $\mathcal{T} \circ \cdots \circ \mathcal{T}$. We have

$$\mathcal{T}^n\mu = \sum_{i_1,\dots,i_n=1}^{N} m_{i_1}\cdots m_{i_n}\, \mu \circ S_{i_1,\dots,i_n}^{-1},$$

support$(\mathcal{T}^n\mu) = \overline{\mathcal{S}^n}(\text{support}\mu)$ if $\mu \in \mathcal{M}^1$, support$(\mathcal{T}^n\mu) = \mathcal{S}^n(\text{support}\mu)$, if $\mu \in \mathcal{M}_c^1$.

We now prove that $\mathcal{T}$ is a contraction in $\mathcal{M}^1$ for the pseudo-distance H:

Lemma 5.1. *We have*

$$H(\mathcal{T}\mu, \mathcal{T}\nu) \le \ell^\gamma H(\mu,\nu),$$

for every $\mu,\nu \in \mathcal{M}^1$, where $\ell = \max\{\ell_1,\dots,\ell_N\}$.

Proof. With notation from Section 2, let $\Phi \in C_b^{0,\gamma}$, with $|\Phi|_{0,\gamma} \le 1$. We put $\phi_i = \ell^{-\gamma}\Phi \circ S_i$. We have $||\phi_i||_\infty \le \ell^{-\gamma}||\Phi||_\infty < \infty$. Moreover, for every $x, y \in X$ we have

$$|\phi_i(x) - \phi_i(y)| = \ell^{-\gamma}|\Phi(S_i(x)) - \Phi(S_i(y))|$$

$$\le \ell^{-\gamma}|\Phi|_{0,\gamma} d(S_i(x), S_i(y))^\gamma \le (\ell_i/\ell)^\gamma d(x,y)^\gamma \le d(x,y)^\gamma,$$

hence $|\phi|_{0,\gamma} \le 1$.

On the other hand,

$$\int \Phi d\mathcal{T}\mu - \int \Phi d\mathcal{T}\nu = \sum_{i=1}^{N} m_i\Big(\int \Phi \circ S_i\, d\mu - \int \Phi \circ S_i\, d\nu\Big)$$

$$= \sum_{i=1}^{N} m_i \ell^\gamma\Big(\int \phi_i\, d\mu - \int \phi_i\, d\nu\Big)$$

therefore,

$$\Big|\int \Phi d\mathcal{T}\mu - \int \Phi d\mathcal{T}\nu\Big| \le \sum_{i=1}^{N} m_i \ell^\gamma \Big|\int \phi_i\, d\mu - \int \phi_i\, d\nu\Big|$$

$$\le \sum_{i=1}^{N} m_i \ell^\gamma H(\mu,\nu) = \ell^\gamma H(\mu,\nu)$$

and this concludes the proof. □

Lemma 5.2. *There exists a compact subset T of X, such that if $\mu_0 \in \mathcal{M}^1$ and $\mu_n = \mathcal{T}^n\mu_0$ for every $n = 1, 2, \dots$, then* support$\mu_n \subset T$ *for every $n \ge \bar{n}$ and some $\bar{n}$.*

Proof. Let K be the compact invariant subset of X for the map $\mathcal{S}$, given by Lemma 4.3. Since X is locally compact, there exists $\varepsilon_0 > 0$ such that $T = \overline{I_{\varepsilon_0} K}$ is compact. We have

$$\mu_n = \mathcal{T}^n \mu = \sum_{i_1,\dots,i_n=1}^{N} m_{i_1} \cdots m_{i_n}\, \mu \circ S^{-1}_{i_1,\dots,i_n},$$

support$(\mathcal{T}^n \mu_0) = \overline{\mathcal{S}^n}($support$\mu)$. By Lemma 4.1, there exists $\bar{n}$ such that for every $n \geq \bar{n}$ we have $\overline{\mathcal{S}^n}($support$\mu) \subset I_{\varepsilon_0} K \subset T$. Therefore, support$(\mathcal{T}^n \mu) \subset T$ for every $n \geq \bar{n}$ and this proves the lemma. □

We are now in a position to prove the theorems stated in the introduction.

Proof of Theorem 1.1. We choose any measure $\mu_0 \in \mathcal{M}^1$ and we consider the measures $\mu_n = \mathcal{T}^n \mu_0$, $n = 1, 2, \dots$. By Lemma 5.2, support$\mu_n \subset T$ for all $n \geq \bar{n}$ and some $\bar{n}$. By Lemma 5.2 and Lemma 3.3, there exists a measure $\mu \in \mathcal{M}^1_c$, with support contained in T, such that $H(\mu_n, \mu) \to 0$ as $n \to \infty$. Moreover, $\mu = \mathcal{T}(\mu)$. In addition, by Lemma 3.2, $\mu_n \rightharpoonup \mu$ as $n \to \infty$. This proves (i) and (ii) of Theorem 1.1 at the same time. Since

$$\mu = \sum_{i=1}^{N} m_i \mu \circ S_i^{-1},$$

and supportμ is compact, then

$$\text{support}\mu = \bigcup_{i=1}^{N} S_i(\text{support}\mu).$$

By Lemma 4.2, support$\mu = K$, where K is the invariant set of $\mathcal{S}$. This proves (iii) and concludes the proof of Theorem 1.1. □

Definition 5.1. A *space of homogeneous type* is a quasi metric space $X \equiv (X, d)$ that satisfies the following homogeneity property: There exists an integer M such that for every $x \in X$ and every $r > 0$ the ball $B(x, r)$ contains at most M points $x_1, \dots x_M$, such that $d(x_i, x_j) > r/2$, $i \neq j$.

It can be easily proved, by recurrence, that the a ball $B(x, r)$, for every $x \in X$ and every $r > 0$, contains at most M^n points x_i, with $d(x_i, x_j) > r/2^n$ if $i \neq j$, where n is an arbitrary integer ≥ 1.

Lemma 5.3. *Let X be a complete quasi-metric space of homogeneous type. Then every ball $B(x, r)$, $x \in X$, $r > 0$, has a compact closure in X. In particular, X is locally compact.*

Proof. In view of a remark in Section 2, it suffices to prove that $B(x, r)$ is totally bounded, that is, given $\varepsilon > 0$, $B = B(x, r)$ can be covered by a finite number of balls $B(x_i, \varepsilon)$ with centers $x_i \in B$. In fact, let $0 < \varepsilon \leq r$ and let $n \geq 1$ be an integer such that $r/2^{n+1} < \varepsilon/2 \leq r/2^n$. Let us choose a maximal set $x_1, x_2, \dots$ in B such that $d(x_i, x_j) > r/2^{n+1}$. By the homogeneity property of X, such a set is finite. The union of all balls $B(x_i, \varepsilon)$ covers B: If not, there exists a point $\bar{x} \in B$ such that $d(\bar{x}, x_i) \geq \varepsilon > \varepsilon/2 \geq r/2^{n+1}$ for all i, in contradiction with the maximality of the set $\{x_i\}$. □

The following result is due to Marchi, [7], see also [13]:

Lemma 5.4. *Let X be a complete quasi-metric space of homogeneous type. Let $S_1, \ldots, S_N$ be contractive similitudes in X satisfying the open set condition and let $\ell_1, \ldots, \ell_N$ be their contraction factors. Let K be the invariant set whose existence is assured by Lemma 4.2. Let d_f be the positive number uniquely determined by the identity $\sum_{i=1}^N \ell_i^{d_f} = 1$ and let $\mathcal{H}^{d_f}$ be the d_f-dimensional Hausdorff measure in the space X. Then,*

$$0 < \mathcal{H}^{d_f}(K) < \infty;$$

in particular, the Hausdorff dimension of K in X equals d_f.

For the construction and the relevant properties of Hausdorff measures in a quasi-metric space we refer to [7].

In view of Theorem 1.1, assuring the existence of the invariant measure μ associated with the weights $m_i = \ell_i^{d_f}$, the preceding result can be also obtained by estimating the upper d_f-dimensional density of μ from above , as in [5], 5.3.(1)(ii). The arguments in [5] go through to our present more general setting, due to the homogeneity property of X.

From the preceding lemma, it follows also easily, see [5], 5.1.(4)(ii), that

$$\mathcal{H}^{d_f}(S_i(K) \cap S_i(K)) = 0, i \neq j,$$

in particular, K is self-similar according to Definition 5.1(1) of [5]. A further immediate consequence of the property above is the following self-similarity identity satisfied by the restriction $\mathcal{H}_{d_f}|K$ of the measure $\mathcal{H}^{d_f}$ to K:

$$\mathcal{H}^{d_f}|K = \sum_{i=1}^{N} \ell_i^{d_f} \left(\mathcal{H}^{d_f}|K\right) \circ S_i^{-1}$$

The proof of Theorem 1.2 of the Introduction is now immediate.

Proof of Theorem 1.2. By Lemma 5.3, X is locally compact, therefore Theorem 1.1 applies and this proves the first part of the theorem.

Let d_f be as in the statement of Theorem 1.2. Let τ be the measure

$$\tau = [\mathcal{H}^{d_f}(K)]^{-1}\mathcal{H}^{d_f}|K.$$

In view of the remarks following Lemma 5.4,

$$\tau = \sum_{i=1}^{N} \ell_i^{d_f} \tau \circ S_i^{-1}.$$

By the uniqueness of the invariant measure μ, established in Theorem 1.1, $\mu = \tau$. This concludes the proof. □

References

[1] M. P. BERNARDI, C. BONDIOLI: *On some dimension problems for self-affine fractals*, Zeitschr. Anal. Anw. **18, 3** (1999), 733-751

[2] R. COIFMAN, G. WEISS: *Analyse harmonique noncommutative sur certains éspaces homogènes*, Lect. Notes Math. **242**, Springer, Berlin 1971

[3] H. FEDERER: *Geometric Measure Theory*, Springer, Berlin 1969

[4] C. L. FEFFERMAN, D. H. PHONG: *Sub-elliptic eigenvalue problems*, in: Harmonic Analysis, Wadsworth Math. Series **2** (1983), 590-606

[5] J. E. HUTCHINSON: *Fractals and self-similarity*, Indiana Univ. Math. J. **30** (1981), 713-747

[6] R. A. MACIAS, C. SEGOVIA: *Lipschitz functions on spaces of homogeneous type*, Adv. Math. **33** (1979), 257-270

[7] M. V. MARCHI: *Self-similarity in spaces of homogeneous type*, Adv. Math. Sci. Appl. **9, 2** (1999), 851-870

[8] P. MATTILA: *Geometry of Sets and Measures in Euclidean Spaces*, Cambridge Univ. Press, Cambridge 1995

[9] U. MOSCO: *Invariant field metrics and dynamical scalings on fractals*, Phys. Rev. Letters **79, 21** (1997), 4067-4070

[10] U. MOSCO: *Lagrangian metrics on fractals*, Amer. Math. Soc. Proc. Symp. Appl. Math. [Ed.: R. SPIGLER, S. VENAKIDES] **54** (1998), p. 301-323

[11] U. MOSCO: *Lagrangian metrics and fractal dynamics*, in: Proc. Int. Conf. Fractal Geometry and Stochastics II, University of Greifswald [Ed.: C. BANDT, S. GRAF, M. ZÄHLE], Birkhäuser, Basel, to appear

[12] L. P. ROTHSCHILD, E. M. STEIN: *Hypoelliptic differential operators and nilpotent groups*, Acta Math. **137** (1976), 247-490

[13] A. SCHIEF: *Self-similar sets in complete metric spaces*, Proc. Amer. Math. Soc. **124** (1996), 481-490

[14] E. M. STEIN: *Harmonic Analysis*, Princeton Univ. Series, Princeton 1994

[15] R. S. STRICHARTZ: *Self-similarity in nilpotent Lie groups*, Contemporary Math. **140** (1992), 123-157

UMBERTO MOSCO, Università di Roma "La Sapienza", Dipartimento di Matematica, Piazzale A. Moro 2, I-00185 Roma, Italy; `mosco@axcasp.caspur.it`

Progress in Nonlinear Differential Equations
and Their Applications, Vol. 40

C^1-Fredholm Maps and Bifurcation for Quasilinear Elliptic Equations on $\mathbb{R}^N$

PATRICK J. RABIER, CHARLES A. STUART

Dedicated to Alfonso Vignoli on the occasion of his 60th birthday

Summary: We discuss a broad class of second order quasilinear elliptic operators on $\mathbb{R}^N$ acting from the Sobolev space $W^{2,p}(\mathbb{R}^N)$ into $L^p(\mathbb{R}^N)$ for $p \in (N, \infty)$. Conditions are given which ensure that such operators are C^1–Fredholm maps of index zero. Then we give additional assumptions which imply that they are proper on the closed bounded subsets of $W^{2,p}(\mathbb{R}^N)$. For operators with these properties the topological degrees developed by Fitzpatrick, Pejsachowicz and Rabier are available. We illustrate their use by deriving results about the bifurcation of connected components of solutions of quasilinear elliptic equations on $\mathbb{R}^N$.

Keywords: bifurcation, topological degree, Fredholm map, global branch, quasilinear elliptic equation, proper map.

Classification: 35B32, 35J60, 55M25, 55C40.

1. Introduction. In this paper we survey some of our recent work on second order quasilinear elliptic operators defined on $\mathbb{R}^N$. In particular, we discuss the global behaviour of some connected sets of solutions (λ, u) of a second order quasilinear elliptic equation

$$(1) \qquad -\sum_{\alpha,\beta=1}^{N} a_{\alpha\beta}(x, u(x), \nabla u(x))\partial_\alpha \partial_\beta u(x) + b(x, u(x), \nabla u(x), \lambda) = 0$$

for $x \in \mathbb{R}^N$. Here λ is a real parameter and the function u is required to satisfy the condition

$$(2) \qquad \lim_{|x|\to\infty} u(x) = 0.$$

In addition to the ellipticity of the matrix $[a_{\alpha\beta}]$ of coefficients, we suppose that $b(x, 0, \lambda) = 0$ for all $(x, \lambda) \in \mathbb{R}^{N+1}$. Thus $u \equiv 0$ is a solution of the problem (1), (2) for every $\lambda \in \mathbb{R}$ and our results deal with components of non-trivial solutions bifurcating from this line of trivial solutions.

As is well known, [2,15-17], topological degree theory is the primary tool for establishing this kind of result and we have shown that the degree for proper Fredholm maps in the form developed by Fitzpatrick, Pejsachowicz and Rabier can be applied to the problem (1), (2) under very natural and rather general assumptions concerning the coefficients.

The first step is to express the problem (1), (2) as the set of zeros of a function $F \in C^1(\mathbb{R} \times X, Y)$ where X and Y are real Banach spaces. We use the standard Sobolev spaces $X_p = W^{2,p}(\mathbb{R}^N)$ and $Y_p = L^p(\mathbb{R}^N)$ where $p \in (N, \infty)$. There are two reasons for choosing these spaces :

(i) all elements of X_p vanish as $|x| \to \infty$, thus ensuring that (2) is satisfied, and

(ii) we can ensure that $F(\lambda, u) \in Y_p$ for all $u \in X_p$ without imposing restrictions on the growth of the functions $a_{\alpha\beta}(x,\xi)$ and $b(x,\xi,\lambda)$ as $|\xi| \to \infty$.

Our results in Section 5, giving explicit conditions for global bifurcation, do not depend on the choice of p within the range (N, ∞).

In this setting we have derived conditions on the functions $a_{\alpha\beta}(x,\xi)$ and $b(x,\xi,\lambda)$ which imply that $F : \mathbb{R} \times X \to Y$ is a C^1–Fredholm operator of index zero which is proper on closed bounded subsets of $\mathbb{R} \times X$. We summarize these criteria in Sections 2 and 3. The remainder of the paper concerns the conclusions about the bifurcation of solutions of (1), (2) which can be obtained within this framework. We recall in Section 4 the abstract bifurcation theorems found in [9] and [7]. The Fredholm property alone is sufficient to guarantee the bifurcation of a branch of solutions from a point across which the parity is equal to -1, so we begin with a result of this kind for (1),(2). The sense in which a branch can fail to be compact can be made more precise if $F : \mathbb{R} \times X \to Y$ is also proper on closed bounded subsets of $\mathbb{R} \times X$. This is a more delicate question and most of our work has been devoted to resolving this issue. The main conclusions about global bifurcation for are presented at the end of Section 5.

2. C^1-Fredholm operators of index zero. Let X and Y be real Banach spaces. A function $F \in C^1(\mathbb{R} \times X, Y)$ is said to be a Fredholm map of index zero on an interval J if the partial derivative $D_u F(\lambda, u) : X \to Y$ is a bounded linear Fredholm operator of index zero for all points $(\lambda, u) \in J \times X$.

In this section we consider the differential operator

$$(3) \qquad F(\lambda,u)(x) = -\sum_{\alpha,\beta=1}^{N} a_{\alpha\beta}(x,u(x),\nabla u(x))\partial_\alpha\partial_\beta u(x) + b(x,u(x),\nabla u(x),\lambda)$$

as a mapping between the spaces $X_p = W^{2,p}(\mathbb{R}^N)$ and $Y_p = L^p(\mathbb{R}^N)$ where $p \in (N, \infty)$. Our aim is to formulate conditions which ensure that it is a C^1-Fredholm map of index zero on an interval J. We must first deal with its smoothness and in this connection we have found the following definition to be convenient when discussing the assumptions on of the coefficients required to ensure that $F \in C^1(\mathbb{R} \times X_p, Y_p)$.

Using the notation

$$f : \mathbb{R}^N \times \mathbb{R}^{N+1} \to \mathbb{R} \text{ with } (x,\eta) = (x,\xi_0,\ldots\xi_N) \mapsto f(x,\xi_0,\ldots\xi_N)$$

and

$$g : \mathbb{R}^N \times \mathbb{R}^{N+2} \to \mathbb{R} \text{ with } (x,\eta) = (x,\xi_0,\ldots\xi_N,\lambda) \mapsto g(x,\xi_0,\ldots\xi_N,\lambda),$$

we see that the variables x and η play markedly different roles when deriving the smoothness properties of the associated Nemytskii operators $u \mapsto f(\cdot,u(\cdot),\nabla u(\cdot))$ and

$(\lambda, u) \mapsto g(\cdot, u(\cdot), \nabla u(\cdot), \lambda)$. The terminology "bundle map" provides a convenient way of handling this distinction where x is the "base" variable and η is the "fiber" variable. Note that since we require smoothness with respect to u and λ it is natural to treat λ as a fiber variable.

Definition 2.1. A function $f : \mathbb{R}^N \times \mathbb{R}^M \to \mathbb{R}$ is called an *equicontinuous C^0-bundle map* if f is continuous and the collection $\left\{f(x,\cdot) : x \in \mathbb{R}^N\right\}$ is equicontinuous at ξ for every $\xi \in \mathbb{R}^M$. For a positive integer k, we say that $f = f(x,\eta)$ is an *equicontinuous C^k_η-bundle map* if the partial derivatives $D^\alpha_\eta f$ exist and are equicontinuous C^0-bundle maps for all multi-indices α with $|\alpha| \leq k$.

Remark 2.1. If $V \in C(\mathbb{R}^N) \cap L^\infty(\mathbb{R}^N)$ and $g \in C^k(\mathbb{R}^M)$ then the function $f(x,\eta) = V(x)g(\eta)$ is an equicontinuous C^k_η-bundle map, as are finite sums of such functions.

Remark 2.2. Equicontinuous C^0-bundle maps are uniformly equicontinuous on compact subsets of $\mathbb{R}^M$ in the following sense. Let $f : \mathbb{R}^N \times \mathbb{R}^M \to \mathbb{R}$ be an equicontinuous C^0-bundle map. Given a compact subset K of $\mathbb{R}^M$ and $\varepsilon > 0$, there exists $\delta(K,\varepsilon) > 0$ such that $|f(x,\xi) - f(x,\eta)| < \varepsilon$ for all $x \in \mathbb{R}^N$ and $\xi, \eta \in K$ with $|\xi - \eta| < \delta(K,\varepsilon)$. See Lemma 2.1 of [11].

We can now formulate the essential smoothness properties of the family of quasilinear second order differential operators defined by (3) where the functions $a_{\alpha\beta} : \mathbb{R}^N \times \mathbb{R}^{N+1} \to \mathbb{R}$ and $b : \mathbb{R}^N \times \mathbb{R}^{N+2} \to \mathbb{R}$ are bundle maps having the following properties.

(B) For $\alpha, \beta = 1, \ldots, N$, the function $a_{\alpha\beta} = a_{\beta\alpha} : \mathbb{R}^N \times \mathbb{R}^{N+1} \to \mathbb{R}$ is an equicontinuous C^1_ξ-bundle map with

$$a_{\alpha\beta}(\cdot, 0) \text{ and } \partial_{\xi_i} a_{\alpha\beta}(\cdot, 0) \in L^\infty(\mathbb{R}^N) \text{ for } i = 0, 1, \ldots, N. \tag{4}$$

The function $b : \mathbb{R}^N \times \mathbb{R}^{N+2} \to \mathbb{R}$ is continuous and its partial derivatives $\partial_{\xi_i} b, \partial_\lambda b, \partial_\lambda \partial_{\xi_i} b$ and $\partial_{\xi_i} \partial_\lambda b$ exist and are continuous on $\mathbb{R}^N \times \mathbb{R}^{N+2}$ for $i = 0, 1, \ldots, N$. For each $\lambda \in \mathbb{R}, b(\cdot, \lambda) : \mathbb{R}^N \times \mathbb{R}^{N+1} \to \mathbb{R}$ is an equicontinuous C^1_ξ-bundle map and $\partial_\lambda \partial_{\xi_i} b : \mathbb{R}^N \times \mathbb{R}^{N+2} \to \mathbb{R}$ is an equicontinuous C^0-bundle map for $i = 0, 1, \ldots, N$. Furthermore, for all $x \in \mathbb{R}^N$ and $\lambda \in \mathbb{R}$,

$$b(x, 0, \lambda) = 0 \tag{5}$$

and

$$\partial_{\xi_i} b(\cdot, 0, \lambda) \text{ and } \partial_{\xi_i} \partial_\lambda b(\cdot, 0, \lambda) \in L^\infty(\mathbb{R}^N) \text{ for } i = 0, \ldots, N. \tag{6}$$

Remark 2.3. The hypothesis (B) ensures that $\partial_\lambda \partial_{\xi_i} b \equiv \partial_{\xi_i} \partial_\lambda b$ for $i = 0, 1, \ldots, N$ and that

$$\partial_\lambda b(x, 0, \lambda) = 0 \text{ for } x \in \mathbb{R}^N \text{ and } \lambda \in \mathbb{R}. \tag{7}$$

Furthermore, it is easy to deduce from (B) that, for each $\lambda \in \mathbb{R}$, $\partial_\lambda b(\cdot, \lambda) : \mathbb{R}^N \times \mathbb{R}^{N+1} \to \mathbb{R}$ is an equicontinuous C^1_ξ-bundle map.

Using this terminology we can formulate the following results giving the requisite smoothness of the differential operator (3). See Theorem 3 in [14].

Theorem 2.2. *Fix $p \in (N, \infty)$ and consider the operator F defined by* (3) *under the hypothesis* (B). *Then*

(i) *$F \in C^1(\mathbb{R} \times X_p, Y_p)$ and the partial derivatives (in the sense of Fréchet) $D_u D_\lambda F$ and $D_\lambda D_u F$ exist and are continuous on $\mathbb{R} \times X_p$.*

(ii) *$F(\cdot, u) : \mathbb{R} \to Y_p$ is equicontinuous with respect to u in bounded subsets of X_p.*

(iii) *In particular, $F(\lambda, 0) = 0$ with*

$$[D_u F(\lambda, 0)v](x) =$$
$$-\sum_{\alpha,\beta=1}^{N} a_{\alpha\beta}(x,0)\partial_\alpha\partial_\beta v(x) + \sum_{i=1}^{N} \partial_{\xi_i} b(x,0,\lambda)\partial_i v(x) + \partial_{\xi_0} b(x,0,\lambda)v(x)$$

and

$$[D_\lambda D_u F(\lambda, 0)v](x) = [D_u D_\lambda F(\lambda, 0)v](x) =$$
$$\sum_{i=1}^{N} \partial_\lambda \partial_{\xi_i} b(x,0,\lambda)\partial_i v(x) + \partial_\lambda \partial_{\xi_0} b(x,0,\lambda)v(x).$$

for all $v \in X_p$.

We now turn to a discussion of the bounded linear operator $D_u F(\lambda, u) : X_p \to Y_p$ with a view to ensuring that it is Fredholm of index zero. For this we shall suppose that the differential operator defined by (3) is elliptic in the following sense.

(E) The operator F is strictly elliptic in the sense that there exists a lower semicontinuous function, $\nu : \mathbb{R}^N \times \mathbb{R}^{N+1} \to (0, \infty)$, such that

$$\sum_{\alpha,\beta=1}^{N} a_{\alpha\beta}(x,\xi)\eta_\alpha\eta_\beta \geq \nu(x,\xi)\,|\eta|^2 \tag{8}$$

for all $\eta \in \mathbb{R}^N$ and $(x, \xi) \in \mathbb{R}^N \times \mathbb{R}^{N+1}$.

From Theorem 3.1 of [11] we now obtain the following result.

Theorem 2.3. *Consider the operator* (3) *under the hypotheses* (B) *and* (E). *For any $p \in (N, \infty)$, $F : \mathbb{R} \times X_p \to Y_p$ is a C^1-Fredholm map of index zero on an interval J if and only if for each $\lambda \in J$, there exists an element $u_\lambda \in X_p$ such that $D_u F(\lambda, u_\lambda) : X_p \to Y_p$ is a Fredholm operator of index zero.*

Remark 2.4. We observe that the choice $u_\lambda = 0$ is particularly attractive since $D_u F(\lambda, 0)$ does not involve any derivatives of the functions $a_{\alpha\beta}$.

3. Properness. Let X and Y be real Banach spaces. A function $G : X \to Y$ is said to be proper on closed bounded subsets of X provided that $G^{-1}(K) \cap W$ is a compact subset of X for every compact subset K of Y and every closed bounded subset W of X. Let $L(X, Y)$ denote the Banach space of all bounded linear operators from X into Y and let the kernel and range of a linear operator T be denoted by $\ker T$ and rge T, respectively. In the case of a bounded linear operator $L : X \to Y$, it is known that L is proper on closed bounded subsets of X if and only if $L \in$

$\Phi_+(X,Y) \equiv \{L \in L(X,Y) : \text{rge}L$ is closed and $\dim\ker L < \infty\}$. This result is due to Yood and appears as on page 78 of [3].

In this section we discuss the properness of the nonlinear differential operator $F(\lambda,\cdot) : X_p \to Y_p$ defined by (3) on closed bounded subsets of X_p. For this we shall suppose that it is asymptotically periodic as x tends to infinity in the following sense.

(A) There exist equicontinuous C^0-bundle maps $a^\infty_{\alpha\beta} = a^\infty_{\beta\alpha} : \mathbb{R}^N \times \mathbb{R}^{N+1} \to \mathbb{R}$ for $\alpha, \beta = 1,\dots,N$ and an equicontinuous C^1_η-bundle map $b^\infty : \mathbb{R}^N \times \mathbb{R}^{N+2} \to \mathbb{R}$ such that $b^\infty(x,0,\lambda) \equiv 0$ and

$$\lim_{|x|\to\infty}[a_{\alpha\beta}(x,\xi) - a^\infty_{\alpha\beta}(x,\xi)] = \lim_{|x|\to\infty}[\partial_{\xi_i} b(x,\xi,\lambda) - \partial_{\xi_i} b^\infty(x,\xi,\lambda)] = 0$$

uniformly for (ξ,λ) in bounded subsets of $\mathbb{R}^{N+2}$, where $1 \le \alpha,\beta \le N$ and $i = 0,1,\dots,N$. Furthermore, $a^\infty_{\alpha\beta}(\cdot,\xi)$ and $b^\infty(\cdot,\xi,\lambda) : \mathbb{R}^N \to \mathbb{R}$ are N-periodic on $\mathbb{R}^N$ in the sense that, for some $T = (T_1,\dots T_N)$ with $T_i > 0$ for all $i = 1,\dots,N$,

$$a^\infty_{\alpha\beta}(x_1,\dots,x_i + T_i,\dots,x_N,\xi) = a^\infty_{\alpha\beta}(x_1,\dots,x_N,\xi)$$

and

$$b^\infty(x_1,\dots,x_i + T_i,\dots,x_N,\xi,\lambda) = b^\infty(x_1,\dots,x_N,\xi,\lambda)$$

for all $(x,\xi,\lambda) \in \mathbb{R}^N \times \mathbb{R}^{N+2}$ and $i = 1,\dots,N$.

Under the assumptions (B) and (A) we define a differential operator, F^∞, by

$$(9) \qquad F^\infty(\lambda,u) = -\sum_{\alpha,\beta=1}^{N} a^\infty_{\alpha\beta}(x,u(x),\nabla u(x))\partial_\alpha\partial_\beta u(x) + b^\infty(x,u(x),\nabla u(x),\lambda).$$

Remark 3.1. The assumptions (B) and (A) imply that $F^\infty(\lambda,u) \in Y_p$ for all $(\lambda,u) \in \mathbb{R} \times X_p$ and that $F^\infty(\lambda,0) = 0$ for all $\lambda \in \mathbb{R}$.

In this context Theorem 6.1 of [11] yields the following result.

Theorem 3.1. *Consider the differential operator defined by* (3) *under the hypotheses* (B), (E) *and* (A) *and let* $p \in (N,\infty)$. *Then* $F(\lambda,\cdot) : X_p \to Y_p$ *is proper on closed bounded subsets of* X_p *if and only if*

$$\begin{cases} \text{(i)} & \textit{there is an element } u_\lambda \in X_p \textit{ such that } D_u F(\lambda,u_\lambda) \in \Phi_+(X_p,Y_p), \\ \text{(ii)} & \textit{the equation } F^\infty(\lambda,u) = 0 \textit{ has no non-trivial solution } u \textit{ in } X_p. \end{cases}$$

4. Abstract bifurcation results. Let X and Y be real Banach spaces. A topological degree for C^1-Fredholm maps of index zero from X to Y has been developed in [7,9]. That degree is based on the notion of the parity, denoted by $\pi(A(\lambda) : \lambda \in [a,b])$, of a continuous path, $\lambda \mapsto A(\lambda)$, of bounded linear Fredholm operators of index zero from X into Y, which was introduced in [6]. For such a path, a parametrix is any continuous function $B : [a,b] \to GL(Y,X)$ such that $B(\lambda)A(\lambda) : X \to X$ is a compact perturbation of the identity for each $\lambda \in [a,b]$. If $A(a)$ and $A(b) \in GL(X,Y)$, the parity of the path A on $[a,b]$ is defined by

$$\pi\,(A(\lambda) : \lambda \in [a,b]) = d_{LS}(B(a)A(a))d_{LS}(B(b)A(b)) \tag{10}$$

where d_{LS} denotes the Leray-Schauder degree. This definition is justified by showing that a parametrix always exists and that $d_{LS}(B(a)A(a))d_{LS}(B(b)A(b))$ is independent of the choice of parametrix B. We observe that for the discussion of parity the Leray-Schauder degree is used only in the very special case of linear isomorphisms of the form I - K where K is compact and thus it coincides with $(-1)^m, m$ being the sum of the multiplicities of the eigenvalues of K which are real and greater than 1.In some circumstances the parity can be expressed in a form which is easier to evaluate directly. In formulating our bifurcation theorems for (1), (2) we shall only use the following criterion.

Proposition 4.1. *Let $A : [a,b] \to L(X,Y)$ be a continuous path of bounded linear operators having the following properties.*
(i) $A \in C^1([a,b], L(X,Y))$.
(ii) *$A(\lambda) : X \to Y$ is a Fredholm operator of index zero for all $\lambda \in [a,b]$.*
(iii) *There exists $\lambda_0 \in (a,b)$ such that*

$$A'(\lambda_0)[\ker\ A(\lambda_0)] \oplus rge\ A(\lambda_0) = Y \tag{11}$$

in the sense of a topological direct sum.
Then there exists $\varepsilon > 0$ such that $[\lambda_0 - \varepsilon, \lambda_0 + \varepsilon] \subset [a,b]$,

$$A(\lambda) \in GL(X,Y) \text{ for all } \lambda \in [\lambda_0 - \varepsilon, \lambda_0) \cup (\lambda_0, \lambda_0 + \varepsilon] \tag{12}$$

and

$$\pi\,(A(\lambda) : \lambda \in [\lambda_0 - \varepsilon, \lambda_0 + \varepsilon]) = (-1)^k \tag{13}$$

where $k = \dim\ker A(\lambda_0)$.

This result is a combination of Proposition 2.1 of [5] and Theorem 6.18 of [6].

Note that for any continuous path $A : [a,b] \to L(X,Y)$ and any $\lambda_0 \in (a,b)$ such that $A(\lambda) \in GL(X,Y)$ for all $\lambda \in [a,b]\setminus\{\lambda_0\}$, the parity $\pi\,(A(\lambda) : \lambda \in [\lambda_0 - \varepsilon, \lambda_0 + \varepsilon])$ is the same for all $\varepsilon > 0$ provided that $[\lambda_0 - \varepsilon, \lambda_0 + \varepsilon] \subset [a,b]$. This quantity is called the *parity of the path A across* λ_0. The preceding proposition provides one way of calculating the parity across λ_0.

We can now state the main result about global bifurcation which can be derived using the above notions.

Theorem 4.2. *Let X and Y be real Banach spaces and consider a function $F \in C^1(J \times X, Y)$ where J is an open interval such that $F(\lambda, 0) = 0$ for all $\lambda \in J$ and F is a Fredholm map of index zero on J. Suppose that $\lambda_0 \in J$ and that there exists $\varepsilon > 0$ such that $[\lambda_0 - \varepsilon, \lambda_0 + \varepsilon] \subset J$,*

$$D_u F(\lambda, 0) \in GL(X,Y) \text{ for } \lambda \in [\lambda_0 - \varepsilon, \lambda_0 + \varepsilon]\setminus\{\lambda_0\}$$

and

$$\pi\left(D_uF(\lambda,0),[\lambda_0-\varepsilon,\lambda_0+\varepsilon]\right)=-1.$$

Let $Z=\{(\lambda,u)\in J\times X: u\neq 0$ and $F(\lambda,u)=0\}$ and let C denote the connected component of $Z\cup\{(\lambda_0,0)\}$ containing $(\lambda_0,0)$. Then C has at least one of the following properties.

(i) C is not a compact subset of $J\times X$.

(ii) The closure of C contains a point $(\lambda_1,0)$ where $\lambda_1\in J\backslash[\lambda_0-\varepsilon,\lambda_0+\varepsilon]$ and $D_uF(\lambda_1,0)\notin GL(X,Y)$.

The above statements refer to Z and C with the metric inherited from $\mathbb{R}\times X$. The basic procedure for proving a result like this is to suppose that C has neither of the properties stated in the conclusion and then to derive a contradiction using the properties of the degree for C^1-Fredholm maps. For the special case $J=\mathbb{R}$, this result appears as Theorem 6.1 of [9], but the same arguments yield the version we have stated.

The ways in which C can fail to be compact can be made more precise provided that F is proper on closed bounded subsets of X.

Theorem 4.3. *Let X and Y be real Banach spaces and let J be an open interval. Consider a function $F\in C^1(J\times X,Y)$ such that the maps $F(\cdot,u):J\to Y$ are equicontinuous for u in bounded subsets of X. Suppose that $F(\lambda,0)=0$ for all $\lambda\in J$ and that F is a Fredholm map of index zero on J, with the property that $F(\lambda,\cdot):X\to Y$ is proper on closed bounded subsets of X for all $\lambda\in J$. Suppose also that $\lambda_0\in J$ and that there exists $\varepsilon>0$ such that $[\lambda_0-\varepsilon,\lambda_0+\varepsilon]\subset J$,*

$$D_uF(\lambda,0)\in GL(X,Y)\ \text{for}\ \lambda\in[\lambda_0-\varepsilon,\lambda_0+\varepsilon]\backslash\{\lambda_0\}$$

and

$$\pi\left(D_uF(\lambda,0),[\lambda_0-\varepsilon,\lambda_0+\varepsilon]\right)=-1.$$

Let $Z=\{(\lambda,u)\in J\times X: u\neq 0$ and $F(\lambda,u)=0\}$ and let C denote the connected component of $Z\cup\{(\lambda_0,0)\}$ containing $(\lambda_0,0)$. Then C has at least one of the following properties.

(i) C is an unbounded subset of $J\times X$.

(ii) The closure of $PC=\{\lambda\in\mathbb{R}:(\lambda,u)\in C\}$ intersects the boundary of J.

(iii) The closure of C contains a point $(\lambda_1,0)$ where $\lambda_1\in J\backslash[\lambda_0-\varepsilon,\lambda_0+\varepsilon]$ and $D_uF(\lambda_1,0)\notin GL(X,Y)$.

Proof. Suppose that C has none of the properties (i), (ii), (iii). We shall show that this implies that C is a compact subset of $J\times X$, contradicting the previous theorem. Let $\{(\lambda_n,u_n)\}\subset C$. For the compactness of C, it is enough to show that this sequence contains a subsequence converging to a point (λ,u) in $J\times X$. Since C is a bounded subset of $\mathbb{R}\times X$ there is a closed bounded subset W of X such that $\{u_n\}\subset W$. Passing to a subsequence we can suppose immediately that $\lambda_n\to\lambda$ and $\lambda\in J$ since $\lambda_n\in J$ and C does not have the property (ii). Furthermore since the functions $F(\cdot,u_n)$ are equicontinuous at λ, $F(\lambda_n,u_n)-F(\lambda,u_n)\to 0$ as $n\to\infty$. Since $F(\lambda_n,u_n)=0$, it follows that $F(\lambda,u_n)\to 0$ and so $K=\{F(\lambda,u_n)\}\cup\{0\}$ is a compact subset of Y. Hence $[F(\lambda,\cdot)^{-1}K]\cap W$ is a compact subset of X. This implies that $\{u_n\}$ contains a convergent subsequence in X which in turn establishes the compactness of C. □

5. Bifurcation for quasilinear elliptic equations on $\mathbb{R}^N$. Throughout this section we fix $p \in (N, \infty)$ and consider the differential operator $F : \mathbb{R} \times X_p \to Y_p$ defined by (3) under the assumption (B). By Theorem 2.2 this already ensures that

$$F \in C^1(\mathbb{R} \times X_p, Y_p) \text{ with } F(\lambda, 0) = 0$$

and

$$D_u F(\lambda, 0)v = -\sum_{\alpha,\beta=1}^{N} a_{\alpha\beta}(\cdot, 0)\partial_\alpha \partial_\beta v + \sum_{i=1}^{N} \partial_{\xi_i} b(\cdot, 0, \lambda)\partial_i v + \partial_{\xi_0} b(\cdot, 0, \lambda)v$$

for all $\lambda \in \mathbb{R}$.

Our next results deal with a situation where the parity of the path $\lambda \mapsto D_u F(\lambda, 0)$ across a value λ_0 can be determined in a relatively explicit way. This is a consequence of the following assumption (L) which ensures that $D_u F(\lambda, 0)$ has a particularly simple form. More general behaviour at $u = 0$ can be handled provided that the asymptotic behaviour required for the discussion of properness is assumed and we shall return to this in due course.

(L) There is a (constant) positive definite matrix $[A_{\alpha\beta}]$ such that

$$a_{\alpha\beta}(x, 0) = A_{\alpha\beta} = A_{\beta\alpha} \text{ for all } x \in \mathbb{R}^N$$

and

$$\partial_{\xi_\alpha} b(x, 0, \lambda) = 0 \text{ for all } x \in \mathbb{R}^N \text{ and } \lambda \in \mathbb{R}$$

for all $\alpha, \beta = 1, \ldots, N$,

Under the hypotheses (B) and (L), it follows from Theorem 2.2 that

$$[D_u F(\lambda, 0)v](x) = -\sum_{\alpha,\beta=1}^{N} A_{\alpha\beta}\partial_\alpha \partial_\beta v(x) + \partial_{\xi_0} b(x, 0, \lambda)v(x)$$

which can be reduced to the form

$$[D_u F(\lambda, 0)v](x) = -\Delta v(x) + \partial_{\xi_0} b(x, 0, \lambda)v(x) \tag{14}$$

by a linear change of the variable x.

By Theorem 2.3, $F : \mathbb{R} \times X_p \to Y_p$ is a C^1-Fredholm map of index zero on an interval J if and only if $D_u F(\lambda, 0) : X_p \to Y_p$ linear Fredholm operator of index zero for all $\lambda \in J$. Using the hypothesis (L) the latter property can be expressed in terms of the linear Schrödinger operator $-\Delta + \partial \xi_0 b(x, 0, \lambda)$ on $L^2(\mathbb{R}^N)$.

We refer to [4] for the notions of spectrum, discrete spectrum and essential spectrum of an unbounded self-adjoint operator acting on a Hilbert space. The discrete spectrum consists of the isolated points in the spectrum which are eigenvalues of finite multiplicity. Those points in the spectrum which do not belong to the discrete spectrum form the essential spectrum.

In [12] we discussed the Fredholm properties of the operator $-\Delta + V$ in $L^p(\mathbb{R}^N)$ for a class of potentials admitting singularities. To deal with (14) it is sufficient to recall the following special case which appears as Theorem 1 in [12].

Theorem 5.1. *Let $V \in L^\infty(\mathbb{R}^N)$. Then $-\Delta + V : W^{2,2}(\mathbb{R}^N) \subset L^2(\mathbb{R}^N) \to L^2(\mathbb{R}^N)$ is a self-adjoint operator whose spectrum and discrete spectrum are denoted by σ and σ_d respectively. For $p \in (1, \infty)$, consider also the operator $S_p : W^{2,p}(\mathbb{R}^N) \to L^p(\mathbb{R}^N)$ defined by*

$$S_p u = (-\Delta + V)u \text{ for } u \in W^{2,p}(\mathbb{R}^N).$$

For every $p \in (1, \infty)$, the following conclusions are valid.
(i) *$S_p - \lambda I : W^{2,p}(\mathbb{R}^N) \to L^p(\mathbb{R}^N)$ is an isomorphism if $\lambda \notin \sigma$,*
whereas, if $\lambda \in \sigma_d$, then
(ii) *$S_p - \lambda I : W^{2,p}(\mathbb{R}^N) \to L^p(\mathbb{R}^N)$ is a Fredholm operator of index zero,*
(iii) *$ker\,(S_p - \lambda I) = ker\,(S_2 - \lambda I)$, and*
(iv) *$L^p(\mathbb{R}^N) = ker\,(S_p - \lambda I) \oplus rge\,(S_p - \lambda I)$ where $\oplus$ denotes a topological direct sum.*

We now use this result in conjunction with Proposition 4 to discuss the parity of the path $\lambda \mapsto D_u F(\lambda, 0)$.

Lemma 5.2. *Suppose that the conditions* (B) *and* (L) *are satisfied and consider a point $\lambda_0 \in \mathbb{R}$ such that 0 does not belong to the essential spectrum of the self-adjoint operator $-\Delta + \partial_{\xi_0} b(x, 0, \lambda_0) : X_2 \subset Y_2 \to Y_2$. Let $p \in (N, \infty)$.*
(i) *Then $D_u F(\lambda_0, 0) : X_p \to Y_p$ is a Fredholm operator of index zero.*
(ii) *Furthermore, if either*

$$\partial_\lambda \partial_{\xi_0} b(\cdot, 0, \lambda_0) \geq 0 \text{ but } \not\equiv 0 \text{ on } \mathbb{R}^N,$$

or

$$\partial_\lambda \partial_{\xi_0} b(\cdot, 0, \lambda_0) \leq 0 \text{ but } \not\equiv 0 \text{ on } \mathbb{R}^N,$$

then there exists $\varepsilon > 0$ such that $D_u F(\lambda, 0) \in GL(X_p, Y_p)$ for all $\lambda \in [\lambda_0 - \varepsilon, \lambda_0 + \varepsilon] \setminus \{\lambda_0\}$ and

$$\pi(D_u F(\lambda, 0), [\lambda_0 - \varepsilon, \lambda_0 + \varepsilon]) = (-1)^k$$

where $k = \dim \ker[-\Delta + \partial_{\xi_0} b(x, 0, \lambda_0)]$.

Remark 5.1. By Theorem 5.1 (iii), the kernel of the linear operator $-\Delta + \partial_{\xi_0} b(x, 0, \lambda) : X_p \to Y_p$ does not depend on the choice of $p \in (1, \infty)$.

We can now formulate our first bifurcation theorem for the problem (1), (2).

Theorem 5.3. *Let the conditions* (B), (E) *and* (L) *be satisfied and let $p \in (N, \infty)$. Consider the operator $F : \mathbb{R} \times X_p \to Y_p$ defined by* (3) *and suppose that J is an open interval having the following properties.*
(a) *For $\lambda \in J$, the essential spectrum of the self-adjoint operator $-\Delta + \partial_{\xi_0} b(x, 0, \lambda) : X_2 \subset Y_2 \to Y_2$ does not contain 0.*
(b) *There is a point $\lambda_0 \in J$ such that dim $ker[-\Delta + \partial_{\xi_0} b(x, 0, \lambda_0)]$ is odd and either*

$$\partial_\lambda \partial_{\xi_0} b(\cdot, 0, \lambda_0) \geq 0 \text{ but } \not\equiv 0 \text{ on } \mathbb{R}^N,$$

or

$$\partial_\lambda \partial_{\xi_0} b(\cdot, 0, \lambda_0) \leq 0 \text{ but } \not\equiv 0 \text{ on } \mathbb{R}^N.$$

Let $Z = \{(\lambda, u) \in J \times X_p : u \neq 0$ *and* $F(\lambda, u) = 0\}$ *and let* C *denote the connected component of* $Z \cup \{(\lambda_0, 0)\}$ *containing* $(\lambda_0, 0)$. *Then* C *has at least one of the following properties.*

(i) C *is a non-compact subset of* $J \times X_p$.

(ii) *The closure of* C *in* $\overline{J} \times X_p$ *contains a point* $(\lambda_1, 0)$ *where* $\lambda_1 \neq \lambda_0$.

Proof. Using Lemma 5.2 (i) and Theorem 2.3, we see that $F : \mathbb{R} \times X_p \to Y_p$ is a Fredholm map of index zero on the interval J. By Lemma 8(ii) its parity across λ_0 is equal to -1. The conclusion follows from Theorem 4.2. □

To resolve the non-compactness property into a global one we suppose that F is asymptotically periodic in the sense of condition (A) and then we discuss the equation $F^\infty(\lambda, u) = 0$ with a view to showing that $u = 0$ is the only solution in X_p. We can also ensure that the operator (3) is a C^1-Fredholm map on an interval by imposing conditions on the linearization of F^∞.

First of all we recall from Section 6 of [11] that, although the assumption (A) does not guarantee the differentiability of the operator $F^\infty : \mathbb{R} \times X_p \to Y_p$, it does imply that $F^\infty(\lambda, \cdot) : X_p \to Y_p$ is differentiable (in the sense of Fréchet) at 0 with

$$D_u F^\infty(\lambda, 0)v = -\sum_{\alpha,\beta=1}^{N} a^\infty_{\alpha\beta}(\cdot, 0)\partial_\alpha\partial_\beta v + \sum_{\alpha=1}^{N} \partial_{\xi_\alpha} b^\infty(\cdot, 0, \lambda)\partial_\alpha v + \partial_{\xi_0} b^\infty(\cdot, 0, \lambda)v$$

for all $v \in X_p$ and $\lambda \in \mathbb{R}$.

We note that $D_u F^\infty(\lambda, 0)$ is a linear second order differential operator with continuous N-periodic coefficients. In [11], Lemma 6.6 and Remark 6.2 describe some situations where it is a Fredholm operator of index zero. The following assumption isolates a particularly agreeable situation.

(L$^\infty$) There is a (constant) positive definite matrix $\left[A^\infty_{\alpha\beta}\right]$ such that

$$a^\infty_{\alpha\beta}(x, 0) = A^\infty_{\alpha\beta} = A^\infty_{\beta\alpha} \text{ for all } x \in \mathbb{R}^N$$

and

$$\partial_{\xi_\alpha} b^\infty(x, 0, \lambda) = 0 \text{ for all } x \in \mathbb{R}^N \text{ and } \lambda \in \mathbb{R}$$

for $1 \leq \alpha, \beta \leq N$.

Remark 5.2. When this condition is satisfied we can assume that $A^\infty_{\alpha\beta} = \delta_{\alpha\beta}$ for $1 \leq \alpha, \beta \leq N$ (by making a linear change of variable) and hence that

$$D_u F^\infty(\lambda, 0)v = -\Delta v + \partial_{\xi_0} b^\infty(\cdot, 0, \lambda)v$$

for $v \in X_p$ and $\lambda \in \mathbb{R}$.

In this context we have the following result which appears as Lemma 11 of [14].

Lemma 5.4. *Suppose that the conditions* (B), (A) *and* (L$^\infty$) *are satisfied and consider* $\lambda_0 \in \mathbb{R}$ *such that the self-adjoint operator,* $S(\lambda_0) : X_2 \subset Y_2 \to Y_2$, *defined by*

$$S(\lambda_0)v = -\Delta v + \partial_{\xi_0} b^\infty(\cdot, 0, \lambda_0)v \text{ for } v \in X_2$$

is an isomorphism. Let $p \in (N, \infty)$.
(i) *Then* $D_u F(\lambda_0, 0) : X_p \to Y_p$ *is a Fredholm operator of index zero.*
(ii) *Let* $\{\varphi_i \in X_p : i = 1, \dots, k\}$ *and* $\{\psi_i \in Y_q : i = 1, \dots, k\}$ *be bases for* $\ker D_u F(\lambda_0, 0)$ *and* $\ker [D_u F(\lambda_0, 0)]^*$ *respectively, with* $k = \dim \ker D_u F(\lambda_0, 0)$ *and* $\frac{1}{p} + \frac{1}{q} = 1$. *Then*

$$\det \left[\int_{\mathbb{R}^N} \psi_i \{ D_\lambda D_u F(\lambda_0, 0) \varphi_j \} \, dx \right] \neq 0 \tag{15}$$

if and only if

$$D_\lambda D_u F(\lambda_0, 0) [\ker D_u F(\lambda_0, 0)] \oplus rge \, D_u F(\lambda_0, 0) = Y_p. \tag{16}$$

When (15) *is satisfied there exists* $\varepsilon > 0$ *such that* $D_u F(\lambda, 0) \in GL(X_p, Y_p)$ *for all* $\lambda \in [\lambda_0 - \varepsilon, \lambda_0 + \varepsilon] \setminus \{\lambda_0\}$ *and*

$$\pi \left(D_u F(\lambda, 0), [\lambda_0 - \varepsilon, \lambda_0 + \varepsilon] \right) = (-1)^k.$$

Remark 5.3. This result gives the same conclusions as Lemma 5.2 without requiring $D_u F(\lambda, 0)$ to be a formally symmetric differential operator. Note that when the conditions (B), (E), (A) and (L) are satisfied then so is (L^∞) with $A^\infty_{\alpha\beta} = A_{\alpha\beta}$. We also observe that, since $\partial_{\xi_0} b^\infty(\cdot, 0, \lambda_0)$ is an N-periodic function, $S(\lambda_0)$ is an isomorphism if and only if 0 does not belong to its essential spectrum. (See Theorem 5.4 of Chapter 3 in [1].) Moreover when there exists a continuous N-periodic function P such that $\partial_{\xi_0} b^\infty(x, 0, \lambda_0) \equiv P(x) - \lambda_0$, $S(\lambda_0) = -\Delta + P - \lambda_0$ and it is an isomorphism if and only if λ_0 does not belong to the spectrum of the N-periodic Schrödinger operator $-\Delta + P$.

Combining the above results we obtain the following rather general global bifurcation theorem for (1), (2). See Theorem 12 in [14].

Theorem 5.5. *Let the conditions* (B),(E),(A) *and* (L^∞) *be satisfied. Choose* $p \in (N, \infty)$ *and consider the operator* $F : \mathbb{R} \times X_p \to Y_p$ *defined by* (3). *Suppose that* J *is an open interval having the following properties.*
(a) *For all* $\lambda \in J$, $\{u \in X_p : F^\infty(\lambda, u) = 0\} = \{0\}$.
(b) *For all* $\lambda \in J$, *the self-adjoint operator* $-\Delta + \partial_{\xi_0} b^\infty(\cdot, 0, \lambda) : X_2 \subset Y_2 \to Y_2$ *is an isomorphism.*
(c) *There is a point* $\lambda_0 \in J$ *such that* $\dim \ker D_u F(\lambda_0, 0)$ *is odd and the condition* (15) *is satisfied.*
Let C *denote the connected component of* $Z \cup \{(\lambda_0, 0)\}$ *containing* $(\lambda_0, 0)$ *where*

$$Z = \{(\lambda, u) \in J \times X_p : u \neq 0 \text{ and } F(\lambda, u) = 0\}$$

and $Z \cup \{(\lambda_0, 0)\}$ *has the metric inherited from* $\mathbb{R} \times X_p$. *Then* C *has at least one of the following properties.*
(i) C *is an unbounded subset of* $J \times X_p$.
(ii) *The closure of* $\{\lambda : (\lambda, u) \in C$ *for some* $u \in X_p\}$ *intersects the boundary of* J.
(iii) *The closure of* C *in* $\overline{J} \times X_p$ *contains a point* $(\lambda_1, 0)$ *where* $\lambda_1 \neq \lambda_0$.

There are several approaches which can be used to check the condition (a) in this result and we shall present three of them. In order to give relatively explicit hypotheses on the operator F we shall strengthen the hypothesis (L).

(LL) The condition (L) is satisfied and there is a constant $c \neq 0$ such that

$$\partial_\lambda \partial_{\xi_0} b(x,0,\lambda) = c \text{ for all } x \in \mathbb{R}^N \text{ and } \lambda \in \mathbb{R}.$$

When (B) and (LL) are satisfied we can and shall suppose (by redefining λ) that there exists a function $V \in C(\mathbb{R}^N) \cap L^\infty(\mathbb{R}^N)$ such that

$$\partial_{\xi_0} b(x,0,\lambda) = V(x) - \lambda \text{ for all } x \in \mathbb{R}^N \text{ and } \lambda \in \mathbb{R}.$$

In this case

$$D_u F(\lambda,0)v = (-\Delta + V)v - \lambda v \tag{17}$$

where $-\Delta + V : X_2 \subset Y_2 \to Y_2$ is a self-adjoint operator. When the condition (LL) is satisfied with $\partial_{\xi_0} b(x,0,\lambda) \equiv V(x) - \lambda$, the condition ($L^\infty$) is also satisfied and $\partial_{\xi_0} b^\infty(x,0,\lambda_0) \equiv P(x) - \lambda$ where P is a continuous N-periodic function. Since $\lim_{|x|\to\infty} \{V(x) - P(x)\} = 0$, it follows that the essential spectrum of the Schrödinger operator $-\Delta + V : X_2 \subset Y_2 \to Y_2$ is equal to the whole spectrum of the N-periodic Schrödinger operator $-\Delta + P : X_2 \subset Y_2 \to Y_2$.

Henceforth we use σ and σ_e to denote its spectrum and essential spectrum, respectively.

5.1. Using the maximum principle. The maximum principle can be used to establish the condition (a) of Theorem 5.5 provided that the functions $a^\infty_{\alpha\beta}$ and b^∞ have the following properties.

(M_λ) There exists a continuous function $\nu : \mathbb{R}^{N+1} \to (0,\infty)$ such that

$$\sum_{\alpha,\beta=1}^{N} a^\infty_{\alpha\beta}(x,s,0)\eta_\alpha \eta_\beta \geq \nu(x,s)\,|\eta|^2 \text{ for all } \eta \in \mathbb{R}^N$$

and for all $(x,s) \in \mathbb{R}^{N+1}$ and

$$b^\infty(x,s,0,\lambda)s > 0 \text{ for all } (x,s) \in \mathbb{R}^{N+1} \text{ with } s \neq 0.$$

Remark 5.4. It follows from this that $\partial_{\xi_0} b^\infty(x,0,\lambda) \geq 0$ for all $x \in \mathbb{R}^N$.

Theorem 5.6. *Let the conditions* (B),(E) *and* (A) *be satisfied and let* $p \in (N,\infty)$. *Suppose that* J *is an open interval such that* (M_λ) *is satisfied for all* $\lambda \in J$.

(i) *Suppose that the condition* (L^∞) *is satisfied and that the operator* $S(\lambda) = -\Delta + \partial_{\xi_0} b^\infty(\cdot,0,\lambda) : X_2 \to Y_2$ *is an isomorphism for all* $\lambda \in J$. *Then the conclusion of Theorem 5.5 is valid for the operator* $F : \mathbb{R} \times X_p \to Y_p$ *defined by* (3) *at any point* $\lambda_0 \in J$ *such that* $\dim\ker D_u F(\lambda_0,0)$ *is odd and the condition* (15) *is satisfied with* $\lambda = \lambda_0$.